AF692566

Chiral Symmetry in Physics

Chiral Symmetry in Physics

Guest Editor

Dubravko Klabučar

MDPI

Basel • Beijing • Wuhan • Barcelona • Belgrade • Novi Sad • Cluj • Manchester

Guest Editor
Dubravko Klabučar
Physics Department
University of Zagreb
Zagreb
Croatia

Editorial Office
MDPI AG
Grosspeteranlage 5
4052 Basel, Switzerland

This is a reprint of the Special Issue, published open access by the journal *Symmetry* (ISSN 2073-8994), freely accessible at: https://www.mdpi.com/journal/symmetry/special_issues/chiral_symmetry_physics.

For citation purposes, cite each article independently as indicated on the article page online and as indicated below:

Lastname, A.A.; Lastname, B.B. Article Title. *Journal Name* **Year**, *Volume Number*, Page Range.

ISBN 978-3-7258-6063-0 (Hbk)
ISBN 978-3-7258-6064-7 (PDF)
https://doi.org/10.3390/books978-3-7258-6064-7

© 2025 by the authors. Articles in this book are Open Access and distributed under the Creative Commons Attribution (CC BY) license. The book as a whole is distributed by MDPI under the terms and conditions of the Creative Commons Attribution-NonCommercial-NoDerivs (CC BY-NC-ND) license (https://creativecommons.org/licenses/by-nc-nd/4.0/).

Contents

About the Editor

Dubravko Klabučar

Dubravko Klabučar is affiliated with the Department of Physics, Faculty of Science, University of Zagreb, Croatia, where he worked as a Full Professor until retirement in October 2023. He received his undergraduate degree in Physics from the University of Zagreb in 1981 and Ph.D. from the State University of New York at Stony Brook in 1986, including a research year at the Niels Bohr Institute, Copenhagen, 1983-84, under mentorship of G.E. Brown. He completed postdoctoral research at the Max-Planck Institute for Nuclear Physics, Heidelberg, 1986-87. Then, he returned to Croatia, to Ruđer Bošković Institute in Zagreb, where he became a research associate in 1991. In 1992, he was elected as assistant professor at the Faculty of Science of the University of Zagreb, associate professor in 1997, and then full professor in 2002. D. Klabučar specializes in theoretical particle physics, especially hadronic and relativistic quark-antiquark bound states, having authored over 70 scientific publications. He was principal investigator of two international projects funded by the U.S. National Science Foundation and European Union and as Croatian coordinator for European networks of ESF (CompStar) and COST (Actions MP1304 NewCompStar, CA15213 THOR, CA16214 PHAROS). He is the Guest Editor for two Special Issues in the journal *Symmetry* and its editorial board member. His teaching contributions include developing courses in Quantum Physics, Relativistic Quantum Physics, and introducing Hadronic Physics courses. He supervised two PhD dissertations and eighteen master's theses. He organized international conferences including the first physics conference in independent Croatia (Brijuni 1994), Jefferson Lab conferences at Dubrovnik (1998, 2003), NewCompStar School (Zadar 2012), and five of the ongoing ACHT conference series. He edited several proceedings of those. He held several leading posts at the Department of Physics and was Croatian delegate to IUPAP 1998-2005.

symmetry

Editorial

Special Issue: Chiral Symmetry in Physics

Dubravko Klabučar

Physics Department, Faculty of Science, University of Zagreb, Bijenička cesta 32, 10000 Zagreb, Croatia; klabucar@phy.hr

The study of symmetry principles has consistently provided excellent guidance in the search to understand the fundamental laws of nature. Symmetries underpin our theoretical frameworks, define conservation laws, and guide the classification of particles and interactions. Among these, chiral symmetry occupies a position of particular significance. It provides profound insights into many areas of physics, notably field theory and elementary particle physics. Rooted in the structure of quantum chromodynamics (QCD), chiral symmetry and its spontaneous (i.e., dynamical) breaking are essential for explaining a considerable portion of particle physics data [1] and understanding a wide array of physical phenomena, from the many properties of hadrons to the collective behavior of matter under extreme conditions.

This Special Issue brings together a diverse set of contributions that explore both the established and emerging roles of chiral symmetry in modern theoretical physics. It assembles seven outstanding papers—five original research articles and two reviews—that employ a range of theoretical tools, from perturbative and nonperturbative field theory methods to lattice QCD simulations and effective field theories. These works collectively offer a broad and insightful survey of how chiral symmetry, and its dynamical breaking or partial restoration, manifest across various physical systems and energy scales. The Table of Contents lists the articles in the chronological order of their acceptance. The first five are research articles; the final two are reviews. Each paper highlights a different aspect of the evolving role of chiral symmetry in modern physics, and together they underscore the enduring relevance of symmetry principles in shaping theoretical inquiry.

Selected Advances in Chiral Symmetry Research

The papers in this Special Issue present some of the key modern developments in our understanding of chiral symmetry. Its spontaneous, viz. dynamical, breaking in quantum chromodynamics (QCD) leads to emergent phenomena in hadron and nuclear physics.

Craig Roberts' exploration of *Empirical Consequences of Emergent Mass* [2] has attracted significant attention and won the Symmetry 2020 Best Paper Award [3] It demonstrates how the relatively simple Lagrangian of QCD generates the remarkable complexity observed in nuclear physics through the mechanism of emergent mass. As Roberts eloquently explains, this emergent mass may be QCD's self-stabilizing mechanism, with observable consequences ranging from pion parton distributions to nucleon electromagnetic form factors. This work provides a crucial framework for understanding how abstract symmetry principles manifest themselves in measurable physical properties. Some subsequent results and related considerations are presented in, e.g., [4–7] and references therein.

Ulf-G. Meißner's highly cited work *Two-Pole Structures in QCD: Facts, Not Fantasy!* [8] challenges conventional wisdom by revealing that certain states in the hadron spectrum—as listed in The Review of Particle Physics [1]—actually represent two distinct states in the complex plane. Beginning with the $\Lambda(1405)$ and extending to excited charm mesons,

https://doi.org/10.3390/sym17081346

Meißner's analysis employs unitarized chiral perturbation theory to demonstrate how this two-pole structure emerges from fundamental symmetry considerations. Here, progress requires an interplay between theoretical methods, experimental data, and lattice QCD simulations. To date, almost a dozen such two-pole structures have now been identified, with those presented in [9–14] being the most recent, establishing a new class of players in hadron spectroscopy. This work exemplifies how careful, symmetry-guided analysis can uncover subtleties missed by more conventional approaches. Related considerations and results are presented in, e.g., refs. [15–17] and references therein.

In a complementary vein, Mannque Rho's investigation titled *Multifarious Roles of Hidden Chiral-Scale Symmetry* [18] has made substantial contributions to our understanding of how the axial current coupling constant impacts nuclear processes. Rho demonstrates that the long-standing puzzle of the "quenched" g_A in nuclear superallowed Gamow–Teller transitions actually encodes the emergence of chiral-scale symmetry hidden in QCD. This insight connects phenomena across vastly different energy scales—from neutrinoless double-beta decays to the equation of state in massive compact stars—illustrating the unifying power of symmetry principles. More recent work on "quenched" g_A can be found in [19–21].

The article by David Blaschke, Denis Devyatyarov, and Olaf Kaczmarek, titled *Quark Cluster Expansion Model for Interpreting Finite-T Lattice QCD Thermodynamics* [22], provides a unified approach to understanding the thermodynamics of hadron–quark–gluon matter. Their generalized Beth-Uhlenbeck approach, with its ansatz for hadronic phase shifts, reveals how the transition from a hadron resonance gas to a quark–gluon plasma occurs in a narrow temperature window of 150–185 MeV, with the Mott dissociation of hadrons triggered by the restoration of chiral symmetry. Their work exemplifies how symmetry considerations can illuminate phase transitions in strongly interacting matter. This approach has triggered follow-up research, as reported in [23], where the ansatz for the hadronic phase shifts has been qualitatively improved and the composition of the hadron–quark–gluon matter has been evaluated, and in [24], where the Beth-Uhlenbeck approach has been extended to the entire QCD phase diagram and the location of the critical end point has been discussed, while also admitting its absence.

Gergely Fejős' contribution, titled *Perturbative RG Analysis of the Condensate Dependence of the Axial Anomaly* [25], employs renormalization group techniques to investigate the thermal behavior of 't Hooft's determinant coupling in the three-flavor linear sigma model. To this end, the standard 't Hooft term is extended by a dimension-5 anomaly operator, resulting in a condensate-dependent effective anomaly coupling. His finding—that mesonic fluctuations enhance the anomaly strength as the chiral condensate decreases at high temperatures—offers critical insight into the intricate relationship between chiral symmetry breaking and the $U_A(1)$ anomaly. Applying formalism to the nuclear liquid–gas transition reveals that the partial restoration of chiral symmetry at the transition point leads to an approximately 20% increase in the effective $U_A(1)$ coupling. This contrasts with many other predictions of a weakening axial anomaly at high temperatures (for example, [26–30], including those based on lattice QCD, e.g., [31,32], and references therein). Hence, this underscores the importance of mesonic fluctuations and highlights the need for further theoretical and numerical investigations into the nonperturbative dynamics of $U_A(1)$ symmetry. Related considerations and results are presented in [33–36].

Especially striking in its bridging of theoretical domains and synthesis of seemingly disparate fields is the mini-review by Ma and Rho, titled *Dichotomy of Baryons as Quantum Hall Droplets and Skyrmions in Compact-Star Matter* [37]. Their exploration of the possible domain-wall structure of compressed baryonic matter in massive compact stars brings together concepts from condensed matter physics, nuclear theory, and astrophysics. By incorporating hidden symmetries—flavor local symmetry and scale symmetry—into an effective nuclear field theory, they offer a fresh perspective on the structure of nuclear matter across density regimes. Their framework suggests a hadron–quark duality in baryonic matter that could revolutionize our approach to nuclear dynamics, from ordinary nuclear matter to the extreme densities found in compact stars. Some examples of papers outlining the broader context of bridging these different fields, and/or presenting recent related results, are [38–46] and references therein. In particular, [21] points out the crucial link between "quenched" g_A, as explained in Ref. [18], and the pseudo-conformal sound velocity in dense compact stars.

The comprehensive pedagogical review *Renormalization Theory and Chiral Gauge Theories in Dimensional Regularization with Non-Anticommuting* γ_5 [47] by Dominik Stöckinger, Amon Ilakovac and collaborators provides an invaluable resource for researchers working with chiral gauge theories, which are the theoretical foundation of our understanding of fundamental interactions of elementary particles. Their detailed exposition of how to handle the spurious breaking of gauge invariance and determine symmetry-restoring counterterms offers both theoretical foundations and practical guidance. The paper's methodical presentation of the BPHZ renormalization, Slavnov–Taylor identities, and BRST formalism makes complex techniques accessible to a broader audience. In particular, it contains the first published pedagogical exposition of the Breitenlohner–Maison–'t Hooft–Veltman (BMHV) scheme of dimensional regularization applied to the renormalization of chiral gauge theories. For related research and results, see [48–51] and references therein.

Summary

Collectively, the contributions to this Special Issue underscore the central role that chiral symmetry and its breaking play across diverse domains of physics—from the substructure of hadrons to the interior of dense astrophysical objects. These papers illustrate the richness of the theoretical landscape and highlight the evolving interplay between analytical methods, numerical simulations, and experimental data. As new tools and perspectives continue to emerge—especially in areas such as lattice QCD, functional renormalization group methods, and holographic duality—the study of chiral symmetry remains a fertile and dynamic field of research. Its unifying nature not only connects very different energy scales but also bridges conceptual frameworks, from quantum field theory to condensed matter analogies. Pedagogical expositions of complex topics are especially valuable for young researchers entering the field. This Special Issue includes such contributions, with the final article in particular offering a clear and comprehensive review of foundational methods in chiral gauge theories. It is our hope that this Special Issue will serve as a resource for both active researchers and those new to the field.

Acknowledgments: The web page for this Special Issue, with links to all included articles, is available at the following address: https://www.mdpi.com/journal/symmetry/special_issues/chiral_symmetry_physics (accessed on 16 July 2025). I sincerely thank all the authors for their valuable contributions and the effort they invested in preparing their work. I also gratefully acknowledge the efficient and professional support of the Symmetry Editorial Team in bringing this issue to fruition.

Conflicts of Interest: The author declares no conflicts of interest.

References

1. Navas, S.; Amsler, C.; Gutsche, T.; Hanhart, C.; Hernández-Rey, J.J.; Lourenço, C.; Masoni, A.; Mikhasenko, M.; Mitchell, R.E.; Patrignani, C.; et al. (Particle Data Group): The Review of Particle Physics 2024. *Phys. Rev. D* **2024**, *110*, 030001. Available online: https://pdg.lbl.gov/ (accessed on 11 October 2024) [CrossRef]
2. Roberts, C.D. Empirical Consequences of Emergent Mass. *Symmetry* **2020**, *12*, 1468. [CrossRef]
3. Available online: https://www.mdpi.com/journal/symmetry/awards/1306 (accessed on 16 July 2025).

4. Roberts, C.D.; Richards, D.G.; Horn, T.; Chang, L. Insights into the emergence of mass from studies of pion and kaon structure *Prog. Part. Nucl. Phys.* **2021**, *120*, 103883. [CrossRef]
5. Ding, M.; Roberts, C.D.; Schmidt, S.M. Emergence of Hadron Mass and Structure. *Particles* **2023**, *6*, 57–120. [CrossRef]
6. Carman, D.S.; Gothe, R.W.; Mokeev, V.I.; Roberts, C.D. Nucleon Resonance Electroexcitation Amplitudes and Emergent Hadron Mass. *Particles* **2023**, *6*, 416–439. [CrossRef]
7. Raya, K.; Bashir, A.; Binosi, D.; Roberts, C.D.; Rodríguez-Quintero, J. Pseudoscalar Mesons and Emergent Mass. *Few Body Syst* **2024**, *65*, 60. [CrossRef]
8. Meißner, U.G. Two-pole structures in QCD: Facts, not fantasy! *Symmetry* **2020**, *12*, 981. [CrossRef]
9. Clymton, S.; Kim, H.C. Two-pole structure of the b1(1235) axial-vector meson. *Phys. Rev. D* **2023**, *108*, 074021. [CrossRef]
10. Clymton, S.; Kim, H.C. Two-pole structure of the h1(1415) axial-vector meson: Resolving the mass discrepancy. *Phys. Rev. D* **2024**, *110*, 114002. [CrossRef]
11. Clymton, S.; Kim, H.C.; Mart, T. Production mechanism of hidden-charm pentaquark states $P_{c\bar{c}s}$ with strangeness $S = -1$. *arXiv* **2025**, arXiv:2504.07693. [CrossRef]
12. Ablikim, M.; Achasov, M.N.; Adlarson, P.; Ai, X.C.; Aliberti, R.; Amoroso, A.; An, M.R.; An, Q.; Bai, Y.; Bakina, O.; et al. Observation of the $Y(4220)$ and $Y(4360)$ in the process $e^+e^- \to \eta J/\psi$. *Phys. Rev. D* **2020**, *102*, 031101. [CrossRef]
13. Molina, R.; Liang, W.H.; Xiao, C.W.; Sun, Z.F.; Oset, E. One or two poles for the $\Xi(1820)$? *PoS* **2025**, *QNP2024*, 034. [CrossRef]
14. Wang, J.Z.; Lin, Z.Y.; Wang, B.; Meng, L.; Zhu, S.L. Double pole structures of $X_1(2900)$ as the P-wave $\bar{D}^*K^*$ resonances. *Phys. Rev. D* **2024**, *110*, 114003. [CrossRef]
15. Zhou, Z.Y.; Xiao, Z. Two-pole structures in a relativistic Friedrichs–Lee-QPC scheme. *Eur. Phys. J. C* **2021**, *81*, 551. [CrossRef]
16. Xie, J.M.; Lu, J.X.; Geng, L.S.; Zou, B.S. Two-pole structures as a universal phenomenon dictated by coupled-channel chiral dynamics. *Phys. Rev. D* **2023**, *108*, L111502. [CrossRef]
17. Meißner, U.G. Chiral dynamics: Quo vadis? *arXiv* **2024**, arXiv:2501.03014.
18. Rho, M. Multifarious Roles of Hidden Chiral-Scale Symmetry: "Quenching" g_A in Nuclei. *Symmetry* **2021**, *13*, 1388. [CrossRef]
19. Rho, M.; Shao, L.Q. The Quenched g_A in Nuclei and Infrared Fixed Point in QCD. *Symmetry* **2024**, *16*, 1704. [CrossRef]
20. Rho, M. The quenched g_A puzzle in nuclei and nuclear matter and "pseudo-conformality" in QCD. *Mod. Phys. Lett. A* **2025**, *40*, 2530004. [CrossRef]
21. Rho, M. Connecting the Quenched g_A in Nuclear Matter To Dense Compact-Star Matter. *arXiv* **2025**, arXiv:2507.04939. [CrossRef]
22. Blaschke, D.; Devyatyarov, K.A.; Kaczmarek, O. Quark cluster expansion model for interpreting finite-T lattice QCD thermodynamics. *Symmetry* **2021**, *13*, 514. [CrossRef]
23. Blaschke, D.; Cierniak, M.; Ivanytskyi, O.; Röpke, G. Thermodynamics of quark matter with multiquark clusters in an effective Beth-Uhlenbeck type approach. *Eur. Phys. J. A* **2024**, *60*, 14. [CrossRef]
24. Blaschke, D. Unified quark-hadron EoS and critical endpoint in the QCD phase diagram. *Acta Phys. Polon. Suppl.* **2021**, *14*, 425–433. [CrossRef]
25. Fejős, G. Perturbative RG analysis of the condensate dependence of the axial anomaly in the three flavor linear sigma model. *Symmetry* **2021**, *13*, 488. [CrossRef]
26. Mitter, M.; Schaefer, B.J. Fluctuations and the axial anomaly with three quark flavors. *Phys. Rev. D* **2014**, *89*, 054027. [CrossRef]
27. Horvatić, D.; Kekez, D.; Klabučar, D. η' and η mesons at high T when the $U_A(1)$ and chiral symmetry breaking are tied. *Phys. Rev. D* **2019**, *99*, 014007. [CrossRef]
28. Horvatić, D.; Kekez, D.; Klabučar, D. Temperature Dependence of the Axion Mass in a Scenario Where the Restoration of Chiral Symmetry Drives the Restoration of the $U_A(1)$ Symmetry. *Universe* **2019**, *5*, 208. [CrossRef]
29. Li, X.; Fu, W.J.; Liu, Y.X. New insight about the effective restoration of $U_A(1)$ symmetry. *Phys. Rev. D* **2020**, *101*, 054034. [CrossRef]
30. Pisarski, R.D.; Rennecke, F. Conjectures about the Chiral Phase Transition in QCD from Anomalous Multi-Instanton Interactions. *Phys. Rev. Lett.* **2024**, *132*, 251903. [CrossRef]
31. Tomiya, A.; Cossu, G.; Aoki, S.; Fukaya, H.; Hashimoto, S.; Kaneko, T.; Noaki, J. Evidence of effective axial U(1) symmetry restoration at high temperature QCD. *Phys. Rev. D* **2017**, *96*, 034509. [CrossRef]
32. Aoki, S.; Aoki, Y.; Cossu, G.; Fukaya, H.; Hashimoto, S.; Kaneko, T.; Rohrhofer, C.; Suzuki, K.; JLQCD collaboration. Study of the axial $U(1)$ anomaly at high temperature with lattice chiral fermions. *Phys. Rev. D* **2021**, *103*, 074506. [CrossRef]
33. Fejös, G.; Patkos, A. Backreaction of mesonic fluctuations on the axial anomaly at finite temperature. *Phys. Rev. D* **2022**, *105*, 096007. [CrossRef]
34. Azcoiti, V. Spectral density of the Dirac-Ginsparg-Wilson operator, chiral U(1)A anomaly, and analyticity in the high temperature phase of QCD. *Phys. Rev. D* **2023**, *107*, 11. [CrossRef]
35. Fejos, G.; Patkos, A. Thermal behavior of effective $U_A(1)$ anomaly couplings in reflection of higher topological sectors. *Phys. Rev. D* **2024**, *109*, 036035. [CrossRef]
36. Giordano, M. Constraints on the Dirac spectrum from chiral symmetry restoration. *Phys. Rev. D* **2024**, *110*, L091504. [CrossRef]

37. Ma, Y.L.; Rho, M. Dichotomy of Baryons as Quantum Hall Droplets and Skyrmions: Topological Structure of Dense Matter. *Symmetry* **2021**, *13*, 1888. [CrossRef]
38. Senthil, T.; Vishwanath, A.; Balents, L.; Sachdev, S.; Fisher, M.P.A. Deconfined Quantum Critical Points. *Science* **2004**, *303*, 1490–1494. [CrossRef]
39. Rho, M. Probing Fractional Quantum Hall Sheets in Dense Baryonic Matter. *arXiv* **2022**, arXiv:2211.14890. [CrossRef]
40. Lin, F.; Ma, Y.L.
Baryons as vortexes on the η' domain wall. *J. High Energy Phys.* **2024**, *5*, 270. [CrossRef]
41. Zhang, L.Q.; Ma, Y.; Ma, Y.L. Peak of sound velocity, scale symmetry and nuclear force in baryonic matter. *arXiv* **2024**, arXiv:2410.04142. [CrossRef]
42. Zhang, L.Q.; Ma, Y.; Ma, Y.L. Nuclear matter properties from chiral-scale effective theory including a dilatonic scalar meson. *arXiv* **2024**, arXiv:2412.19023. [CrossRef]
43. Lin, F.; Ma, Y.L. Confined monopoles in a chiral bag. *Phys. Rev. D* **2025**, *112*, 014009. [CrossRef]
44. Lin, F.; Ma, Y.L. Baryon Construction with η' Meson Field. *Symmetry* **2025**, *17*, 477. [CrossRef]
45. Cheng, M.; Musser, S.; Raz, A.; Seiberg, N.; Senthil, T. Ordering the topological order in the fractional quantum Hall effect. *arXiv* **2025**, arXiv:2505.14767. [CrossRef]
46. Sheng, B.K.; Ma, Y.L. Connecting dilaton thermal fluctuation with the Polyakov loop at finite temperature. *arXiv* **2025**, arXiv:2506.13549. [CrossRef]
47. Bélusca-Maïto, H.; Ilakovac, A.; Kühler, P.; Mađor-Božinović, M.; Stöckinger, D.; Weißwange, M. Introduction to Renormalization Theory and Chiral Gauge Theories in Dimensional Regularization with Non-Anticommuting γ_5. *Symmetry* **2023**, *15*, 622. [CrossRef]
48. Bélusca-Maïto, H.; Ilakovac, A.; Kühler, P.; Mađor-Božinović, M.; Stöckinger, D. Two-loop application of the Breitenlohner-Maison/'t Hooft-Veltman scheme with non-anticommuting γ_5: Full renormalization and symmetry-restoring counterterms in an abelian chiral gauge theory. *J. High Energy Phys.* **2021**, *11*, 159. [CrossRef]
49. Kühler, P.; Stöckinger, D.; Weißwange, M. Advances at the γ_5-Frontier. *PoS* **2024**, *LL2024*, 022. [CrossRef]
50. Ebert, P.L.; Kühler, P.; Stöckinger, D.; Weißwange, M. Shedding light on evanescent shadows — Exploration of non-anticommuting γ_5 in Dimensional Regularisation. *J. High Energy Phys.* **2025**, *1*, 114. [CrossRef]
51. Kühler, P.; Stöckinger, D. Two-Loop Renormalization of a Chiral $SU(2)$ Gauge Theory in Dimensional Regularization with Non-Anticommuting γ_5. *arXiv* **2025**, arXiv:2504.06080. [CrossRef]

Disclaimer/Publisher's Note: The statements, opinions and data contained in all publications are solely those of the individual author(s) and contributor(s) and not of MDPI and/or the editor(s). MDPI and/or the editor(s) disclaim responsibility for any injury to people or property resulting from any ideas, methods, instructions or products referred to in the content.

symmetry

MDPI

Article

Empirical Consequences of Emergent Mass

Craig D. Roberts [1,2]

[1] School of Physics, Nanjing University, Nanjing 210093, Jiangsu, China; cdroberts@nju.edu.cn
[2] Institute for Nonperturbative Physics, Nanjing University, Nanjing 210093, Jiangsu, China

Received: 28 July 2020; Accepted: 28 August 2020; Published: 7 September 2020

Abstract: The Lagrangian that defines quantum chromodynamics (QCD), the strong interaction piece of the Standard Model, appears very simple. Nevertheless, it is responsible for an astonishing array of high-level phenomena with enormous apparent complexity, e.g., the existence, number and structure of atomic nuclei. The source of all these things can be traced to emergent mass, which might itself be QCD's self-stabilising mechanism. A background to this perspective is provided, presenting, inter alia, a discussion of the gluon mass and QCD's process-independent effective charge and highlighting an array of observable expressions of emergent mass, ranging from its manifestations in pion parton distributions to those in nucleon electromagnetic form factors.

Keywords: confinement of gluons and quarks; dynamical chiral symmetry breaking; Dyson-Schwinger equations; emergence of hadronic mass; hadron elastic form factors; hadron spectroscopy and structure; Higgs mechanism; parton distribution amplitudes and functions; strong (non-perturbative) QCD

1. Introduction

One might define *emergent phenomena* as those features of nature which do not readily admit an explanation solely in terms of known or conjectured mathematical rules. The concept is at least as old as Aristotle (384–322 BC), who argued that a compound item can have (emergent) properties in the whole which are not explicable merely through the independent actions of the item's constituent parts. His view is often represented by the statement "The whole is more than the sum of its parts." In this sense, *emergence* has its origins in the Greek "sunergos": "together" plus "working", which is the origin of our current concept of synergy, viz. things working *together* more effectively than could be anticipated from their independent actions in isolation. (Etymologically, the word "emergence" entered English in the mid 17th century, meaning "unforeseen occurrence" and derived from the medieval Latin "emergentia", itself from the Latin "emergere", meaning "bring to light" (Source: Oxford English Dictionary). Herein, "emergence" is seen as a larger notion than is typically expected from this literal connection.).

This perspective is typically contrasted with that described as *reductionism*; namely, the view that everything in nature can ultimately be viewed as no more complex in principle than, e.g., a (very good) watch, which is clearly a complex object; but, equally clearly, not more than the sum of its parts.

In developing such a contrast through debate, hard lines are sometimes drawn, with individuals deciding or being forced to choose one side or the other. This is taking the argument too far, however. During each epoch in history, there has always been a line dividing physics from metaphysics; but the location of that border is neither fixed nor impermeable. As time and humanity have progressed, more aspects of nature have seeped into the pool of physics. Notwithstanding such progress, it does not follow that fewer questions have been left beyond physics. Typically, as mathematics has succeeded in explaining more phenomena, new discoveries have been made, often emerging from attempts to test the newly formulated theories. (One may think here of the discovery of the neutron and then the plethora of other so-called elementary particles, to which the introduction of order demanded the development of quantum field theory.) Therefore, the tension remains. If nature is bounded, then this might change at some future time; but today it would be gross hubris to maintain such a position.

The question "Can all objects that have emerged in nature be explained by a finite collection of rules?" is at least as old as human thought, and we cannot know the answer until all that nature can produce has been discovered. Plainly, the debate must continue, and, perhaps, its greatest merit is the spur that each side provides the other as we seek to understand our place in *the scheme of things*.

2. Strong Interactions in the Standard Model

2.1. Natural Mass Scale

A significant part of the ongoing debate centres on the character of mass and its consequences in the Standard Model of Particle Physics (SM), especially as it emerges from the strong interaction sector; namely, quantum chromodynamics (QCD). To introduce this problem, it is worth recalling the Nobel Prize acceptance speech given by H. D. Politzer; in particular, the following remarks [1].

> *The establishment by the mid-1970's of QCD as the correct theory of the strong interactions completed what is now known prosaically as the Standard Model. It offers a description of all known fundamental physics except for gravity, and gravity is something that has no discernible effect when particles are studied a few at a time. However, the situation is a bit like the way that the Navier-Stokes equation accounts for the flow of water. The equations are at some level obviously correct, but there are only a few, limited circumstances in which their consequences can be worked out in any detail. Nevertheless, many leading physicists were inclined to conclude in the late 1970's that the task of basic physics was nearly complete, and we'd soon be out of jobs. A famous example was the inaugural lecture of Stephen Hawking as Lucasian Professor of Mathematics, a chair first held by Isaac Barrow at Cambridge University. Hawking titled his lecture, 'Is the End in Sight for Theoretical Physics?' And he argued strongly for 'Yes'.*

Concerning the character of mass, many might believe that the answer was found in 2012 with the discovery of the Higgs boson [2,3] and the subsequent Nobel Prize awarded in equal share to F. Englert and P. Higgs [4,5], with the citation "for the theoretical discovery of a mechanism that contributes to our understanding of the origin of mass of subatomic particles ... ". Nevertheless, while discovery of the Higgs was a watershed, it should be placed in context; something achieved nicely and informally on The Guardian's live blog: theguardian.com/science/2011/dec/13/higgs-boson-seminar-god-particle

> *The Higgs field is often said to give mass to everything. That is wrong. The Higgs field only gives mass to some very simple particles. The field accounts for only one or two percent of the mass of more complex things, like atoms, molecules, and everyday objects, from your mobile phone to your pet llama. The vast majority of mass comes from the energy needed to hold quarks together inside atoms.*

These remarks implicitly highlight QCD, the quantum field theory formulated in four spacetime dimensions which defines what is arguably the SM's most important chapter. QCD is supposed to describe all of nuclear physics through the interactions between quarks (matter fields) that are mediated by gluons (gauge bosons). Yet, fifty years after the discovery of quarks [6–8], science is only just beginning to grasp how QCD moulds the basic bricks for nuclei: pions, neutrons, protons, etc., and it is far from understanding how QCD produces nuclei.

The natural scale for nuclear physics (strong interactions) is characterised by the proton mass:

$$m_p \approx 1\,\text{GeV} \approx 2000\, m_e\,, \tag{1}$$

where m_e is the electron mass, i.e., $m_p = 1.783 \times 10^{-27}$ kg. In the SM, m_e is correctly attributed to the Higgs boson, but what is the cause of the prodigious enhancement required to produce m_p? Followed logically to its source, this question leads to an appreciation that our Universe's existence depends critically on, inter alia, the following empirical facts. (*i*) The proton mass is large, i.e., the mass-scale for QCD is very much larger than that of electromagnetism; (*ii*) the proton does not decay, despite being a compound state built from three valence-quarks; and (*iii*) the pion, which carries long-range interactions between nucleons (neutrons and protons), is abnormally light (not massless), having a lepton-like mass even though it is a strongly interacting object built from a valence-quark and valence antiquark. (The μ-lepton, discovered in 1936 [9], was initially mistaken for the pion. The pion was only found a decade later [10].) These qualities of Nature transport us to a reductionist definition of emergence. Namely, assuming it does describe strong interactions, then the one-line Lagrangian of QCD—a very simple low-level rule—must somehow produce high-level phenomena with enormous apparent complexity.

At this point, it is worth studying the Lagrangian of chromodynamics, which appeared as the culmination of a distillation process applied to a large array of distinct ideas and discoveries [11,12]:

$$\mathcal{L}_{\rm QCD} = \sum_{j=u,d,s,\ldots} \bar{q}_j[\gamma_\mu D_\mu + m_j]q_j + \tfrac{1}{4}G^a_{\mu\nu}G^{a\mu\nu}, \tag{2a}$$

$$D_\mu = \partial_\mu + ig\tfrac{1}{2}\lambda^a A^a_\mu, \quad G^a_{\mu\nu} = \partial_\mu A^a_\nu + \partial_\nu A^a_\mu - \underline{g f^{abc} A^b_\mu A^c_\nu}. \tag{2b}$$

Here, $\{q_j\}$ are the quark fields, with j their flavour label and m_j their Higgs-generated current-quark masses, and $\{A^a_\mu, a = 1,\ldots,8\}$ are the gluon fields, with $\{\frac{1}{2}\lambda^a\}$ the generators of the SU(3) (chromo/colour) gauge-group in the fundamental representation. Comparing with quantum electrodynamics (QED), the solitary difference is the piece describing gluon self-interactions, marked as the underlined term in Equation (2b). Somehow, the origin, mass and extent of almost all visible matter in the Universe is attributable to $\mathcal{L}_{\rm QCD}$—one line plus two definitions. That being true, then ...

QCD is quite possibly the most remarkable fundamental theory ever invented.

The only apparent energy scales in Equation (2) are the Higgs-generated current-quark masses, but focusing on the u (up) and d (down) quarks that define nucleons, this scale is more-than one-hundred-times smaller than m_p [13]. No amount of "staring" at $\mathcal{L}_{\rm QCD}$ can reveal the source of that enormous amount of "missing mass"; yet, it must be there. (This is a stark contrast to QED wherein, e.g., the scale in the spectrum of the hydrogen atom is set by m_e, which is a prominent feature of $\mathcal{L}_{\rm QED}$ that is generated by the Higgs boson.).

Models and effective field theories (EFTs) for nuclear physics typically assume existence of the $m_p \approx 1\,$GeV mass-scale and build upon it. They also assume the reality of effectively pointlike nuclear constituents (nucleons) and force carriers (pions and, perhaps, other meson-like entities). Their task is not to elucidate the internal structure of such objects. Instead, they aim to develop systematically improvable techniques that can describe the number and properties of atomic nuclei. This may be seen as reductionism built on an emergent plateau. The basic reductionist question generated by this approach is "Can the plateau upon which the nuclear model/EFT paradigm is built be constructed from QCD?" If the answer is "yes", then all parameters used and fitted in such theories of nuclear structure will (some day) be confronted with ab initio predictions in a profound test of the SM.

Today there is a good case to be made for an affirmative answer to this question. Amongst the numerous supporting examples, one may list QCD-connected computations of the hadron spectrum. Depicted in Figure 1, one sees that two disparate approaches to solving QCD [14,15] produce a spectrum of ground-state hadrons in good agreement with experiment.

Importantly, however, neither calculation presented in Figure 1 was able to predict the size of the proton mass. Apart from the Higgs-generated current-quark masses, each has an undetermined mass-scale parameter, denoted hereafter as $\Lambda_{\rm QCD}$. Its value is chosen to fit one experimental number, e.g., the pion's leptonic decay constant, but after $\Lambda_{\rm QCD}$ is fixed, all other results are predictions.

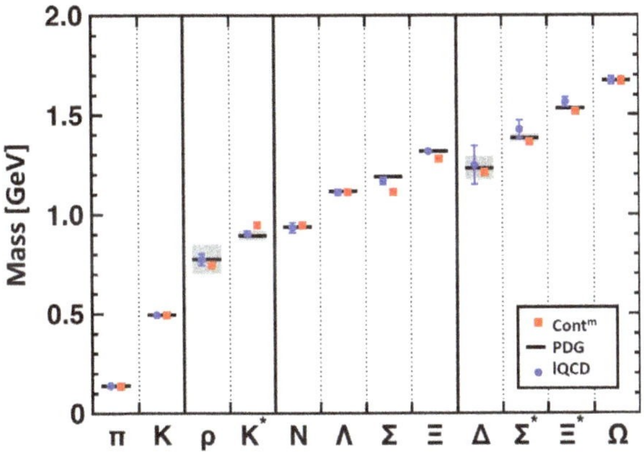

Figure 1. Masses of pseudoscalar and vector mesons, and ground-state positive-parity octet and decuplet baryons calculated using continuum (Cont^{m}—squares, red) [14] and lattice (lQCD—circles, blue) [15] methods in QCD compared with experiment [13] (PDG—black bars, with decay-widths of unstable states shaded in grey). The continuum study did not include isospin symmetry breaking effects, which are evidently small, as highlighted by the empirically determined Σ-Λ mass difference (<7%).

2.2. *Whence Mass?*

Herein, it is worth explaining the need for such a scale-setting procedure. Regarded classically, chromodynamics is a local non-Abelian gauge field theory, and defined in four spacetime dimensions, there is no mass-scale if Lagrangian masses for the fermions are omitted. (The absence of such masses defines the chiral limit.) Scale invariant theories do not support dynamics, only kinematics. Therefore, bound states are impossible; so, the Universe cannot exist.

Spontaneous symmetry breaking, as generated by the Higgs mechanism, does not resolve this issue. The masses of the neutron and proton, which lie at the heart of all visible matter, are two orders-of-magnitude greater than the Higgs-generated current-masses of the u- and d-quarks, which are the defining constituents of protons and neutrons.

Consequently, questions like "How does a mass-scale appear?" and "Why does this scale have its observed value?" are consanguineous with the question "How did the Universe form?"

Modern quantum field theories are built upon Poincaré invariance. In this connection, consider the energy-momentum tensor in classical chromodynamics, $T_{\mu\nu}$. In field theory, conservation of energy and momentum is a consequence of spacetime translational invariance, one of the family of Poincaré transformations. Consequently,

$$\partial_\mu T_{\mu\nu} = 0\,. \tag{3}$$

Consider now a global scale transformation in the Lagrangian of classical chromodynamics:

$$x \to x' = \mathrm{e}^{-\sigma} x\,, \quad A^a_\mu(x) \to A^{a\prime}_\mu(x') = \mathrm{e}^{-\sigma} A^a_\mu(\mathrm{e}^{-\sigma} x)\,, \quad q(x) \to q'(x') = \mathrm{e}^{-(3/2)\sigma} q(\mathrm{e}^{-\sigma} x)\,, \tag{4}$$

where σ is spacetime-independent. The associated Noether current is

$$\mathcal{D}_\mu = T_{\mu\nu} x_\nu\,, \tag{5}$$

viz. the dilation current. In the absence of fermion masses, the classical action is invariant under Equation (4), i.e., the theory is scale invariant; therefore,

$$\partial_\mu \mathcal{D}_\mu = 0 = [\partial_\mu T_{\mu\nu}] x_\nu + T_{\mu\nu}\delta_{\mu\nu} = T_{\mu\mu}\,, \tag{6}$$

where the last equality follows from Equation (3). Evidently, the energy-momentum tensor must be traceless in a scale invariant theory.

Classical chromodynamics is not a meaningful physical framework for many reasons; amongst them the fact, illustrated by Figure 1, that strong interactions are empirically known to be characterised by a large mass-scale, $m_p \approx 1\,\text{GeV}$. In quantising the theory, a mass-scale is introduced by regularisation and renormalisation of ultraviolet divergences. This is "dimensional transmutation": all quantities, including the field operators themselves, become dependent on a mass-scale. This entails the violation of Equation (6), i.e., the appearance of the chiral-limit "trace anomaly" [16]:

$$T_{\mu\mu} = \beta(\alpha(\zeta))\tfrac{1}{4}G^a_{\mu\nu}G^a_{\mu\nu} =: \Theta_0\,, \tag{7}$$

where $\beta(\alpha(\zeta))$ is QCD's β-function, $\alpha(\zeta)$ is the associated running-coupling, and ζ is the renormalisation scale. Equation (7) indicates that a mass-scale linked to the resolving power of a given measurement is engendered via quantisation; to wit, the scale *emerges* as an indispensable part of the theory's quantum definition.

This mass is exhibited in the gauge-boson vacuum polarization. In QED, the photon vacuum polarization does not possess an infrared mass-scale, and dimensional transmutation acts simply to generate the gentle evolution of the QED coupling, i.e., any dynamical breaking of QED's conformal features is small; therefore, the associated trace anomaly is normally negligible. In contrast, gauge sector dynamics drives a Schwinger mechanism in QCD [17–34], so that the QCD trace anomaly expresses a mass-scale which is, empirically, very significant. This is discussed in Section 4.2.

There is another aspect of chromodynamics that should be highlighted, namely, the classical Lagrangian still defines a non-Abelian local gauge theory. Accordingly, the concept of local gauge invariance persists; but without a mass-scale, there is no notion of confinement. For instance, one can compose a colour-singlet combination of three quarks and colour rotations will preserve its colour neutrality; but the participating quarks need not be close together. In fact, it is meaningless to discuss proximity because, in a scale invariant theory, all lengths are equivalent. Accordingly, the question of "Whence mass?" is indistinguishable from "Whence a mass-scale?", is indistinguishable from "Whence a confinement scale?".

Evidently, one does not learn much from knowing that a trace anomaly exists. It only means there is a mass-scale. The central concern is whether the magnitude of that scale can be computed and understood.

The magnitude of the scale anomaly can definitely be measured, and this simply by considering the in-proton expectation value of the energy-momentum tensor (see, e.g., in [35]):

$$\langle p(P)|T_{\mu\nu}|p(P)\rangle = -P_\mu P_\nu\,, \tag{8}$$

where the equations-of-motion for a one-particle proton state were used to obtain the right-hand-side. Now, in the chiral limit

$$\langle p(P)|T_{\mu\mu}|p(P)\rangle = -P^2 = m_p^2 = \langle p(P)|\Theta_0|p(P)\rangle\,. \tag{9}$$

Therefore, there is a sound position from which one may conclude that gluons generate all the proton's mass: the measured value of the trace anomaly is large; and, logically, that feature owes to gluon self-interactions, the agent behind asymptotic freedom.

This is a valid deduction because, ultimately, what else could account for a mass-scale in QCD? Gluon self-interactions are QCD's definitive feature, and it is these interactions that might enable a rigorous (non-perturbative) definition of the matrix element in Equation (9). Nevertheless, it is only reasonable to conclude this when the operator and wave function are computed at a resolving scale $\zeta \gg m_p$, i.e., when one employs a parton-model basis [36].

There is another issue, too, which can be exposed by returning to Equation (8) and replacing the proton by the pion

$$\langle \pi(q) | T_{\mu\nu} | \pi(q) \rangle = -q_\mu q_\nu \quad \Rightarrow \quad \langle \pi(q) | \Theta_0 | \pi(q) \rangle = m_\pi^2 \overset{\text{chiral limit}}{=} 0 \tag{10}$$

because the chiral-limit pion is a massless Nambu–Goldstone (NG) mode [37,38]. It is possible that Equation (10) means the scale anomaly is trivially zero in the pion, i.e., gluon self-interactions have no effect in the pion because each term required to express the operator vanishes separately. Yet, such a conclusion would sit uncomfortably with known QCD dynamics, which expresses both attraction and repulsion, but never passive inactivity. More likely, then, the final identity in Equation (10) results from cancellations between different terms in the complete operator matrix element. Naturally, such precise cancellation could not be accidental. It would require that some symmetry is broken in a very particular manner. (The mechanism is explained in Section 4.5.)

Equations (9) and (10) present a quandary. They stress that any understanding of the proton's mass is incomplete unless it simultaneously explains Equation (10). Moreover, any discussion of confinement, fundamental to the proton's stability, is unreasonable before this conundrum is resolved. As will become clear, at least some of these features of Nature have a reductionist explanation grounded in the dynamics responsible for the emergence of m_p as the natural mass-scale for nuclear physics; and one of the most important goals in modern science is to explain and elucidate the entire array of empirical consequences of this dynamics [39–43].

3. Confinement

It is a textbook result, with its origin in the Nobel Prize for the discovery of asymptotic freedom [1,44,45], that QCD is characterised by an interaction which becomes stronger as the participants try to separate. Remaining within those results that can be established using perturbation theory, one is led to contemplate some unusual possibilities: if the coupling strength rises rapidly with separation, then perhaps it is not bounded, and perhaps an infinite amount of energy is required to remove a gluon or quark from the interior of a hadron? Such thinking has produced

> *The Confinement Hypothesis*: Colour-charged particles cannot be isolated and therefore cannot be directly observed. They clump together in colour-neutral bound-states.

Confinement seems to be an empirical reality, but a credited mathematical proof is lacking. Partly as a result, the Clay Mathematics Institute proffered a "Millennium Problem" prize of $1 million for a proof that pure-glue QCD is mathematically well defined [46]. One necessary part of such a proof would be to establish whether the confinement postulate is correct in pure-gauge QCD i.e., the theory obtained from Equation (2) after omitting all terms containing quark fields.

There is a pitfall here, however: no reader of this material can be described within pure-glue QCD. Light quarks are essential to understanding all known visible matter. Thus, a proof of confinement in pure-glue QCD is chiefly irrelevant to our Universe. Life exists because nature has supplied two light quark flavours, combinations of which form the pion, and the pion is unnaturally light, thus very easily produced and capable of propagating over nuclear-size distances. Therefore, as noted previously by others [47,48], no explanation of SM confinement is empirically relevant unless it also describes the link between confinement and the emergence of mass, and so the existence and role of pions, i.e., pseudo-NG modes with $m_\pi \ll m_p$.

One piece of the Yang–Mills millennium problem [46] is to prove that a mass-gap, $\Delta > 0$, exists in pure-glue QCD. This conjecture is supported by some strong evidence, e.g., numerical studies of lattice-regularised QCD (lQCD) find $\Delta \gtrsim 1.5\,\text{GeV}$ [49]. This sharpens the conundrum described above: with $\Delta^2/m_\pi^2 \gtrsim 100$, can the mass-gap in pure Yang–Mills theory really play any role in understanding confinement when the emergence of mass, driven by kindred dynamics, ensures that our Universe supports almost-massless strongly-interacting excitations? Skirting the question, one can respond that any explanation of confinement must simultaneously describe its link to pion properties. From this position, pions are viewed as playing a critical role in any explanation of SM confinement, and a discussion that omits reference to pions is *practically irrelevant*.

These observations indicate that the potential between infinitely-heavy quarks computed in numerical simulations of quenched lQCD—the static potential [50], often associated with formation of an incredibly strong flux tube between the colour source and sink [51]—is detached from the question of confinement in our Universe. In fact, as light–particle annihilation and creation effects are essentially non-perturbative in QCD, it is impossible to calculate a quantum mechanical potential between two light quarks [52–54]. It follows that there is no discernible flux tube in a Universe with light quarks; so, the flux tube is not the correct paradigm for confinement.

As highlighted already, the emergence of mass is key here. It ensures the existence of pseudo-NG modes, and no flux tube linking a static colour source and sink can have an observable existence in the presence of these modes. To verify this statement, suppose that such a tube is stretched between a source and sink. The potential energy stored within the tube may only increase until it reaches the amount required to produce a particle–antiparticle pair of the system's pseudo-NG modes. Simulations of lQCD demonstrate [52,53] that the flux tube then disappears instantaneously, leaving two separated colour-singlet systems. The length-scale characterising this effect is $r_{\not{\sigma}} \simeq (1/3)$ fm; so, if any such string forms, it would disintegrate well within a hadron's interior.

An alternative realisation associates confinement with marked changes in the analytic properties of coloured propagators and vertices, driven by QCD dynamics. That leads such coloured n-point functions to violate the axiom of reflection positivity, thereby eliminating the associated excitations from the Hilbert space associated with asymptotic states [55]. This is certainly a sufficient condition for confinement [25,31,56–71].

It should be highlighted, however, that the appearance of such modifications when analysing some simplification of a given theory does not signify that the theory itself is truly confining: uncommon spectral properties can be introduced by approximations, yielding a truncated version of a theory which expresses confinement even though the complete theory does not, see, e.g., in [72,73]. Notwithstanding exceptions like these, in a veracious treatment of QCD the computed violation of reflection positivity by coloured functions does express confinement.

4. Strong QCD

4.1. Dyson–Schwinger Equations

The appearance and size of the natural scale for nuclear physics ($m_p \approx 1$ GeV) and the confinement of gluons and quarks are emergent phenomena. They are not apparent in the QCD Lagrangian, yet they determine the character of QCD's spectrum, the structure of bound states and so forth. Given Equation (2), the natural question to ask is whether one can understand these features reductively, i.e., directly in terms of the degrees-of-freedom used to formulate QCD, or does the complexity of strong interaction phenomena make prediction and explanation impractical (impossible)? For instance, is it pointless to attempt the QCD-connected prediction of any nucleon structural property on a domain that is not yet empirically accessible?

If a reductive explanation is impossible, then science must rely on a tower of EFTs, each level developed for a different energy domain, in order, e.g., to express and understand the consequences of the emergence of mass and contingent effects, such as confinement, without identifying their source.

On the other hand, if a reductive approach is possible, then non-perturbative calculational methods must be developed to define and solve QCD. Prominent amongst such techniques today are (*i*) the numerical simulation of lQCD [74–76] and (*ii*) continuum Schwinger function methods (CSMs), viz.ã collection of models and schemes, each with varying degrees of connection to Equation (2). Currently, each of these two approaches has strengths and weaknesses, so the best way forward is to combine them to the fullest extent that is reasonably possible and exploit the synergies that emerge.

Among CSMs, the Dyson–Schwinger equations (DSEs) have proven useful [63,77–86]. These quantum field theory generalisations of the Euler–Lagrange equations provide a continuum method for calculating Schwinger functions; namely, the same Euclidean-space Green functions that are computed using lQCD. Consequently, there are many opportunities for cross-fertilisation between DSE and lQCD studies, and this has been exploited to increasingly good effect during the past twenty years, especially at the level of propagators (2-point functions) and vertices (3-point functions) defined using QCD's elementary degrees-of-freedom.

The challenge to DSE analyses is found in the fact that the equation for any given n-point function is coupled with those for some higher-n-point functions, e.g., the gap equation for the quark 2-point function is coupled to those for the gluon 2-point function and the gluon-quark 3-point function. Therefore, truncations are necessary in order to define a tractable problem. Systematic, symmetry-preserving schemes have been developed [87–93], so that today DSE predictions can be distributed into three classes: (A) model-independent results in QCD; (B) illustrations of such results, using well-constrained model elements and possessing a recognisable connection to QCD; and (C) analyses that can reasonably be described as QCD-based, but whose elements have not been calculated using a truncation that maintains a systematically-improvable connection with QCD. Regarding the last two classes, comparisons between schemes and orders within schemes can be used to identify robust outcomes. Results can also be compared with lQCD predictions, when available, capitalising on the overlap domain of these two approaches, and, of course, predictions can be tested against experiment, which is, as always, the final arbiter in physics.

4.2. Gluon Mass

It is now possible and appropriate to return to *the confinement hypothesis*, introduced in the opening paragraph of Section 3 and expressed in the "Millennium Problem". Beginning with pioneering efforts roughly forty years ago [17,18], continuum and lattice studies of QCD's gauge sector have been growing in sophistication and reliability. The current state of understanding can be traced from an array of sources, see, e.g., in [17–34]. Of specific interest is the property that the gluon propagator saturates at infrared momenta; to wit,

$$\Delta(k^2 \simeq 0) = 1/m_g^2, \tag{11}$$

where $\Delta(k^2)$ is the scalar function that characterises the dressed gluon propagator. This entails that the long-range propagation characteristics of gluons are markedly altered by their self-interactions. Significantly, one may associate a renormalisation-group-invariant (RGI) gluon mass-scale with this effect: $m_0 \approx m_p/2$, and summarise a large body of work by recording that gluons, although behaving as massless entities on the perturbative domain, actually possess a running mass, with m_0 characterising its value at infrared momenta.

Asymptotic freedom guarantees that QCD's ultraviolet behaviour is tractable; but the emergence of a gluon mass reveals a new SM physics frontier because the existence of a running gluon mass, sizeable at infrared momenta, influences all analyses of the continuum bound-state problem. For instance, it could be a harbinger of gluon saturation [94,95].

Furthermore, $m_0 > 0$ entails that QCD dynamically generates its own infrared cut-off, so that gluons with wavelength $\lambda \gtrsim \sigma := 1/m_0 \approx 0.5\,\text{fm}$ decouple from the strong interaction, hinting at a dynamical realisation of confinement [25,31,56–71]. In this picture, once a gluon or quark is produced, it starts to propagate in spacetime, but following each "step" of average length σ an interaction occurs and the parton loses its identity, sharing it with others. Ultimately a parton cloud is produced, which fuses into colour-singlet final states. This physics is embodied in parton fragmentation functions (PFFs), which describe how QCD partons, (nearly) massless when produced in a high-energy event, transform into a cascade of massive hadrons. PFFs express the emergence of hadrons with mass from massless partons [96]. Such observations suggest that PFFs are the cleanest expression of dynamical confinement in QCD. This perspective can be explored at modern and anticipated facilities.

4.3. Effective Charge

Among the many other consequences of QCD's intricate non-perturbative gauge-sector dynamics is the generation of a process-independent (PI) running coupling, $\hat{\alpha}(k^2)$, see, e.g., in [32–34]. Depicted as the solid (black) curve in Figure 2, this is a new type of effective charge. It is an analogue of the Gell–Mann–Low effective coupling in QED [97] because it is completely determined by the gauge-boson vacuum polarisation, when the problem is approached using the pinch technique [17,98–100] and background field method [101]. The result in Figure 2 is a parameter-free Class-A DSE prediction, capitalising on analyses of QCD's gauge sector undertaken using continuum methods and informed by numerical simulations of lQCD.

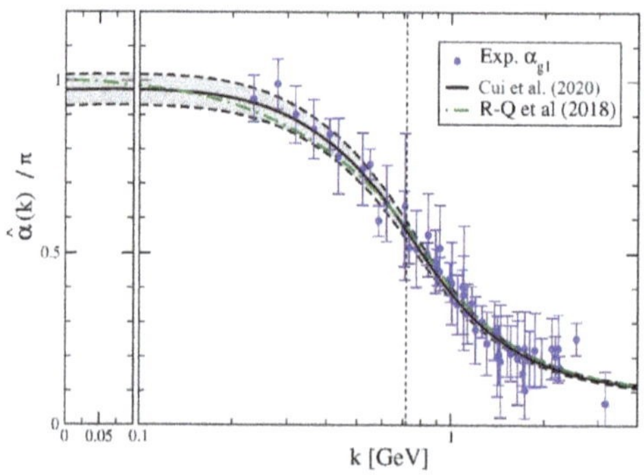

Figure 2. Solid black curve within grey band—RGI PI running-coupling, $\hat{\alpha}(k^2)/\pi$, computed in [34] (Cui et al. 2020), and dot-dashed green curve—earlier result (R-Q et al. 2018) [33]. (The grey band bordered by dashed curves indicates the uncertainty in the result arising from that in both continuum and lattice-QCD inputs and is detailed in [34].) For comparison, world data on the process-dependent charge, α_{g_1}, defined via the Bjorken sum rule, are also depicted. (The data sources are listed elsewhere [34]. For additional details, see e.g., in [102–104].) The k-axis scale is linear to the left of the vertical partition, and logarithmic otherwise. The vertical line, $k = m_G$, marks the gauge sector screening mass, Equation (12).

The data in Figure 2 represent empirical information on α_{g_1}, a process-*dependent* effective-charge [105,106] determined from the Bjorken sum rule, a basic constraint on our knowledge of nucleon spin structure. Solid theoretical reasons underpin the almost precise agreement between $\hat{\alpha}$ and α_{g_1} [32–34], so that the Bjorken sum serves as a window through which to gain empirical insight into QCD's effective charge.

Figure 2 shows that QCD's unique effective coupling is everywhere finite, i.e., there is no Landau pole and the theory plausibly possesses an infrared-stable fixed point. Apparently, QCD is infrared-finite because a gluon mass-scale is dynamically generated. (A theory is said to possess a Landau pole at k_L^2 if the effective charge diverges at that point. In QCD perturbation theory, such a pole exists at $k_L^2 = \Lambda_{\rm QCD}^2$. Were such a pole to persist in a complete treatment of QCD, it would signal an infrared failure of the theory. On the other hand, the absence of a Landau pole supports a view that QCD is alone amongst four-dimensional quantum field theories in being defined and internally consistent at all energy scales. This might have implications for attempts to develop an understanding of physics beyond the SM based upon non-Abelian gauge theories [71,107–113]). In this case, the value of the PI charge at $k^2 = \Lambda_{\rm QCD}^2$ defines a screening mass [114,115]: $m_G \approx 1.4\,\Lambda_{\rm QCD} \approx 0.71\,{\rm GeV}$. As evident in Figure 2, m_G marks a boundary: the running coupling alters character at $k \simeq m_G$ so that modes with $k^2 \lesssim m_G^2$ are screened from interactions and the theory enters a practically conformal domain. Evidently, the line $k = m_G$ draws a natural border between soft and hard physics; therefore, defines the "hadronic scale":

$$\zeta_H = m_G \,. \tag{12}$$

This is the scale at which all the properties of a hadron are expressed by the dressed quasiparticles that form the DSE kernels and emerge as the self-consistent solutions of the associated equations.

As a unique PI effective charge, $\hat{\alpha}$ appears in each of QCD's dynamical equations of motion, including the gauge- and matter-sector gap equations, setting the strength of all interactions. It therefore plays a critical role in settling the fate of chiral symmetry; to wit, the dynamical origin of light-quark masses in the SM even in the absence of a Higgs coupling.

4.4. *Dynamical Chiral Symmetry Breaking*

Just as a gluon mass-scale emerges dynamically in QCD, massless current-quarks become massive dressed-quarks through a phenomenon known as dynamical chiral symmetry breaking (DCSB) [116]. This effect is another of the critical emergent phenomena in QCD. It is expressed in hadron wave functions, not in vacuum condensates [81,117–120], and modern theory indicates that more than 98% of the visible mass in the Universe can be attributed to DCSB. As classical massless-QCD is a scale-invariant theory (Section 2.2), this means that DCSB is fundamentally connected with the *origin of mass from nothing*.

DCSB is most readily apparent in the dressed-quark propagator

$$S(p) = 1/[i\gamma \cdot pA(p^2) + B(p^2)] = Z(p^2)/[i\gamma \cdot p + M(p^2)] \,, \tag{13}$$

which is obtained as the solution of a gap equation whose kernel is critically dependent upon $\hat{\alpha}$. $M(p^2)$ in Equation (13) is the dressed-quark mass-function, depicted and explained in Figure 3.

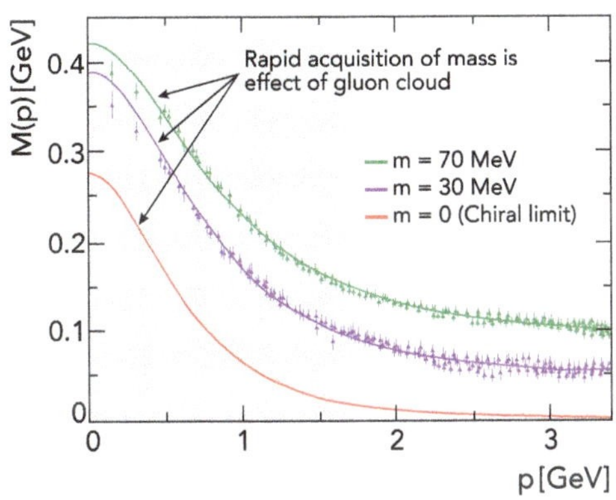

Figure 3. Renormalisation-group-invariant dressed-quark mass function, $M(p)$ in Equation (13): *solid curves*—gap equation solutions [121,122], "data"—numerical simulations of lQCD [123], available for current-quark masses $m = 30, 70$ MeV. QCD's current-quark evolves into a constituent-quark as its momentum becomes smaller. The constituent-quark mass arises from a cloud of low-momentum gluons attaching themselves to the current-quark. This is DCSB, the essentially non-perturbative effect that generates a quark *mass from nothing*; namely, it occurs even in the chiral limit. Notably, the size of $M(0)$ is a measure of the magnitude of the QCD scale anomaly in $n = 1$-point Schwinger functions [124]. Moreover, experiments on $Q^2 \in [0, 12]\,\text{GeV}^2$ at the modern Thomas Jefferson National Accelerator Facility (JLab) will be sensitive to the momentum dependence of $M(p)$ within a domain that is here indicated approximately by the shaded region.

One must insist that chiral symmetry breaking in the absence of a Higgs mechanism is dynamical. It is distinct from Higgs-induced spontaneous symmetry breaking because (*a*) nothing is added to

QCD to catalyse this remarkable outcome and (*b*) no simple change of variables in the QCD action can reveal it. Instead, following quantisation of classical chromodynamics, with its massless gluons and quarks, a large mass-scale emerges in both the gauge- and matter-sectors.

DCSB is empirically revealed very clearly in properties of the pion, whose structure in QCD is described by a Bethe–Salpeter amplitude:

$$\Gamma_\pi(k;P) = \gamma_5 \left[iE_\pi(k;P) + \gamma \cdot P F_\pi(k;P) + \gamma \cdot k\, G_\pi(k;P) + \sigma_{\mu\nu} k_\mu P_\nu H_\pi(k;P) \right], \tag{14}$$

where k is the relative momentum between the pion's valence-quark and -antiquark (defined such that the scalar functions in Equation (14) are even under $k \cdot P \to -k \cdot P$) and P is their total momentum. $\Gamma_\pi(k;P)$ is directly related to an entity that, if a nonrelativistic limit were appropriate, would become the pion's Schrödinger wave function.

In chiral-limit QCD, if, and only if, chiral symmetry is dynamically broken, then [89,91,93]

$$f_\pi^0 E_\pi(k;0) = B(k^2), \tag{15}$$

where f_π^0 is the pion's leptonic decay constant evaluated in the chiral-limit. This identity is exceptional. It is independent of the renormalisation scheme and true in any covariant gauge, and it entails that the two-body problem is solved, nearly completely, once the solution to the one body problem is known. Furthermore, Equation (15) has many corollaries, e.g., it ensures that chiral-QCD generates a massless pion in the absence of a Higgs mechanism; predicts $m_\pi^2 \propto m$ on $m \simeq 0$, where m is the current-quark mass; and entails that the chiral-limit leptonic decay constant vanishes for all excited-state pseudoscalar mesons with nonzero isospin [125,126]. It is also the keystone that supports the success of chiral effective field theories in nuclear physics.

Equation (15), which may be described as a quark-level Goldberger–Treiman relation, is the most basic statement in QCD of the Nambu–Goldstone theorem [37,38]. (Additional considerations applying to the η-η' complex are described elsewhere [127,128].)

> *The Nambu–Goldstone theorem is fundamentally an expression of equivalence between the one-body problem and the two-body problem in QCD's colour-singlet pseudoscalar channel.*

It means that pion properties are practically a direct measure of the dressed-quark mass function rendered in Figure 3. Thus, the qualities of the nearly-massless pion are, enigmatically, the cleanest expression of the mechanism that is responsible for almost all visible mass in the Universe.

Reinstating the Higgs mechanism, so that the light-quarks possess their small current-masses, roughly commensurate with the electron mass, then DCSB is responsible for, inter alia, the physical size of the pion mass ($m_\pi \approx 0.15\, m_p$); the large mass-splitting between the pion and its valence-quark spin-flip partner, the ρ-meson ($m_\rho > 5\, m_\pi$); and the natural scale of nuclear physics, $m_p \approx 1\,\text{GeV}$. Curious things also happen to the kaon. Similar to a pion, except that a strange quark replaces one of the light quarks, the kaon comes to possess a mass $m_K \approx 0.5\,\text{GeV}$. In this case, a competition is taking place between dynamical and Higgs-driven mass generation [114,115,128–131].

Expanding upon these observations, it is worth highlighting that the physical size of m_π is actually much bigger than that linked with the Higgs mechanism for light quarks. Empirically, the scale of the Higgs effect in the light-quark sector is $\sim$1 MeV [13]. As remarked above, $m_\pi = 0$ without a Higgs mechanism, but the current-masses of the light quarks in the pion are the same as they are in nucleons. Therefore, the simple Higgs mechanism result is $m_\pi \approx (m_u + m_d)$, yielding a value which is only 5% of the physical mass. This physical mass emerges as the result of a hefty DCSB-induced enhancement factor, which multiplies the current-quark mass contribution [42]. However, the scale of DCSB is $\sim m_p/3$, i.e., the size of a typical u or d constituent-quark mass, and the special NG-character of the pion means that although it *should* have a mass $\sim (2/3)m_p \approx m_\rho$, most of that mass is cancelled by gluon binding effects owing to constraints imposed by DCSB [124]. The mechanism will now be explained.

4.5. Pion and the Trace Anomaly

At this point, it is possible to resolve the dichotomy expressed by Equations (9) and (10). These statements hold equally on a measurable neighbourhood of the chiral limit because each hadron's mass is a continuous function of the current-masses of its valence-quarks/antiquarks. So consider that for any meson, H_2, constituted from a valence-quark with current-mass m_q and a valence-antiquark with mass $m_{\bar{q}}$,

$$s_{H_2}(0) = \langle H_2(q)|m_\Sigma \bar{\psi}\psi|H_2(q)\rangle = m_\Sigma \frac{\partial m_{H_2}^2}{\partial m_\Sigma}\,, \tag{16}$$

where $m_\Sigma = (m_q + m_{\bar{q}})$, viz. the scalar form factor at zero momentum transfer measures the reaction of the meson's mass-*squared* to a variation in current-quark mass. It is merely a misleading *convention* to define the meson's σ-term as $\sigma_{H_2} = s_{H_2}(0)/[2m_{H_2}]$. Notably, the pion (and any other NG mode) possesses the peculiar property that

$$s_\pi(0) \stackrel{m_\Sigma \simeq 0}{=} m_\Sigma \frac{\partial m_\pi^2}{\partial m_\Sigma} = 1 \times m_\pi^2, \tag{17}$$

which is the statement that in the vicinity of the chiral limit, 100% of the pion mass-squared owes to the existence of the current-mass in $\mathcal{L}_{\rm QCD}$. One should compare this result with that for the pion's valence-quark spin-flip partner, i.e., the ρ-meson [132]:

$$s_\rho(0) \approx 0.06\, m_\rho^2, \tag{18}$$

indicating that just 6% of the ρ-meson's mass-squared is generated by the current-mass term in $\mathcal{L}_{QCD}$. The remainder arises largely as a consequence of EHM [42], as suggested by the fact that $m_\rho \approx 2M(0)$, where $M(k^2)$ is the dressed-quark mass function in Figure 3. The key to understanding Equation (10) is Equation (15) and three associated Goldberger–Treiman-like relations, which are exact in chiral QCD. Utilising these identities when working with those DSEs needed to describe a pseudoscalar meson, an algebraic proof of the following statement can be constructed [89,91,93]: at every order in a symmetry-preserving analysis, the masses generated for the valence-quarks that constitute the system are exactly cancelled by the attraction produced by interactions between them. This cancellation ensures that the initial two-valence-parton system, which began massless, grows into a complex system, with a bound-state wave function tied to a pole in the scattering matrix at $P^2 = 0$. Stated simply, Equation (10) is obtained through cancellations between one-body dressing and two-body binding effects:

$$M_{\rm quark}^{\rm dressed} + M_{\rm antiquark}^{\rm dressed} + U_{\rm quark-antiquark\ interaction}^{\rm dressed} \stackrel{\rm chiral\ limit}{\equiv} 0\,, \tag{19}$$

with the sum being exactly zero if, and only if, chiral symmetry is broken dynamically in the Nambu pattern. (A full discussion is provided in Sec. 3.3 of Ref. [124].) Away from the chiral limit for NG modes and always in other channels, the cancellation is incomplete.

It is important to remark that such a transparent resolution of the conundrum expressed by Equations (9) and (10) is impossible if one insists on using a parton model basis, in which the trace anomaly operator is given simply by Equation (7). One must employ a modern quasiparticle formalism. Then, with dressed-particle propagators and bound-state wave functions, obtained at a low renormalisation scale, $\zeta \lesssim m_p$, as solutions of coupled integral equations, each of which sums a countable infinity of diagrams, Equation (10) can be re-expressed:

$$\langle \pi(q)|\Theta_0|\pi(q)\rangle \stackrel{\zeta \gg m_p}{=} \langle \pi(q)|\tfrac{1}{4}\beta(\alpha(\zeta))G^a_{\mu\nu}G^a_{\mu\nu}|\pi(q)\rangle \stackrel{\zeta \lesssim m_p}{\to} \langle \pi(q)|\mathcal{D}_1 + \mathcal{I}_2|\pi(q)\rangle\,, \tag{20a}$$

$$\mathcal{D}_1 = \sum_{f=u,d} M_f(\zeta)\, \bar{\mathcal{Q}}_f(\zeta)\mathcal{Q}_f(\zeta)\,, \quad \mathcal{I}_2 = \tfrac{1}{4}[\beta(\alpha(\zeta))\mathcal{G}^a_{\mu\nu}\mathcal{G}^a_{\mu\nu}]_{\rm 2PI}\,. \tag{20b}$$

Equations (20) describe the transfigurement of the chiral-limit parton-basis expression of the trace anomaly's expectation value in the pion into a new form, written in terms of a non-perturbatively-dressed quasi-particle basis, with $\mathcal{Q}$ denoting dressed-quarks and $\mathcal{G}$ the dressed-gluon field strength tensor. Here, the first term expresses the one-body-dressing content of the trace anomaly. It is positive. Patently, a massless valence-quark (antiquark) gaining a large mass by way of interactions with its own gluon field is an effect of the trace-anomaly in what may be described as the one-quasiparticle subspace of a complete pion wave function. The second term expresses the two-particle-irreducible (2PI) interaction content of the forward scattering process represented by this trace anomaly matrix element. It is negative and acquires a scale because the gluon- and quark-propagators and the couplings in the 2PI processes have all gained a mass-scale.

The discussion of Equations (19) and (20) and their connection with Equation (10) bring a well-known adage to mind, quoted here from S. Weinberg in [133]: "You may use any degrees of freedom you like to describe a physical system, but if you use the wrong ones, you'll be sorry!"

5. Empirical Manifestations of Emergent Mass

5.1. Pion Wave Function

Empirical signals of the emergence of mass are ubiquitous, but their appearance typically changes from one system to another. This means that it is essential to study a wide variety of observables because each one is apt to expose different aspects of the underlying mechanisms. It is useful, therefore, to begin with the most obvious, whose importance derives from Equation (15), i.e., the pion's leading-twist two-particle parton distribution amplitude (PDA), φ_π, the simplest component of its light-front wave function. Any framework that is capable of delivering a hadron's Poincaré-covariant bound-state amplitude can also be used to calculate its PDA, and in this case, φ_π is given by a light-front projection of the pion's Bethe–Salpeter wave function [134]

$$f_\pi \, \varphi_\pi(x;\zeta) = \text{tr}_{\text{CD}} Z_2 \int_{dq}^{\Lambda} \delta(n \cdot q_+ - x\, n \cdot P)\, \gamma_5 \gamma \cdot n\, S(q_+) \Gamma(q;P) S(q_-)\,. \tag{21}$$

In Equation (21), the trace is over colour and spinor indices; $\int_{dq}^{\Lambda}$ is a Poincaré-invariant regularisation of the four-dimensional integral, with Λ the ultraviolet regularization mass-scale; $Z_2(\zeta,\Lambda)$ is the quark wavefunction renormalisation constant; n is a light-like four-vector, $n^2 = 0$; and P is the pion's four-momentum, $P^2 = -m_\pi^2$ and $n \cdot P = -m_\pi$ in the pion's rest frame.

Two distinctively different truncations of QCD's DSEs [134] have been used to calculate the amplitude in Equation (21). Both agree: compared with the asymptotic profile, which is valid on $\Lambda_{\text{QCD}}/\zeta \simeq 0$, there is a marked broadening of $\varphi_\pi(x;\zeta)$, which owes exclusively to DCSB, i.e., the emergence of mass as exhibited in Figure 3. This causal connection may be asserted because the PDA is calculated at a low renormalisation scale in the chiral limit, in which case the quark mass function owes entirely to DCSB via Equation (15). Moreover, the PDA's dilation is related to the rate at which a dressed-quark approaches the asymptotic bare-parton limit. The prediction determined using the most sophisticated kernel [114,115] is depicted in Figure 4:

$$\varphi_\pi(x;\zeta_H) = 20.227\, x(1-x)[1 - 2.5088\sqrt{x(1-x)} + 2.0250\, x(1-x)]\,, \tag{22}$$

It can be verified empirically at JLab 12, e.g., in measurements of the pion's electromagnetic form factor (see Section 5.2 below).

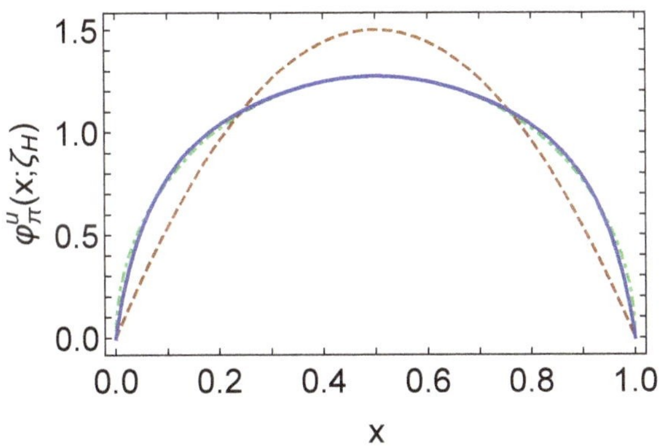

Figure 4. Twist-two pion PDA at the hadronic scale, ζ_H. Solid blue curve, Equation (22)—determined using the most sophisticated available symmetry preserving DSE kernels; dot-dashed green curve—original prediction from the work in [134]. These PDA results are consistent with contemporary lQCD results [135,136]. The dashed orange curve is $\varphi_\pi^{\rm asy}(x) = 6x(1-x)$, the limiting form under QCD evolution [137–139]. The PDAs are dimensionless.

It is worth remarking here that the authors of [134] chose to reconstruct the pion's DA from its Mellin moments using an order-α Gegenbauer expansion. This is useful as a first step because the procedure converges rapidly, so enables the qualitative feature of broadening driven by EHM in the SM to be exposed. Such broadening, however, need not and should not disturb the DA's endpoint behaviour, which QCD predicts to be linear in the neighbourhoods $x \simeq 0, 1$. Therefore, as a second step, the authors of [114,115] re-expressed the result from the work in [134] as the function in Equation (22). The first eleven moments agree at the level of 1.6 (1.4)%, i.e., well within any sensible estimate of uncertainty in the computation of high-order moments. Pointwise, the curves are practically indistinguishable, as highlighted in Figure 4.

It is important to emphasise that the computed PDAs in Figure 4 are concave functions. Such pointwise behaviour contrasts markedly with the "bimodal" or "double-humped" distributions that have been favoured in phenomenological applications by some authors [140]. It should be understood in this connection that a double-humped profile for the twist-two PDA places it in the class of distributions produced by a Bethe–Salpeter wave function which is suppressed at zero relative momentum, instead of maximal thereat. No ground-state solution of the pseudoscalar or vector meson Bethe–Salpeter equation exhibits corresponding behaviour [141–145]. Thus, a bimodal distribution cannot be an accurate pointwise rendering of the PDA for a ground-state meson. Notwithstanding that, if one employs such a distribution in a phenomenological application for which only its lowest Mellin moments are important, then some carefully-constrained bimodal distributions may supply useful approximations to the moments of a broad, concave PDA and thereby prove practically useful.

A question of more than thirty-years standing can be answered using Figure 4, namely, when does $\varphi_\pi^{\rm asy}(x)$ provide a good approximation to the pion PDA? Plainly, not at ζ_H. The ERBL evolution equation [137–139] describes the ζ-evolution of $\varphi_\pi(x;\zeta)$, and applied to $\varphi_\pi(x;\zeta)$ in Figure 4, one finds [135,136] that $\varphi_\pi^{\rm asy}(x)$ is a poor approximation to the true result even at $\zeta = 10\,$GeV. Thus at energy scales accessible to experiment, the twist-two PDAs of ground-state hadrons are "fat and squat". Evidence supporting this picture had long been accumulating [146–149], and the dilation is confirmed by simulations of lQCD [135,136].

5.2. Pion Electromagnetic Form Factor

The cross sections for many hard exclusive hadronic reactions, i.e., processes involving a highly energetic probe that strikes a target and leaves it intact, can be written in terms of the PDAs of the hadrons involved. An example is the elastic electromagnetic form factor of the pion, for which the prediction can be stated thus [137–139]:

$$\exists Q_0 > \Lambda_{\rm QCD} \mid Q^2 F_\pi(Q^2) \overset{Q^2>Q_0^2}{\approx} 16\pi\alpha(Q^2) f_\pi^2 w_\varphi^2, \tag{23a}$$

$$w_\varphi = \frac{1}{3}\int_0^1 dx\, \frac{1}{x}\varphi_\pi(x), \tag{23b}$$

where $\alpha(Q^2)$ is the running coupling, which is practically equivalent to $\hat{\alpha}(Q^2)$ in Figure 2 on any domain within which perturbation theory is valid. The value of Q_0 is not predicted by perturbative QCD (pQCD), but it is computable in any non-perturbative framework that veraciously expresses EHM.

It was anticipated that JLab would verify the fundamental prediction in Equation (23). Indeed, the first publication by the F_π Collaboration [150] heralded the beginning of a new era in mapping the pion's internal structure. Ensuing measurements [151–155] confirmed the data's trend, leading to a common perception that, at $Q^2 = 2.45\,\text{GeV}^2$, one remains far from the resolution field wherein the pion acts like an elementary partonic quark–antiquark pair, i.e., far from validating Equation (23). This conclusion was based on the assumption that inserting $\varphi_\pi^{\rm asy}(x)$ into Equation (23) delivers a valid approximation at $\zeta^2 = Q^2 = 2.45\,\text{GeV}^2$, so that

$$Q^2 F_\pi(Q^2) \overset{Q^2=4\,\text{GeV}^2}{\approx} 0.15\,. \tag{24}$$

The result in Equation (24) is a factor of 2.7 smaller than the empirical value quoted at $Q^2 = 2.45\,\text{GeV}^2$ [154]: $0.41^{+0.04}_{-0.03}$, and a factor of three smaller than that computed at $Q^2 = 4\,\text{GeV}^2$ in [156]. At the time, the authors of [156] supplied the lone prediction for the Q^2-dependence of F_π that was both applicable on the full spacelike region mapped reliably by experiment and confirmed thereby.

Here, the perception of a mismatch and a real discrepancy are not equivalent because, as elucidated above, $Q^2 = 4\,\text{GeV}^2$ is not within the domain $\Lambda_{\rm QCD}^2/Q^2 \simeq 0$ upon which Equation (23) used with $\varphi_\pi^{\rm asy}(x)$ provides a valid approximation. This being so, and given the successful prediction in [156], it was natural to ask whether DSE methods could address the issue of the ultimate validity of Equation (23).

Developments within the past decade have made that possible and now a computation of the pion's electromagnetic form factor is available to arbitrarily large-Q^2 [131,157,158]. The result is the solid (black) curve in Figure 5. In addition, the study enables that result to be correlated with Equation (23) via the modern PDA computed in the same framework, which is the dashed (blue) curve in Figure 5. On the depicted domain, this leading-order, leading-twist QCD prediction, computed with a pion valence-quark PDA evaluated at a scale fitting the experiment, underestimates the full DSE result by a fairly uniform 15%. The mismatch is explained by a mix of higher-order, higher-twist corrections to Equation (23) in pQCD on the one hand and, on the other hand, shortcomings in the leading-order DSE truncation used in [131,157,158], which predicts the right power-law behaviour for the form factor but not precisely the correct anomalous dimension (logarithm exponent) in the strong-coupling calculation. It is now anticipated that the upgraded JLab facility will reveal a maximum in $Q^2 F_\pi(Q^2)$ at $Q^2 \approx 6\,\text{GeV}^2$ and an experiment at $Q^2 = 9\,\text{GeV}^2$ will see an indication of parton model scaling and scaling violations for the first time in a hadron elastic form factor. While JLab's grip on these things might be tenuous, the reach of an electron ion collider would certainly enable validation of these predictions [39].

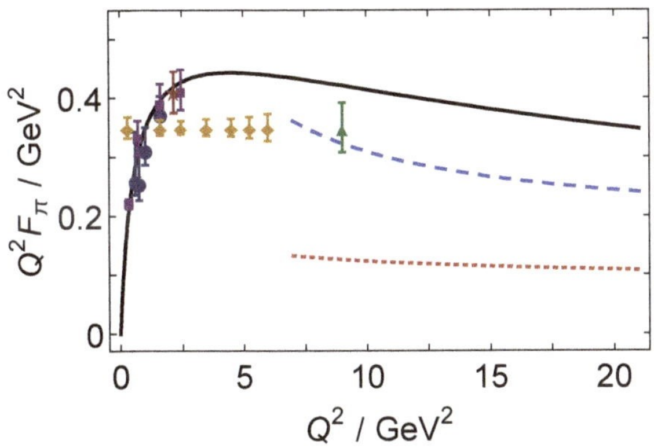

Figure 5. $Q^2F_\pi(Q^2)$. Solid curve (black)—theoretical prediction [131,157,158]; dashed curve (blue)—pQCD prediction computed with the modern, dilated pion PDA described in Section 5.1; and dotted (red) curve—pQCD prediction computed with the asymptotic profile, $\varphi_\pi^{\rm asy}(x)$, which had previously been used to guide expectations for the asymptotic behaviour of $Q^2F_\pi(Q^2)$. The filled-circles and -squares represent existing JLab data [154] and the filled diamonds and triangle, whose normalisations are arbitrary, indicate the projected Q^2-reach and accuracy of forthcoming experiments [159,160].

5.3. Valence-Quark Distributions in the Pion

Given the pions' simple valence-quark content, i.e., one quark and one antiquark, a basic quantity in any discussion of their structure is the associated parton distribution function (PDF), $q^\pi(x;\zeta)$. This density charts the probability that a valence q-quark in a pion carries a light-front fraction x of the system's total momentum [161], and one of the earliest predictions of pQCD is [162–164]:

$$q^\pi(x;\zeta=\zeta_H) \sim (1-x)^2 . \tag{25}$$

Moreover, the exponent evolves as ζ increases beyond ζ_H, becoming $2+\gamma$, where $\gamma \gtrsim 0$ is an anomalous dimension that increases logarithmically with ζ [165–168]. (In the limit of exact $\mathcal{G}$-parity symmetry, which is a good approximation in the SM, $u^{\pi^+}(x) = \bar{d}^{\pi^+}(x)$, etc. Therefore, it is only necessary to discuss one unique distribution.)

As described in connection with Equations (19) and (20), for nature's pion there is near-complete cancellation between (a) strong-mass-generating dressing of the valence-quark and -antiquark and (b) binding attraction. Such effects distinguish this system from the more massive kaon, within which the cancellation is far less efficient because the s-quark current mass generated by the Higgs is so much larger than that of the u- and d-quarks. Consequently, high-precision measurements of the valence-parton distributions in the pion and kaon are a high priority at existing and anticipated facilities [39,169–171].

Such efforts are driven by ongoing progress in theory. Marking one significant class of advances, lQCD is beginning to yield results for the pointwise behaviour of the pion's valence-quark distribution [172–176], offering promise for information beyond the lowest few moments [177,178].

Extensions of the continuum analysis in [179] are also yielding new insights [180], leading to the first parameter-free predictions of the valence, glue and sea distributions within the pion [181,182], unifying them with, inter alia, electromagnetic pion elastic and transition form factors [128,131,157,158,183,184]. The analysis reveals that, like the pion's PDA in Figure 4, the valence-quark distribution function is hardened by DCSB, producing the following apportioning of light-front momentum at the scale $\zeta = \zeta_2 = 2\,\text{GeV}$:

$$\langle x_{\text{valence}} \rangle = 0.48(3)\,, \quad \langle x_{\text{glue}} \rangle = 0.41(2)\,, \quad \langle x_{\text{sea}} \rangle = 0.11(2)\,. \tag{26}$$

A similar valence-quark momentum fraction was obtained in Ref. [185] by analysing data on π-nucleus Drell–Yan and leading neutron electroproduction [185]: $\langle 2x \rangle_q^\pi = 0.49(1)$ at $\zeta = \zeta_2$. Even though this phenomenological PDF yields a compatible momentum fraction, its x-profile is different. In fact, the phenomenological PDF conflicts with the QCD constraint, Equation (25). Significantly, the analysis in [185] ignored threshold resummation effects, which are known to have a material impact at large x [186,187]. (Similar remarks apply to the analysis in [188].)

Importantly, as illustrated in Figure 6, after evolution [165–168] to $\zeta = \zeta_5 = 5.2\,\text{GeV}$, using $\hat{\alpha}(k^2)$ in Figure 2 to integrate the evolution equations, the continuum prediction for $u^\pi(x)$ from the works in [181,182] matches that obtained using lQCD [175]. Given that no parameters were varied in order to procure this or any other outcome in [181,182], a remarkable, modern confluence has been reached, which suggests that real strides are being made toward understanding pion structure and its relation to the emergence of mass. (Predictions for kaon PDFs are described in [114,115].)

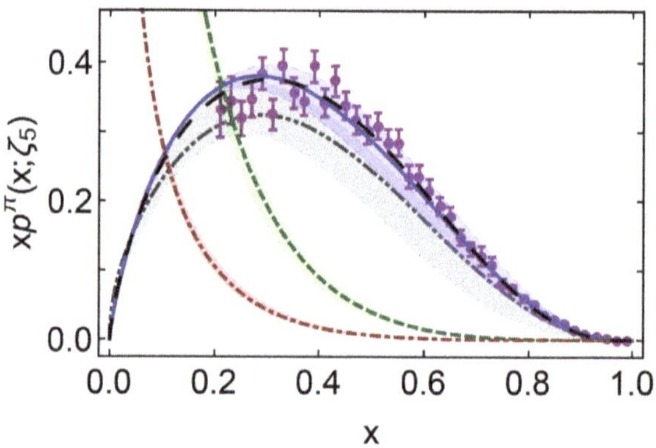

Figure 6. Pion valence-quark momentum distribution function, $xu^\pi(x;\zeta_5)$: dot-dot-dashed (grey) curve within shaded band—lQCD result [175]; long-dashed (black) curve—early continuum analysis [179]; and solid (blue) curve within shaded band—modern, continuum calculation [181,182]. From the works in [181,182]: gluon momentum distribution in pion, $xg^\pi(x;\zeta_5)$—dashed (green) curve within shaded band; and sea-quark momentum distribution, $xS^\pi(x;\zeta_5)$—dot-dashed (red) curve within shaded band. (The shaded bands indicate the size of calculation-specific uncertainties, detailed in the source material [175,181,182].) Data (purple) from in [189], rescaled according to the analysis in [186].

5.4. *Emergence of Diquark Correlations*

The emergence of mass is also expressed in the properties of baryons, something which can readily be seen when the three valence-quark bound-state problem is studied with the same level of sophistication that is now typical for mesons. In this connection, DCSB, as displayed in the momentum-dependence of the dressed-quark mass—Figure 3, is as important to baryons as it is to mesons. In fact, one important consequence of DCSB is that any interaction able to create composite pseudo–NG modes from a light dressed-quark and -antiquark, and reproduce the observed value of their leptonic decay constants, will also generate tight colour-antitriplet correlations between any two of a nucleon's dressed-quarks. This conclusion is based upon evidence gathered in thirty years of studying two- and three-body bound-states in hadron physics [190].

The properties of such diquark correlations are known. As colour-carrying correlations, diquarks are confined [88,191]. Additionally, owing to properties of charge-conjugation, a diquark with spin-parity J^P may be viewed as a partner to the analogous J^{-P} meson [192]. It follows that the strongest diquark correlations in the nucleon are scalar isospin-zero, $[ud]_{0^+}$, and pseudovector, isospin-one, $\{uu\}_{1^+}$, $\{ud\}_{1^+}$, $\{dd\}_{1^+}$. Furthermore, although no pole-mass exists, the following mass-scales, expressing the range and strength of the correlation, can be associated with these diquarks [192–199] (in GeV),

$$m_{[ud]_{0^+}} \approx 0.7 - 0.8\,,\; m_{\{uu\}_{1^+}} \approx 0.9 - 1.1\,, \tag{27}$$

with $m_{\{dd\}_{1^+}} = m_{\{ud\}_{1^+}} = m_{\{uu\}_{1^+}}$ in the isospin symmetric limit. Notably, the nucleon contains both scalar-isoscalar and pseudovector-isovector correlations. Neither can safely be ignored and their presence has many observable consequences [200–203].

Realistic diquark correlations are also soft and interacting. They all carry charge, scatter electrons, and have an electromagnetic size which is similar to that of the kindred mesonic system, see, e.g., in [194,204,205]:

$$r_{[ud]_{0^+}} \gtrsim r_\pi, \quad r_{\{uu\}_{1^+}} \gtrsim r_\rho, \tag{28}$$

with $r_{\{uu\}_{1^+}} > r_{[ud]_{0^+}}$. As in the meson sector, these scales are set by that associated with DCSB.

It should be emphasised that such fully dynamical diquark correlations are vastly different from the static, pointlike "diquarks" introduced originally [206] in an attempt to solve the so-called "missing resonance" problem [207], viz. the fact that quark models predict a far greater number of baryon states than were observed in the previous millennium [208]. Moreover, their existence enforces certain distinct interaction patterns for the singly- and doubly-represented valence-quarks within the proton, as exhibited elsewhere [81,84,199–202,209–213].

The existence of these strong correlations between two dressed quarks is the path to converting the three valence-quark bound-state problem into the simpler Faddeev equation drawn in Figure 7, without loss of dynamical information [214]. The three-gluon vertex, a definitive feature of QCD's non-Abelian nature, is not an explicit part of the scattering kernel in this picture. Instead, one profits from the following features, phase-space factors materially enhance two-body interactions over $n \geq 3$-body interactions and the primary three-body force, produced by a three-gluon vertex attaching once, and only once, to each valence quark, vanishes when projected into the colour-singlet channel; and subsequently capitalises on diquark dominance in the two-body subsystems. Then, while iterated, overlapping three-body terms might alter fine details of baryon structure, the primary effect of non-Abelian multi-gluon vertices is manifested in the formation of diquark correlations. (On the other hand, the three-gluon vertex appears to play a material and distinguishable role in the formation of hybrid hadrons [215] and glueballs [216]). Accordingly, the active kernel describes binding in the baryon through diquark breakup and reconstitution, mediated by exchange of a dressed-quark. The properties and interactions of such a baryon are chiefly determined by the quark+diquark structure evident in Figure 7.

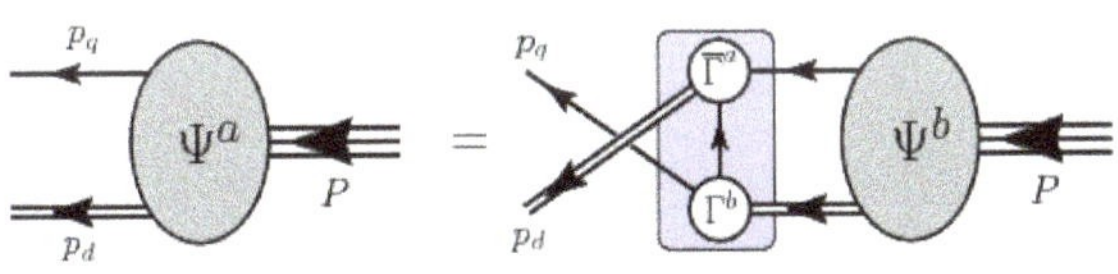

Figure 7. Poincaré covariant Faddeev equation: a homogeneous linear integral equation for Ψ, the matrix-valued Faddeev amplitude for a baryon of total momentum $P = p_q + p_d$, which expresses the relative momentum correlation between the dressed-quarks and -diquarks within the baryon. The shaded rectangle demarcates the kernel of the Faddeev equation: *single line*, dressed-quark propagator; Γ, diquark correlation amplitude; and *double line*, diquark propagator.

The spectrum of baryons produced by the Faddeev equation [197,198,211,217] is like that found in the three-constituent quark model and consistent with lQCD analyses [218]. Notably, modern data and recent analyses have reduced the number of missing resonances [219–224].

5.5. Proton Wave Function

After solving the Faddeev equation in Figure 7 for the proton's Faddeev amplitude, one can then compute the proton's dressed-quark leading-twist PDA [203]. The result for this proton "wave function" is depicted in Figure 8.

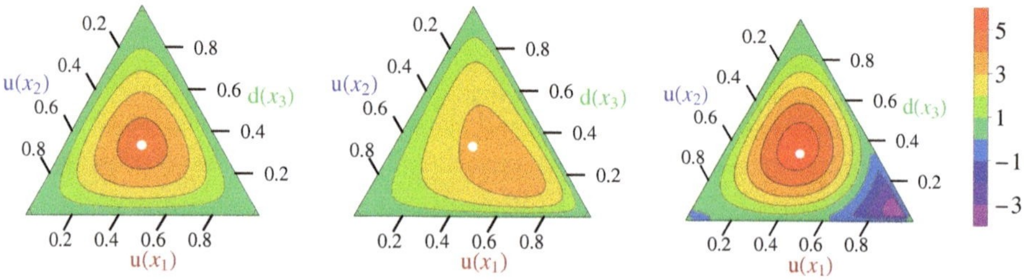

Figure 8. Baryon leading-twist dressed-quark distribution amplitudes depicted using barycentric plots. *Left panel*—asymptotic profile, baryon PDA, $\varphi_N^{\text{asy}}([x]) = 120x_1x_2x_3$; *middle panel*—computed proton PDA evolved to $\zeta = 2\,\text{GeV}$, which peaks at $([x]) = (0.55, 0.23, 0.22)$; and *right panel*—Roper resonance PDA. The white circle in each panel marks the centre of mass for $\varphi_N^{\text{asy}}([x])$, whose peak lies at $([x]) = (1/3, 1/3, 1/3)$. Here, x_1, x_2, x_3 indicate the fraction of the bound-state's light-front momentum carried by the associated quark; naturally, $x_1 + x_2 + x_3 = 1$. The amplitudes are dimensionless; hence, the height is simply a real number.

Table 1 lists the proton PDA's four lowest-order moments. They reveal important insights, e.g., when the proton is pictured as solely a quark+scalar-diquark correlation, $\langle x_2\rangle_u = \langle x_3\rangle_d$, because these two form the scalar quark+quark correlation, and the system is very asymmetric, with the PDA's peak being shifted strongly in favour of $\langle x_1\rangle_u > \langle x_2\rangle_u$. This outcome conflicts with lQCD results [225,226]. On the other hand, as explained above, realistic Faddeev equation calculations suggest that pseudovector diquark correlations are an essential piece of the proton's wave function. When these $\{uu\}$ and $\{ud\}$ diquarks are included, momentum is shared more evenly, shifting from the spectator $u(x_1)$ quark into $u(x_2), d(x_3)$. Adding these diquarks with the known weighting, the PDA's peak shifts back toward the centre, relocating to $([x]) = (0.55, 0.23, 0.22)$, and the computed values of the first moments line up with those computed using lQCD. This convergence delivers a more complete understanding of the lQCD simulations, which by that means are seen to confirm a picture of the proton as a bound-state with both tight scalar *and* pseudovector diquark correlations, wherein the scalar diquarks are responsible for $\approx 65\%$ of the Faddeev amplitude's canonical normalisation.

Importantly, like ground-state S-wave mesons [134,136,227–230], the leading-twist PDA of the ground-state nucleon is both broader than $\varphi_N^{\text{asy}}([x])$ (defined by Table 1, Row 1) and decreases monotonically away from its maximum in all directions, i.e., the PDAs of these ground-state S-wave systems possess endpoint enhancements, but neither humps nor bumps. Models which produce such humped structures were previously considered reasonable [140]. However, it is now clear that pointwise behaviour of this type is in conflict with QCD. The models may nonetheless be viewed as possessing a qualitative truth, insofar as they represent a way by which endpoint enhancements can be expressed in hadron PDAs if one limits oneself to the polynomial basis characterising QCD in the neighbourhood $\Lambda_{\text{QCD}}/\zeta \simeq 0$.

As with the pion prediction described in Section 5.2, the veracity of this result for the proton PDA can be tested in future experiments. For instance, it can be used to provide the first realistic evaluation of the scale at which exclusive measurements involving proton targets may reasonably be compared with predictions based on pQCD hard scattering formulae. Analogous to the pion case, the value of this mass-scale is an empirical manifestation of the emergence of mass, here within the three-valence-quark proton bound-state.

Table 1. Computed values of the first four moments of the proton and Roper-resonance PDAs. The error on f_N, a dynamically-determined quantity which measures the proton's "wave function at the origin", reflects a nucleon scalar diquark content of 65 ± 5%, and values in rows marked with "$\not{\!\partial}$ av" were obtained assuming the baryon is constituted solely from a scalar diquark. Including axial-vector diquark correlations, the mean absolute relative difference between continuum and lattice results improves by 36%. (All results listed at a renormalisation scale $\zeta = 2\,\text{GeV}$.)

	$10^3 f_N/\text{GeV}^2$	$\langle x_1\rangle_u$	$\langle x_2\rangle_u$	$\langle x_3\rangle_d$
asymptotic PDA		0.333	0.333	0.333
lQCD [225]	2.84 (33)	0.372 (7)	0.314 (3)	0.314 (7)
lQCD [226]	3.60 (6)	0.358 (6)	0.319 (4)	0.323 (6)
DSE proton [203]	3.78 (14)	0.379 (4)	0.302 (1)	0.319 (3)
DSE proton $\not{\!\partial}$ av	2.97	0.412	0.295	0.293
DSE Roper [203]	5.17 (32)	0.245 (13)	0.363 (6)	0.392 (6)
DSE Roper $\not{\!\partial}$ av	2.63	0.010	0.490	0.500

5.6. *Proton's First Radial Excitation*

Almost immediately after discovery of the Roper resonance [231–235], the proton's first positive-parity excitation, questions were asked regarding the nature of like-parity excitations of ground-state positive-parity baryons. A lucid picture is now emerging following [84], (*i*) the acquisition and analysis of a large amount of very precise nucleon-resonance electroproduction data, involving single- and double-pion final states on a large domain of energy and momentum-transfer; (*ii*) development of an advanced dynamical reaction theory, able to simultaneously describe all partial waves extracted from available data; (*iii*) formulation and extensive application of the DSE approach to the continuum bound state problem in relativistic quantum field theory, which expresses diverse local and global impacts of EHM; and and (*iv*) the improvement of constituent quark models so that they can also qualitatively incorporate these features of strong QCD. In this picture, such states are primarily radial excitations of the associated ground-state baryon, comprised of a well-defined dressed-quark core supplemented by a meson cloud.

Using the Faddeev equation framework sketched above, properties of the Roper-resonance's dressed-quark core have been exposed [195,199]: it is found that the scalar functions in the Roper's Faddeev amplitude have a zero at quark+diquark relative momentum $\sqrt{\ell^2} \approx 0.4\,\text{GeV} \approx 1/[0.5\,\text{fm}]$. Working with this input, the authors of [203] delivered the associated leading-twist PDA, depicted in the rightmost panel of Figure 8 and whose first four moments are listed in Table 1. The prediction reveals some interesting features, e.g., the PDA of this excitation is not positive definite and there is a conspicuous locus of zeros in the lower-right section of the barycentric plot, both of which duplicate aspects of the wave function describing the first radial excitation of a quantum mechanical system that are also seen in the PDAs of meson radial excitations [236,237], and here the influence of pseudovector diquarks is contrary to that in the ground-state, viz. they shift momentum into $u(x_1)$ from $u(x_2), d(x_3)$.

These observations highlight that the ground state is just one isolated member of a set of Hamiltonian eigenvectors with infinitely many elements: many Hamiltonians can possess practically equivalent ground states and yet lead to excited-state spectra that are vastly different. Moreover, masses alone, as infrared-dominated quantities, contain relatively little information. Different Hamiltonians may adequately reproduce known hadron spectra; but these same Hamiltonians can yield predictions that disagree markedly when used to compute structural properties. Such properties—like wave functions and the Q^2-dependence of elastic and transition form factors—possess the greatest discriminating power. Therefore, study of the structure of nucleon resonances is a critical complement to that of ground-state nucleons and mesons because it is capable of revealing additional novel features of strong QCD. Modern theory must be deployed to compute observable properties of all these systems. Aspects of this effort are sketched elsewhere [41,43].

5.7. Emergent Features of Nucleon Form Factors

The character of emergent mass and the diquark correlations within baryons it induces are visible in baryon elastic and transition form factors, and particular examples of contemporary significance are neutron and proton elastic form factors. Nucleons are the most basic elements of nuclear physics; and these form factors are manifestations of the nature of the nucleons' constituents and the dynamics holding them together.

Paradigmatic shifts in our understanding of these things are being driven by new, precise form factor data. This is nowhere more evident than in analyses of experimental data acquired in the past vicennium, which have established a new ideal. Namely, despite its elementary valence-quark content, the nucleon's internal structure is very complex. For instance, there are measurable differences between the distributions of charge and magnetisation throughout the interiors [238] and between the way these qualities are carried by the different quark flavours [202,209,239]. The aim now is to explain the observations in terms of emergent features of the strong interaction.

In this connection, it is here worth highlighting that the behaviour of the proton's electric form factor on $Q^2 \gtrsim 4\,\text{GeV}^2$ is particularly sensitive to the running of the dressed-quark mass (displayed in Figure 3), especially the rate at which the dressed-quark mass runs between the infrared and ultraviolet domains [209,213,240].

The proton's momentum-space charge and magnetisation distributions are expressed in combinations of the two Poincaré-invariant elastic form factors that are needed to express the proton's electromagnetic current:

$$ie\,\bar{u}(p')\,[\gamma_\mu F_1(Q^2) + \frac{Q_\nu}{2m_N}\,\sigma_{\mu\nu}\,F_2(Q^2)]\,u(p)\,, \tag{29}$$

where m_N is the nucleon mass; $Q = p' - p$, $u(p)$ and $\bar{u}(p')$ are, respectively, spinors describing the incident and scattered proton; and $F_{1,2}(Q^2)$ are the proton's Dirac and Pauli form factors. The combinations that feature in the electron-proton elastic scattering cross section are the charge and magnetisation distributions [241]

$$G_E(Q^2) = F_1(Q^2) - (Q^2/[4m_N^2])F_2(Q^2)\,, \quad G_M(Q^2) = F_1(Q^2) + F_2(Q^2)\,. \tag{30}$$

Data available before the year 1999 led to a view that

$$\mu_p \left.\frac{G_E^p(Q^2)}{G_M^p(Q^2)}\right|_{\text{Rosenbluth}} \approx 1\,, \tag{31}$$

$\mu_p\, G_M^p(Q^2=0) = 1$; therefore, a conclusion that the distributions of charge and magnetisation within the proton are approximately identical [242]. At the time, the proton was viewed as a simple bound state, wherein quark+quark correlations and attendant orbital angular momentum played little role. Equation (31) is consistent with this picture.

The situation changed dramatically when the combination of high-energy, -current and -polarisation at JLab enabled a new type of experiment to be performed, viz. polarisation-transfer reactions [238], which are directly proportional to $G_E(Q^2)/G_M(Q^2)$ [243,244]. A series of these experiments [238,245–248] has determined that

$$\mu_p \left.\frac{G_E^p(Q^2)}{G_M^p(Q^2)}\right|_{\text{JLab PT}} \approx 1 - \text{constant} \times Q^2\,, \tag{32}$$

where the constant is such that the ratio might become negative for $Q^2 \gtrsim 8\,\text{GeV}^2$. This behaviour contrasts starkly with Equation (31); and since the proton's magnetic form factor is reliably known on a spacelike domain that extends to $Q^2 \approx 30\,\text{GeV}^2$ [249,250], the Q^2-dependence of G_E^p/G_M^p exposes novel features of the proton's charge distribution, as expressed in $G_E^p(Q^2)$.

An understanding of the behaviour in Equation (32) is provided by the analyses in [209,213,240], and the answer lies largely with the proton's Pauli form factor. F_2^p is a gauge of the proton's magnetisation distribution. Ultimately, this magnetisation is borne by the dressed-quarks and influenced by the k^2-dependence of (*i*) their propagators and (*ii*) the correlations amongst them. Both are expressed in the Faddeev wavefunction.

Thus, for the sake of argument, suppose that dressed-quarks are described by a momentum-independent mass-function, e.g., as obtained using a symmetry-preserving regularisation of a vector $\times$ vector contact interaction [251]. They then behave as Dirac particles with constant Dirac values for their magnetic moments. Consequently, the composite proton possesses a hard Pauli form factor and this produces a zero in G_E^p at $Q^2 \approx 4\,\text{GeV}^2$ [240,252].

Alternatively, suppose that the dressed-quarks have a momentum-dependent mass-function, like that depicted in Figure 3, which is large at small momenta but vanishes as their momentum increases. At infrared momenta the dressed-quarks will then act as constituent-like particles with a large magnetic moment, but as the probe momentum grows, their mass and magnetic moment will drop toward zero. (*N.B.* Massless fermions do not have a measurable magnetic moment [253], so that any significant magnetic moment for a constituent-like quark is an emergent feature.) Such dressed-quarks will produce a proton Pauli form factor that is large on $Q^2 \simeq 0$ but drops rapidly on the domain of transition between the infrared and ultraviolet domains, to give a very small result at large-Q^2. The proton's Dirac form factor is far less sensitive to spin-related effects; hence the interplay between the Dirac and Pauli form factors, expressed in Equation (30), entails that $G_E(Q^2)$ must have a zero at larger values of Q^2 when calculated with a running mass function than we computed with momentum-independent dressed-quark masses. The precise form of the Q^2-dependence will depend on the evolving nature of the angular momentum correlations between the dressed-quarks [209,213].

The Class-C DSE analyses in [209,213] implement a dressed-quark mass function that is distinguished by a particular transition rate between the non-perturbative and perturbative domains. If that rate were increased, then the transformation to partonic quarks would become more rapid; hence the proton's Pauli form factor would drop even more quickly to zero. In this event, the quark angular momentum correlations, embodied in the diquark structure, remain but the separate dressed-quark magnetic moments diminish markedly. Thus a quicker transition will push the zero in $\mu_p G_E^p/G_M^p$ to larger values of Q^2. Moreover, there will be a rate of transformation beyond which the zero disappears completely: there is no zero in a theory whose mass function rapidly becomes partonic. (For instance, pQCD analyses cannot produce a zero.) These expectations are realised in explicit calculations, as illustrated in Figure 4 of Ref. [209].

It follows that the possible existence and location of a zero in the ratio of proton elastic form factors $\mu_p G_E^p(Q^2)/G_M^p(Q^2)$ are a fairly direct measure of the character of EHM in the SM. Consequently, in pushing experimental measurements of this ratio, and thereby the proton's electric form factor, to larger momenta, i.e., in reliably determining the proton's charge distribution, there is an exceptional opportunity for a positive dialogue between experiment and theory. That feedback should assist greatly with contemporary efforts to reveal the character of the SM's strong interaction and its emergent phenomena.

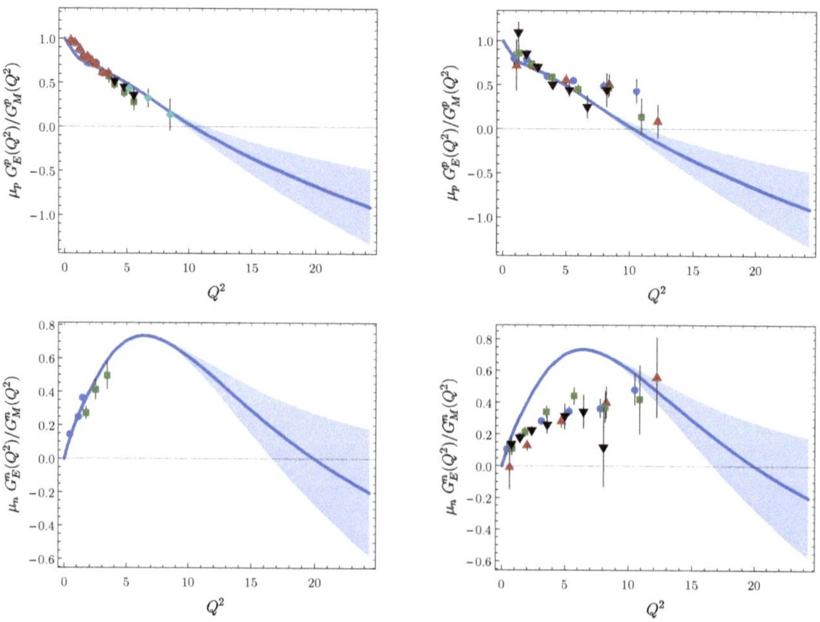

Figure 9. Ratios of Sach's form factors , $\mu_N G_E^N(Q^2)/G_M^N(Q^2)$. *Upper panels*— Proton. Left , [213] calculation compared with data (red up-triangles [238]; green squares [245]; blue circles [246]; black down-triangles [247]; and cyan diamonds [248]); right, compared with available lQCD results, drawn from Ref. [254]. *Lower panels*—Neutron. Left, comparison with data (blue circles [255] and green squares [256]); right, with available lQCD results, drawn from the work in [254]. In all panels, the 1σ band for the SPM approximants is shaded in light blue.

Given such potential, the analysis in [209] was recently revisited [213], using improved algorithms to calculate the form factors on $Q^2 \gtrsim 10\, m_p^2$. The new study employed a statistical implementation of the Schlessinger point method (SPM) [31,210,257–260] for the interpolation and extrapolation of smooth functions to deliver predictions for form factors on $Q^2 > 9\, m_p^2$ with a quantified uncertainty estimate. Results for the form factor ratios discussed herein are presented in Figure 9.

For the proton, the new analysis predicts

$$\mu_p \frac{G_E^p(Q_{z_p}^2)}{G_M^p(Q_{z_p}^2)} = 0\,, \quad Q_{z_p}^2 = 10.3^{+1.1}_{-0.7}\,\text{GeV}^2\,. \tag{33}$$

This value is compatible with, although a little larger than, that obtained earlier [209]: $Q_{z_p}^2 \approx 9.5\,\text{GeV}^2$; and a more recent inference based on ρ-meson elastic form factors [261]: $Q_{z_p}^2 \approx 9.4(3)\,\text{GeV}^2$.

Regarding the neutron, the authors of [209] predicted a peak in this ratio at $Q^2 \approx 6\,\text{GeV}^2$, which is reproduced in [213]. Furthermore, it located a zero at $Q_{z_n}^2 \approx 12\,\text{GeV}^2$. With the statistical SPM method for reaching to large-Q^2:

$$\mu_n \frac{G_E^n(Q_{z_n}^2)}{G_M^n(Q_{z_n}^2)} = 0\,, \quad Q_{z_n}^2 = 20.1^{+10.6}_{-\ 3.5}\,\text{GeV}^2\,, \tag{34}$$

viz. at 1σ SPM confidence level, this ratio is likely to exhibit a zero, but it probably lies beyond the reach of 12 GeV beams at JLab. On the other hand, the prediction of a peak in $R_{EM}^n = \mu_n G_E^n / G_M^n$, which is a harbinger of the zero in this ratio, can be tested at the 12 GeV JLab.

The properties of the dressed-quark propagators and bound-state amplitudes which influence the appearance of a zero in R^n_{EM} are qualitatively the same as those described in connection with R^p_{EM}. However, because of the different electric charge weightings attached to the quark contributions in the neutron (1 valence u-quark and 2 valence d-quarks), the quantitative effect is opposite to that for the proton (2 valence u-quarks and 1 valence d-quark). Namely, when the transformation from dressed-quark to parton is accelerated, the zero occurs at smaller Q^2. On the other hand, a model which generates a momentum-independent dressed-quark mass typically produces no zero in the neutron ratio [240].

In order to understand the source of these features, consider that the strange quark contributes very little to nucleon electromagnetic form factors [262,263], in which case one can write

$$G_E^p = e_u G_E^{p,u} - |e_d| G_E^{p,d}\,,\; G_E^n = e_u G_E^{n,u} - |e_d| G_E^{n,d} \tag{35}$$

($e_u = 2/3, e_d = -1/3$), where the contribution from each quark flavour is made explicit. Consider next that charge-symmetry is almost exact in QCD; to wit,

$$G_E^{n,d} = G_E^{p,u}\,,\quad G_E^{n,u} = G_E^{p,d}\,. \tag{36}$$

Therefore, to a very good level of approximation,

$$G_E^n = e_u G_E^{n,u} - |e_d| G_E^{n,d} = e_u G_E^{p,d} - |e_d| G_E^{p,u}\,. \tag{37}$$

Now, with a zero in G_E^p at $Q^2 = Q^2_{z_p} =: s_z^p$, one has $G_E^{p,d}(s_z) = 2\,G_E^{p,u}(s_z)$ and hence $G_E^n(s_z) = G_E^{p,u}(s_z) > 0$. This shows that although the behaviours of $G_E^{p,u}$ and G_E^p are qualitatively similar, the zero in $G_E^{p,u}$ occurs at a larger value of Q^2 than that in G_E^p itself. Under these conditions, any zero in G_E^n must occur at a larger value of Q^2 than the zero in G_E^p, a prediction confirmed in Figure 9.

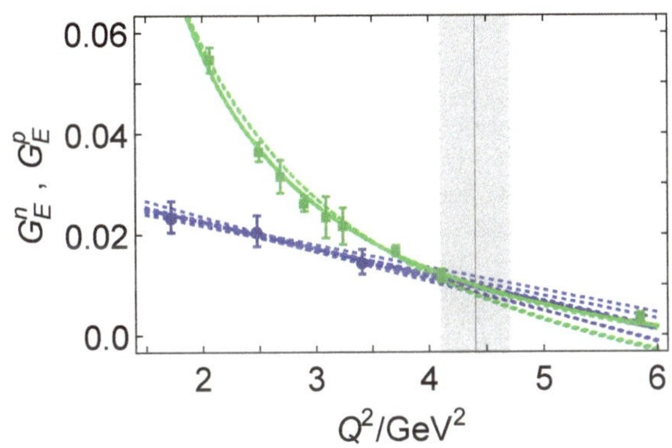

Figure 10. Data: neutron electric form factor—blue circles [256], and proton electric form factor—green squares [264]. Associated curves depict least-squares fits obtained via a jackknife analysis of the depicted G_E^n, G_E^p data. Vertical black line within grey bands marks the boundary of the domain described by Equation (38).

A curious effect follows from the combination of faster-than-dipole decrease of the proton's electric form factor (and probable appearance of a zero) and steady increase of R^n_{EM} to a peak at $Q^2 \approx 6\,\text{GeV}^2$. Namely, there must be a domain of larger Q^2 upon which the neutron's electric form factor is greater than that of the proton's; therefore, at some value of Q^2, the electric form factor of the neutral compound fermion is actually larger than that of its positive counterpart. This was predicted in [209] and confirmed by the updated analyses in [213]. It is therefore worth testing with available data [256,264]. Figure 10 depicts G_E^n, G_E^p data along with a set of associated least-squares fits produced using a jackknife analysis. On the domain currently covered by data, $G_E^n < G_E^p$; but extrapolation of the fits suggests that

$$G_E^n(Q^2) \overset{Q^2>Q^2_{n>p}}{>} G_E^p(Q^2)\,, \quad Q^2_{n>p} = 4.4(3)\,\mathrm{GeV}^2. \tag{38}$$

With 12 GeV operations under way at JLab, neutron electric form factor data to $Q^2 = 10.2\,\mathrm{GeV}^2$ will soon be available [265], enabling this prediction to be tested.

It is worth reiterating that the existence and location of a zero in G_E^p—similarly for G_E^n—are sensitive to the rate at which the dressed-quark mass function transits from the strong to perturbative QCD domains: denote this rate by R_M^{sp}. Note, too, that in contrast to $G_E^{p,u}$, $G_E^{p,d}$ evolves more slowly with changes in R_M^{sp}. This inertia derives from the d-quark being preferentially contained within a (soft) scalar diquark. Subject to these insights, consider Equation (37): with the location of a zero in $G_E^{p,d}$ shifting slowly to larger values of Q^2, but that in $G_E^{p,u}$ moving rapidly (as noted above, the zero in $G_E^{p,u}$ disappears if R_M^{sp} is sufficiently quick) one is subtracting from $G_E^{p,d}(Q^2)$ a function whose domain of positive support is increasing in size. That operation typically shifts the zero in G_E^n to smaller values of Q^2, potentially enabling a zero in G_E^n even when that in G_E^p has disappeared.

6. Epilogue

The Lagrangian that defines QCD, the strong interaction sector of the Standard Model (SM), appears very simple, yet it is responsible for a remarkable array of high-level phenomena with enormous apparent complexity. One can argue that the foundation for all such effects is laid by the emergence of hadronic mass (EHM), whose consequences are variously expressed in a wide range of empirical observables. This emergence itself is currently beyond a reductionist explanation, unless one is content to find that in the need for ultraviolet renormalisation of four-dimensional quantum field theory. Even supposing the latter, then the size of the associated mass-scale, Λ_{QCD}, is not something which can be predicted from within the SM.

The simplest expression of EHM is to be found in the generation of a running mass for the gluon. Mechanically, this is driven by gluon self-interactions, and the scale of the effect is known once Λ_{QCD} is fixed. It is interesting theoretically to explore the impact on observables induced by reducing/increasing Λ_{QCD}. Perhaps there is a critical value (or set of values) at which the Universe would appear very different?

Gluon mass generation in the SM entails that quarks, which are massless in the absence of a Higgs mechanism, also acquire a running mass, whose scale at infrared momenta is roughly one-third of the proton mass. This effect is known as dynamical chiral symmetry breaking (DCSB). One of its corollaries is the emergence of pseudoscalar Nambu–Goldstone bosons—most recognisably, the pions; and their appearance and properties have a profound impact on the character of the physical Universe. Nuclear physics, e.g., the formation of the elements and the number of elements that may be formed, seems to be very sensitive to the pion mass; and the pion mass is a quantity which is uniquely sensitive to the interplay between emergent and explicit (Higgs) mass generating mechanisms.

A non-perturbative framework that can unify the emergence of gluon and quark masses, and express their impacts on solutions of the associated bound-state equations in quantum field theory and also in the matrix elements which describe complex observable processes, can provide reductive explanations for physical phenomena. No single such framework exists, but an amalgam of high-level approaches is bearing fruit.

Funding: This work is supported in part by Jiangsu Province Hundred Talents Plan for Professionals.

Acknowledgments: These notes are based on results obtained and insights developed through collaborations with many people, to all of whom I am greatly indebted.

Conflicts of Interest: The author declares no conflict of interest.

Abbreviations

The following abbreviations are used in this manuscript.

CSM	continuum Schwinger-function method
DA (PDA)	(parton) distribution amplitude
DF (PDF)	(parton) distribution function
DCSB	dynamical chiral symmetry breaking
DSE	Dyson–Schwinger Equation
EFT	effective field theory
EHM	emergent hadronic mass
ERBL	Efremov–Radyushkin–Brodsky–Lepage
JLab	Thomas Jefferson National Accelerator Facility
lQCD	lattice-regularised quantum chromodynamics
PDG	Particle Data Group
PI	process independent
PFF	parton fragmentation function
pQCD	perturbative quantum chromodynamics
QED	quantum electrodynamics
QCD	quantum chromodynamics
RGI	renormalisation group invariant
SM	Standard Model (of Particle Physics)
SPM	Schlessinger point method
2PI	two-particle irreducible

References

1. Politzer, H.D. The dilemma of attribution. *Proc. Nat. Acad. Sci. USA* **2005**, *102*, 7789–7793. [CrossRef] [PubMed]
2. ATLAS Collaboration. Observation of a new particle in the search for the Standard Model Higgs boson with the ATLAS detector at the LHC. *Phys. Lett. B* **2012**, *716*, 1–29. [CrossRef]
3. CMS Collaboration. Observation of a New Boson at a Mass of 125 GeV with the CMS Experiment at the LHC. *Phys. Lett. B* **2012**, *716*, 30–61. [CrossRef]
4. Englert, F. Nobel Lecture: The BEH mechanism and its scalar boson. *Rev. Mod. Phys.* **2014**, *86*, 843. [CrossRef]
5. Higgs, P.W. Nobel Lecture: Evading the Goldstone theorem. *Rev. Mod. Phys.* **2014**, *86*, 851. [CrossRef]
6. Taylor, R.E. Deep inelastic scattering: The Early years. *Rev. Mod. Phys.* **1991**, *63*, 573–595. [CrossRef]
7. Kendall, H.W. Deep inelastic scattering: Experiments on the proton and the observation. *Rev. Mod. Phys.* **1991**, *63*, 597–614. [CrossRef]
8. Friedman, J.I. Deep inelastic scattering: Comparisons with the quark model. *Rev. Mod. Phys.* **1991**, *63*, 615–629. [CrossRef]
9. Neddermeyer, S.H.; Anderson, C.D. Note on the Nature of Cosmic-Ray Particles. *Phys. Rev.* **1937**, *51*, 884–886. [CrossRef]
10. Lattes, C.M.G.; Muirhead, H.; Occhialini, G.P.S.; Powell, C.F. Processes involving charged mesons. *Nature* **1947**, *159*, 694–697. [CrossRef]
11. Marciano, W.J.; Pagels, H. Quantum Chromodynamics: A Review. *Phys. Rept.* **1978**, *36*, 137. [CrossRef]
12. Marciano, W.J.; Pagels, H. Quantum Chromodynamics. *Nature* **1979**, *279*, 479–483. [CrossRef]
13. Particle Data Group. Review of Particle Properties. *Prog. Theor. Exp. Phys.* **2020**, *083C01*.
14. Qin, S.X.; Roberts, C.D.; Schmidt, S.M. Spectrum of light- and heavy-baryons. *Few Body Syst.* **2019**, *60*, 26. [CrossRef]
15. Durr, S.; Fodor, Z.; Frison, J.; Hoelbling, C.; Hoffmann, R.; Katz, S.D.; Krieg, S.; Kurth, T.; Lellouch, L.; Lippert, T.; et al. Ab-Initio Determination of Light Hadron Masses. *Science* **2008**, *322*, 1224–1227. [CrossRef]
16. Pascual, P.; Tarrach, R. *QCD: Renormalization for the Practitioner*; Springer: Berlin, Germany, 1984; Lecture Notes in Physics 194.
17. Cornwall, J.M. Dynamical Mass Generation in Continuum QCD. *Phys. Rev. D* **1982**, *26*, 1453. [CrossRef]
18. Gribov, V.N. The theory of quark confinement. *Eur. Phys. J. C* **1999**, *10*, 91–105. [CrossRef]

19. Dudal, D.; Verschelde, H.; Gracey, J.A.; Lemes, V.E.R.; Sarandy, M.S.; Sobreiro, R.F.; Sorella, S.P. Dynamical gluon mass generation from $\langle A^2(\mu)\rangle$ in linear covariant gauges. *JHEP* **2004**, *01*, 044. [CrossRef]
20. Bowman, P.O.; Heller, U.M.; Leinweber, D.B.; Parappilly, M.B.; Williams, A.G. Unquenched gluon propagator in Landau gauge. *Phys. Rev. D* **2004**, *70*, 034509. [CrossRef]
21. Luna, E.; Martini, A.; Menon, M.; Mihara, A.; Natale, A. Influence of a dynamical gluon mass in the pp and p anti-p forward scattering. *Phys. Rev. D* **2005**, *72*, 034019. [CrossRef]
22. Aguilar, A.; Binosi, D.; Papavassiliou, J. Gluon and ghost propagators in the Landau gauge: Deriving lattice results from Schwinger-Dyson equations. *Phys. Rev. D* **2008**, *78*, 025010. [CrossRef]
23. Rodríguez-Quintero, J. On the massive gluon propagator, the PT-BFM scheme and the low-momentum behaviour of decoupling and scaling DSE solutions. *JHEP* **2011**, *1101*, 105. [CrossRef]
24. Boucaud, P.; Leroy, J.P.; Le Yaouanc, A.; Micheli, J.; Pene, O.; Rodríguez-Quintero, J. The Infrared Behaviour of the Pure Yang-Mills Green Functions. *Few Body Syst.* **2012**, *53*, 387–436. [CrossRef]
25. Strauss, S.; Fischer, C.S.; Kellermann, C. Analytic structure of the Landau gauge gluon propagator. *Phys. Rev. Lett.* **2012**, *109*, 252001. [CrossRef]
26. Binosi, D.; Chang, L.; Papavassiliou, J.; Roberts, C.D. Bridging a gap between continuum-QCD and ab initio predictions of hadron observables. *Phys. Lett. B* **2015**, *742*, 183–188. [CrossRef]
27. Aguilar, A.C.; Binosi, D.; Papavassiliou, J. The Gluon Mass Generation Mechanism: A Concise Primer. *Front. Phys. China* **2016**, *11*, 111203. [CrossRef]
28. Siringo, F. Analytical study of Yang–Mills theory in the infrared from first principles. *Nucl. Phys. B* **2016**, *907*, 572–596. [CrossRef]
29. Cyrol, A.K.; Fister, L.; Mitter, M.; Pawlowski, J.M.; Strodthoff, N. Landau gauge Yang-Mills correlation functions. *Phys. Rev. D* **2016**, *94*, 054005. [CrossRef]
30. Gao, F.; Qin, S.X.; Roberts, C.D.; Rodríguez-Quintero, J. Locating the Gribov horizon. *Phys. Rev. D* **2018**, *97*, 034010. [CrossRef]
31. Binosi, D.; Tripolt, R.A. Spectral functions of confined particles. *Phys. Lett. B* **2020**, *801*, 135171. [CrossRef]
32. Binosi, D.; Mezrag, C.; Papavassiliou, J.; Roberts, C.D.; Rodríguez-Quintero, J. Process-independent strong running coupling. *Phys. Rev. D* **2017**, *96*, 054026. [CrossRef]
33. Rodríguez-Quintero, J.; Binosi, D.; Mezrag, C.; Papavassiliou, J.; Roberts, C.D. Process-independent effective coupling. From QCD Green's functions to phenomenology. *Few Body Syst.* **2018**, *59*, 121. [CrossRef]
34. Cui, Z.F.; Zhang, J.L.; Binosi, D.; de Soto, F.; Mezrag, C.; Papavassiliou, J.; Rodríguez-Quintero, J.; Roberts, C.D.; Segovia, J.; Zafeiropoulos, S. Effective charge from lattice QCD. *Chin. Phys. C* **2020**, *44*, 083102. [CrossRef]
35. Kharzeev, D. Quarkonium interactions in QCD. *Proc. Int. Sch. Phys. Fermi* **1996**, *130*, 105–131.
36. Close, F. The Quark Parton Model. *Rept. Prog. Phys.* **1979**, *42*, 1285–1335. [CrossRef]
37. Nambu, Y. Quasiparticles and Gauge Invariance in the Theory of Superconductivity. *Phys. Rev.* **1960**, *117*, 648–663. [CrossRef]
38. Goldstone, J. Field Theories with Superconductor Solutions. *Nuovo Cim.* **1961**, *19*, 154–164. [CrossRef]
39. Aguilar, A.C.; Ahmed, Z.; Aidala, C.; Ali, S.; Andrieux, V.; Arrington, J.A.; Bashir, A.; Berdnikov, V.; Binosi, D.; Chang, L.; et al. Pion and Kaon Structure at the Electron-Ion Collider. *Eur. Phys. J. A* **2019**, *55*, 190. [CrossRef]
40. Roberts, C.D. *Resonance Electroproduction and the Origin of Mass*—arXiv:1909.11102 [nucl-th]. In Proceedings of the 12th International Workshop on the Physics of Excited Nucleons (NSTAR 2019), Bonn, Germany, 10–14 June 2019.
41. Brodsky, S.J.; Burkert, V.D.; Carman, D.S.; Chen, J.P.; Cui, Z.F.; Döring, M.; Dosch, H.G.; Draayer, J.; Elouadrhiri, L.; Glazier, D.I.; et al. Strong QCD from Hadron Structure Experiments—arXiv:2006.06802 [hep-ph]. *Intern. J. Mod. Phys. E* **2020**, in press.
42. Roberts, C.D.; Schmidt, S.M. Reflections upon the Emergence of Hadronic Mass. *arXiv* **2020**, arXiv:2006.08782.
43. Aznauryan, I.; Bashir, A.; Braun, V.; Brodsky, S.J.; Burkert, V.D.; Chang, L.; Chen, C.; El-Bennich, B.; Cole, P.; Edwards, R.G.; et al. Studies of Nucleon Resonance Structure in Exclusive Meson Electroproduction. *Int. J. Mod. Phys. E* **2013**, *22*, 1330015. [CrossRef]
44. Gross, D.J. The discovery of asymptotic freedom and the emergence of QCD. *Proc. Nat. Acad. Sci. USA* **2005**, *102*, 9099–9108. [CrossRef] [PubMed]

45. Wilczek, F. Asymptotic freedom: From paradox to paradigm. *Proc. Nat. Acad. Sci. USA* **2005**, *102*, 8403–8413. [CrossRef] [PubMed]
46. Jaffe, A.M. The Millennium Grand Challenge in Mathematics. *Not. Am. Math. Soc.* **2006**, *53*, 652–660.
47. Casher, A. Chiral Symmetry Breaking in Quark Confining Theories. *Phys. Lett. B* **1979**, *83*, 395. [CrossRef]
48. Banks, T.; Casher, A. Chiral Symmetry Breaking in Confining Theories. *Nucl. Phys. B* **1980**, *169*, 103. [CrossRef]
49. McNeile, C. Lattice status of gluonia/glueballs. *Nucl. Phys. Proc. Suppl.* **2009**, *186*, 264–267. [CrossRef]
50. Wilson, K.G. Confinement of quarks. *Phys. Rev. D* **1974**, *10*, 2445–2459. [CrossRef]
51. Isgur, N.; Paton, J.E. A Flux Tube Model for Hadrons. *Phys. Lett. B* **1983**, *124*, 247–251. [CrossRef]
52. Bali, G.S.; Neff, H.; Duessel, T.; Lippert, T.; Schilling, K. Observation of string breaking in QCD. *Phys. Rev. D* **2005**, *71*, 114513. [CrossRef]
53. Prkacin, Z.; Bali, G.S.; Dussel, T.; Lippert, T.; Neff, H.; Schilling, K. Anatomy of string breaking in QCD. *PoS* **2006**, *LAT2005*, 308.
54. Chang, L.; Cloet, I.C.; El-Bennich, B.; Klähn, T.; Roberts, C.D. Exploring the light-quark interaction. *Chin. Phys. C* **2009**, *33*, 1189–1196.
55. Glimm, J.; Jaffee, A. *Quantum Physics. A Functional Point of View*; Springer: New York, NY, USA, 1981.
56. Munczek, H.J.; Nemirovsky, A.M. The Ground State $q\bar{q}$ Mass Spectrum in QCD. *Phys. Rev. D* **1983**, *28*, 181–186. [CrossRef]
57. Cahill, R.T.; Roberts, C.D. Soliton Bag Models of Hadrons from QCD. *Phys. Rev. D* **1985**, *32*, 2419. [CrossRef] [PubMed]
58. Stingl, M. Propagation Properties and Condensate Formation of the Confined Yang-Mills Field. *Phys. Rev. D* **1986**, *34*, 3863–3881; Erratum in **1987**, *36*, 651. [CrossRef]
59. Roberts, C.D.; Williams, A.G.; Krein, G. On the implications of confinement. *Int. J. Mod. Phys. A* **1992**, *7*, 5607–5624. [CrossRef]
60. Burden, C.J.; Roberts, C.D.; Williams, A.G. Singularity structure of a model quark propagator. *Phys. Lett. B* **1992**, *285*, 347–353. [CrossRef]
61. Hawes, F.T.; Roberts, C.D.; Williams, A.G. Dynamical chiral symmetry breaking and confinement with an infrared vanishing gluon propagator. *Phys. Rev. D* **1994**, *49*, 4683–4693. [CrossRef]
62. Maris, P. Analytic structure of the full fermion propagator in quenched and unquenched QED. *Phys. Rev. D* **1994**, *50*, 4189–4193. [CrossRef]
63. Roberts, C.D.; Williams, A.G. Dyson-Schwinger equations and their application to hadronic physics. *Prog. Part. Nucl. Phys.* **1994**, *33*, 477–575. [CrossRef]
64. Bhagwat, M.; Pichowsky, M.; Tandy, P.C. Confinement phenomenology in the Bethe-Salpeter equation. *Phys. Rev. D* **2003**, *67*, 054019. [CrossRef]
65. Roberts, C.D. Hadron Properties and Dyson-Schwinger Equations. *Prog. Part. Nucl. Phys.* **2008**, *61*, 50–65. [CrossRef]
66. Bashir, A.; Raya, A.; Sánchez-Madrigal, S.; Roberts, C.D. Gauge invariance of a critical number of flavours in QED3. *Few Body Syst.* **2009**, *46*, 229–237. [CrossRef]
67. Bashir, A.; Raya, A.; Rodríguez-Quintero, J. QCD: Restoration of Chiral Symmetry and Deconfinement for Large N_f. *Phys. Rev. D* **2013**, *88*, 054003. [CrossRef]
68. Qin, S.X.; Rischke, D.H. Quark Spectral Function and Deconfinement at Nonzero Temperature. *Phys. Rev. D* **2013**, *88*, 056007. [CrossRef]
69. Lowdon, P. Conditions on the violation of the cluster decomposition property in QCD. *J. Math. Phys.* **2016**, *57*, 102302. [CrossRef]
70. Lucha, W.; Schöberl, F.F. Analytic Bethe-Salpeter Description of the Lightest Pseudoscalar Mesons. *Phys. Rev. D* **2016**, *93*, 056006. [CrossRef]
71. Binosi, D.; Roberts, C.D.; Rodríguez-Quintero, J. Scale-setting, flavour dependence and chiral symmetry restoration. *Phys. Rev. D* **2017**, *95*, 114009. [CrossRef]
72. Krein, G.; Nielsen, M.; Puff, R.D.; Wilets, L. Ghost poles in the nucleon propagator: Vertex corrections and form-factors. *Phys. Rev. C* **1993**, *47*, 2485–2491. [CrossRef]
73. Bracco, M.E.; Eiras, A.; Krein, G.; Wilets, L. Selfconsistent solution of the Schwinger-Dyson equations for the nucleon and meson propagators. *Phys. Rev. C* **1994**, *49*, 1299–1308. [CrossRef]
74. Gattringer, C.; Lang, C.B. Quantum chromodynamics on the lattice. *Lect. Notes Phys.* **2010**, *788*, 1–343.

75. Philipsen, O. The QCD equation of state from the lattice. *Prog. Part. Nucl. Phys.* **2013**, *70*, 55–107. [CrossRef]
76. Bañuls, M.C.; Cichy, K. Review on Novel Methods for Lattice Gauge Theories. *Rept. Prog. Phys.* **2020**, *83*, 024401. [CrossRef]
77. Roberts, C.D.; Schmidt, S.M. Dyson-Schwinger equations: Density, temperature and continuum strong QCD. *Prog. Part. Nucl. Phys.* **2000**, *45*, S1–S103. [CrossRef]
78. Maris, P.; Roberts, C.D. Dyson-Schwinger equations: A tool for hadron physics. *Int. J. Mod. Phys. E* **2003**, *12*, 297–365. [CrossRef]
79. Chang, L.; Roberts, C.D.; Tandy, P.C. Selected highlights from the study of mesons. *Chin. J. Phys.* **2011**, *49*, 955–1004.
80. Roberts, C.D. Strong QCD and Dyson-Schwinger Equations. *IRMA Lect. Math. Theor. Phys.* **2015**, *21*, 356–458.
81. Roberts, C.D. Three Lectures on Hadron Physics. *J. Phys. Conf. Ser.* **2016**, *706*, 022003. [CrossRef]
82. Horn, T.; Roberts, C.D. The pion: An enigma within the Standard Model. *J. Phys. G* **2016**, *43*, 073001. [CrossRef]
83. Eichmann, G.; Sanchis-Alepuz, H.; Williams, R.; Alkofer, R.; Fischer, C.S. Baryons as relativistic three-quark bound states. *Prog. Part. Nucl. Phys.* **2016**, *91*, 1–100. [CrossRef]
84. Burkert, V.D.; Roberts, C.D. Roper resonance: Toward a solution to the fifty year puzzle. *Rev. Mod. Phys.* **2019**, *91*, 011003. [CrossRef]
85. Fischer, C.S. QCD at finite temperature and chemical potential from Dyson–Schwinger equations. *Prog. Part. Nucl. Phys.* **2019**, *105*, 1–60. [CrossRef]
86. Qin, S.X.; Roberts, C.D. Impressions of the Continuum Bound State Problem in QCD. *arXiv* **2020**, arXiv:2008.07629.
87. Munczek, H.J. Dynamical chiral symmetry breaking, Goldstone's theorem and the consistency of the Schwinger-Dyson and Bethe-Salpeter Equations. *Phys. Rev. D* **1995**, *52*, 4736–4740. [CrossRef] [PubMed]
88. Bender, A.; Roberts, C.D.; von Smekal, L. Goldstone Theorem and Diquark Confinement Beyond Rainbow-Ladder Approximation. *Phys. Lett. B* **1996**, *380*, 7–12. [CrossRef]
89. Maris, P.; Roberts, C.D.; Tandy, P.C. Pion mass and decay constant. *Phys. Lett. B* **1998**, *420*, 267–273. [CrossRef]
90. Chang, L.; Roberts, C.D. Sketching the Bethe-Salpeter kernel. *Phys. Rev. Lett.* **2009**, *103*, 081601. [CrossRef]
91. Qin, S.X.; Roberts, C.D.; Schmidt, S.M. Ward-Green-Takahashi identities and the axial-vector vertex. *Phys. Lett. B* **2014**, *733*, 202–208. [CrossRef]
92. Williams, R.; Fischer, C.S.; Heupel, W. Light mesons in QCD and unquenching effects from the 3PI effective action. *Phys. Rev. D* **2016**, *93*, 034026. [CrossRef]
93. Binosi, D.; Chang, L.; Qin, S.X.; Papavassiliou, J.; Roberts, C.D. Symmetry preserving truncations of the gap and Bethe-Salpeter equations. *Phys. Rev. D* **2016**, *93*, 096010. [CrossRef]
94. Accardi, A.; Albacete, J.L.; Anselmino, M.; Armesto, N.; Aschenauer, E.C.; Bacchetta, A.; Boer, D.; Brooks, W.K.; Burton, T.; Chang, N.B.; et al. Electron Ion Collider: The Next QCD Frontier. *Eur. Phys. J. A* **2016**, *52*, 268. [CrossRef]
95. Brodsky, S.J.; Deshpande, A.L.; Gao, H.; McKeown, C.A.; Meziani, Z.E.; Milner, R.G.; Qiu, J.W.; Richards, D.G.; Roberts, C.D. QCD and Hadron Physics. *arXiv* **2015**, arXiv:1502.05728.
96. Field, R.D.; Feynman, R.P. A Parametrization of the Properties of Quark Jets. *Nucl. Phys. B* **1978**, *136*, 1–76. [CrossRef]
97. Gell-Mann, M.; Low, F.E. Quantum electrodynamics at small distances. *Phys. Rev.* **1954**, *95*, 1300–1312. [CrossRef]
98. Cornwall, J.M.; Papavassiliou, J. Gauge Invariant Three Gluon Vertex in QCD. *Phys. Rev. D* **1989**, *40*, 3474. [CrossRef] [PubMed]
99. Pilaftsis, A. Generalized pinch technique and the background field method in general gauges. *Nucl. Phys. B* **1997**, *487*, 467–491. [CrossRef]
100. Binosi, D.; Papavassiliou, J. Pinch Technique: Theory and Applications. *Phys. Rept.* **2009**, *479*, 1–152. [CrossRef]
101. Abbott, L.F. The Background Field Method Beyond One Loop. *Nucl. Phys. B* **1981**, *185*, 189. [CrossRef]
102. Deur, A.; Burkert, V.; Chen, J.P.; Korsch, W. Experimental determination of the effective strong coupling constant. *Phys. Lett. B* **2007**, *650*, 244–248. [CrossRef]
103. Deur, A.; Burkert, V.; Chen, J.P.; Korsch, W. Determination of the effective strong coupling constant $\alpha_{g_1}(s)$ from CLAS spin structure function data. *Phys. Lett. B* **2008**, *665*, 349–351. [CrossRef]

104. Deur, A.; Brodsky, S.J.; de Teramond, G.F. The QCD Running Coupling. *Prog. Part. Nucl. Phys.* **2016**, *90*, 1–74. [CrossRef]
105. Grunberg, G. Renormalization Scheme Independent QCD and QED: The Method of Effective Charges. *Phys. Rev. D* **1984**, *29*, 2315. [CrossRef]
106. Dokshitzer, Y.L. *Perturbative QCD theory (includes our knowledge of $\alpha(s)$)*—hep-ph/9812252. In Proceedings of the 29th International Conference on High Energy Physics: ICHEP '98, Vancouver, CU, Canada, 23–29 July 1998; Volumes 1, 2, pp. 305–324.
107. Appelquist, T.; Terning, J.; Wijewardhana, L.C.R. The zero temperature chiral phase transition in SU(N) gauge theories. *Phys. Rev. Lett.* **1996**, *77*, 1214–1217. [CrossRef] [PubMed]
108. Sannino, F. Conformal Dynamics for TeV Physics and Cosmology. *Acta Phys. Polon. B* **2009**, *40*, 3533–3743.
109. LSD Collaboration. Toward TeV Conformality. *Phys. Rev. Lett.* **2010**, *104*, 071601. [CrossRef] [PubMed]
110. Hayakawa, M.; Ishikawa, K.I.; Osaki, Y.; Takeda, S.; Uno, S.; Yamada, N. Running coupling constant of ten-flavor QCD with the Schrődinger functional method. *Phys. Rev. D* **2011**, *83*, 074509. [CrossRef]
111. Cheng, A.; Hasenfratz, A.; Petropoulos, G.; Schaich, D. Scale-dependent mass anomalous dimension from Dirac eigenmodes. *JHEP* **2013**, *07*, 061. [CrossRef]
112. LatKMI Collaboration. Walking signals in $N_f = 8$ QCD on the lattice. *Phys. Rev. D* **2013**, *87*, 094511. [CrossRef]
113. DeGrand, T. Lattice tests of beyond Standard Model dynamics. *Rev. Mod. Phys.* **2016**, *88*, 015001. [CrossRef]
114. Cui, Z.F.; Ding, M.; Gao, F.; Raya, K.; Binosi, D.; Chang, L.; Roberts, C.D.; Rodríguez-Quintero, J.; Schmidt, S.M. Kaon parton distributions: Revealing Higgs modulation of emergent mass. *arXiv* **2020**, arXiv:2006.14075.
115. Cui, Z.F.; Ding, M.; Gao, F.; Raya, K.; Binosi, D.; Chang, L.; Roberts, C.D.; Rodríguez-Quintero, J.; Schmidt, S.M. Kaon and pion parton distributions. **2020**, in progress.
116. Nambu, Y. From BCS to NJL: An old story retold. *AIP Conf. Proc.* **2011**, *1388*, 86–92.
117. Brodsky, S.J.; Shrock, R. Condensates in Quantum Chromodynamics and the Cosmological Constant. *Proc. Natl. Acad. Sci. USA* **2011**, *108*, 45–50. [CrossRef]
118. Brodsky, S.J.; Roberts, C.D.; Shrock, R.; Tandy, P.C. New perspectives on the quark condensate. *Phys. Rev. C* **2010**, *82*, 022201(R). [CrossRef]
119. Chang, L.; Roberts, C.D.; Tandy, P.C. Expanding the concept of in-hadron condensates. *Phys. Rev. C* **2012**, *85*, 012201(R). [CrossRef]
120. Brodsky, S.J.; Roberts, C.D.; Shrock, R.; Tandy, P.C. Confinement contains condensates. *Phys. Rev. C* **2012**, *85*, 065202. [CrossRef]
121. Bhagwat, M.S.; Pichowsky, M.A.; Roberts, C.D.; Tandy, P.C. Analysis of a quenched lattice QCD dressed quark propagator. *Phys. Rev. C* **2003**, *68*, 015203. [CrossRef]
122. Bhagwat, M.S.; Tandy, P.C. Analysis of full-QCD and quenched-QCD lattice propagators. *AIP Conf. Proc.* **2006**, *842*, 225–227.
123. Bowman, P.O.; Heller, U.M.; Leinweber, D.B.; Parappilly, M.B.; Williams, A.G.; Zhang, Z.B. Unquenched quark propagator in Landau gauge. *Phys. Rev. D* **2005**, *71*, 054507. [CrossRef]
124. Roberts, C.D. Perspective on the origin of hadron masses. *Few Body Syst.* **2017**, *58*, 5. [CrossRef]
125. Höll, A.; Krassnigg, A.; Roberts, C.D. Pseudoscalar meson radial excitations. *Phys. Rev. C* **2004**, *70*, 042203(R). [CrossRef]
126. Höll, A.; Krassnigg, A.; Maris, P.; Roberts, C.D.; Wright, S.V. Electromagnetic properties of ground and excited state pseudoscalar mesons. *Phys. Rev. C* **2005**, *71*, 065204. [CrossRef]
127. Bhagwat, M.S.; Chang, L.; Liu, Y.X.; Roberts, C.D.; Tandy, P.C. Flavour symmetry breaking and meson masses. *Phys. Rev. C* **2007**, *76*, 045203. [CrossRef]
128. Ding, M.; Raya, K.; Bashir, A.; Binosi, D.; Chang, L.; Chen, M.; Roberts, C.D. $\gamma^*\gamma \to \eta, \eta'$ transition form factors. *Phys. Rev. D* **2019**, *99*, 014014. [CrossRef]
129. Ding, M.; Gao, F.; Chang, L.; Liu, Y.X.; Roberts, C.D. Leading-twist parton distribution amplitudes of S-wave heavy-quarkonia. *Phys. Lett. B* **2016**, *753*, 330–335. [CrossRef]
130. Chen, C.; Chang, L.; Roberts, C.D.; Wan, S.; Zong, H.S. Valence-quark distribution functions in the kaon and pion. *Phys. Rev. D* **2016**, *93*, 074021. [CrossRef]
131. Gao, F.; Chang, L.; Liu, Y.X.; Roberts, C.D.; Tandy, P.C. Exposing strangeness: Projections for kaon electromagnetic form factors. *Phys. Rev. D* **2017**, *96*, 034024. [CrossRef]

132. Flambaum, V.V.; Höll, A.; Jaikumar, P.; Roberts, C.D.; Wright, S.V. Sigma terms of light-quark hadrons. *Few Body Syst.* **2006**, *38*, 31–51. [CrossRef]
133. Guth, A.H.; Huang, K.; Jaffe, R.L. (Eds.) *Asymptotic Realms of Physics*; MIT Press: Cambridge, MA, USA, 1983; 262p.
134. Chang, L.; Cloet, I.C.; Cobos-Martinez, J.J.; Roberts, C.D.; Schmidt, S.M.; Tandy, P.C. Imaging dynamical chiral symmetry breaking: Pion wave function on the light front. *Phys. Rev. Lett.* **2013**, *110*, 132001. [CrossRef]
135. Segovia, J.; Chang, L.; Cloet, I.C.; Roberts, C.D.; Schmidt, S.M.; Zong, H.S. Distribution amplitudes of light-quark mesons from lattice QCD. *Phys. Lett. B* **2014**, *731*, 13–18. [CrossRef]
136. Shi, C.; Chang, L.; Roberts, C.D.; Schmidt, S.M.; Tandy, P.C.; Zong, H.S. Flavour symmetry breaking in the kaon parton distribution amplitude. *Phys. Lett. B* **2014**, *738*, 512–518. [CrossRef]
137. Lepage, G.P.; Brodsky, S.J. Exclusive Processes in Quantum Chromodynamics: Evolution Equations for Hadronic Wave Functions and the Form-Factors of Mesons. *Phys. Lett. B* **1979**, *87*, 359–365. [CrossRef]
138. Efremov, A.V.; Radyushkin, A.V. Factorization and Asymptotical Behavior of Pion Form- Factor in QCD. *Phys. Lett. B* **1980**, *94*, 245–250. [CrossRef]
139. Lepage, G.P.; Brodsky, S.J. Exclusive Processes in Perturbative Quantum Chromodynamics. *Phys. Rev. D* **1980**, *22*, 2157–2198. [CrossRef]
140. Chernyak, V.L.; Zhitnitsky, A.R. Asymptotic Behavior of Exclusive Processes in QCD. *Phys. Rept.* **1984**, *112*, 173. [CrossRef]
141. Maris, P.; Roberts, C.D. π and K meson Bethe-Salpeter amplitudes. *Phys. Rev. C* **1997**, *56*, 3369–3383. [CrossRef]
142. Maris, P.; Tandy, P.C. Bethe-Salpeter study of vector meson masses and decay constants. *Phys. Rev. C* **1999**, *60*, 055214. [CrossRef]
143. Bhagwat, M.S.; Maris, P. Vector meson form factors and their quark-mass dependence. *Phys. Rev. C* **2008**, *77*, 025203. [CrossRef]
144. Krassnigg, A. Survey of J=0,1 mesons in a Bethe-Salpeter approach. *Phys. Rev. D* **2009**, *80*, 114010. [CrossRef]
145. Qin, S.X.; Chang, L.; Liu, Y.x.; Roberts, C.D.; Wilson, D.J. Investigation of rainbow-ladder truncation for excited and exotic mesons. *Phys. Rev. C* **2012**, *85*, 035202. [CrossRef]
146. Mikhailov, S.; Radyushkin, A. Nonlocal condensates and QCD sum rules for pion wave function. *JETP Lett.* **1986**, *43*, 712–715. [CrossRef] [PubMed]
147. Petrov, V.Y.; Polyakov, M.V.; Ruskov, R.; Weiss, C.; Goeke, K. Pion and photon light cone wave functions from the instanton vacuum. *Phys. Rev. D* **1999**, *59*, 114018. [CrossRef]
148. Braun, V.; Gockeler, M.; Horsley, R.; Perlt, H.; Pleiter, D.; Rakow, P.E.L.; Schierholz, G.; Schiller, A.; Schroers, W.; Stuben, H.; et al. Moments of pseudoscalar meson distribution amplitudes from the lattice. *Phys. Rev. D* **2006**, *74*, 074501. [CrossRef]
149. Brodsky, S.J.; de Teramond, G.F. Hadronic spectra and light-front wavefunctions in holographic QCD. *Phys. Rev. Lett.* **2006**, *96*, 201601, [CrossRef] [PubMed]
150. Jefferson Lab F_π Collaboration. Measurement of the Charged Pion Electromagnetic Form-Factor. *Phys. Rev. Lett.* **2001**, *86*, 1713–1716. [CrossRef]
151. Jefferson Lab F_π-2 Collaboration. Determination of the Charged Pion Form Factor at $Q^2 = 1.60$ and $2.45\,(\mathrm{GeV/c})^2$. *Phys. Rev. Lett.* **2006**, *97*, 192001. [CrossRef]
152. Jefferson Lab F_π Collaboration. Determination of the pion charge form factor for $Q^2 = 0.60 - 1.60\,\mathrm{GeV}^2$. *Phys. Rev. C* **2007**, *75*, 055205. [CrossRef]
153. Horn, T.; Qian, X.; Arrington, J.A.; Asaturyan, R.; Benmokthar, F.; Boeglin, W.; Bosted, P.; Bruell, A.; Christy, M.E.; Chudakov, E.; et al. Scaling study of the pion electroproduction cross sections and the pion form factor. *Phys. Rev. C* **2008**, *78*, 058201. [CrossRef]
154. Jefferson Lab F_π Collaboration. Charged pion form-factor between $Q^2 = 0.60\,\mathrm{GeV}^2$ and $2.45\,\mathrm{GeV}^2$. II. Determination of, and results for, the pion form-factor. *Phys. Rev. C* **2008**, *78*, 045203. [CrossRef]
155. Jefferson Lab F_π Collaboration. Charged pion form factor between $Q^2 = 0.60$ and $2.45\,\mathrm{GeV}^2$. I. Measurements of the cross section for the $^1\mathrm{H}(e,e'\pi^+)n$ reaction. *Phys. Rev. C* **2008**, *78*, 045202. [CrossRef]
156. Maris, P.; Tandy, P.C. The π, K^+, and K^0 electromagnetic form factors. *Phys. Rev. C* **2000**, *62*, 055204. [CrossRef]
157. Chang, L.; Cloet, I.C.; Roberts, C.D.; Schmidt, S.M.; Tandy, P.C. Pion electromagnetic form factor at spacelike momenta. *Phys. Rev. Lett.* **2013**, *111*, 141802. [CrossRef] [PubMed]

158. Chen, M.; Ding, M.; Chang, L.; Roberts, C.D. Mass-dependence of pseudoscalar meson elastic form factors. *Phys. Rev. D* **2018**, *98*, 091505. [CrossRef]
159. Huber, G.M.; Gaskell, D.; Papandreou, Z.; Bosted, P.; Bruell, A.; Ent, R.; Fenker, H.C.; Gaskel, D.; Horn, T.; Jones, M.K.; et al. *Measurement of the Charged Pion Form Factor to High Q^2*; Jefferson Lab Experiment E12-06-101; 2006. Available online: http://www.jlab.org/exp_prog/proposals/06/PR12-06-101.pdf (accessed on 7 July 2006).
160. Horn, T.; Huber, G.M. Jefferson Lab Experiment E12-07-105. 2007. Available online: https://www.jlab.org/exp_prog/experiments/summaries/E12-07-105_summary.pdf (accessed on 22 August 2020).
161. Ellis, R.K.; Stirling, W.J.; Webber, B.R. *QCD and Collider Physics*; Cambridge University Press: Cambridge, UK, 2011.
162. Ezawa, Z.F. Wide-Angle Scattering in Softened Field Theory. *Nuovo Cim. A* **1974**, *23*, 271–290. [CrossRef]
163. Farrar, G.R.; Jackson, D.R. Pion and Nucleon Structure Functions Near x=1. *Phys. Rev. Lett.* **1975**, *35*, 1416. [CrossRef]
164. Berger, E.L.; Brodsky, S.J. Quark Structure Functions of Mesons and the Drell-Yan Process. *Phys. Rev. Lett.* **1979**, *42*, 940–944. [CrossRef]
165. Gribov, V.N.; Lipatov, L.N. Deep inelastic e p scattering in perturbation theory. *Sov. J. Nucl. Phys.* **1972**, *15*, 438–450.
166. Lipatov, L.N. The parton model and perturbation theory. *Sov. J. Nucl. Phys.* **1975**, *20*, 94–102.
167. Altarelli, G.; Parisi, G. Asymptotic Freedom in Parton Language. *Nucl. Phys. B* **1977**, *126*, 298. [CrossRef]
168. Dokshitzer, Y.L. Calculation of the Structure Functions for Deep Inelastic Scattering and e+ e- Annihilation by Perturbation Theory in Quantum Chromodynamics. *Sov. Phys. JETP* **1977**, *46*, 641–653. (In Russian).
169. Keppel, C.; Wojtsekhowski, B.; King, P.; Dutta, D.; Annand, J.; Zhang, J. *Measurement of Tagged Deep Inelastic Scattering (TDIS)*; Jefferson Lab experiment PR12-15-006; 2015, approved. Available online: https://www.jlab.org/exp_prog/proposals/15/PR12-15-006.pdf (accessed on 22 August 2020).
170. Park, K.; Montgomery, R.; Horn, T. *Measurement of Kaon Structure Function through Tagged Deep Inelastic Scattering (TDIS)*; Jefferson Lab experiment C12-15-006A; 2017, approved. Available online: https://www.jlab.org/exp_prog/proposals/17/C12-15-006A.pdf (accessed on 22 August 2020).
171. COMPASS++/AMBER Collaboration. Letter of Intent (Draft 2.0): A New QCD facility at the M2 beam line of the CERN SPS. *arXiv* **2018**, arXiv:1808.00848.
172. Xu, S.S.; Chang, L.; Roberts, C.D.; Zong, H.S. Pion and kaon valence-quark parton quasidistributions. *Phys. Rev. D* **2018**, *97*, 094014. [CrossRef]
173. Zhang, J.H.; Chen, J.W.; Jin, L.; Lin, H.W.; Schäfer, A.; Zhao, Y. First direct lattice-QCD calculation of the x-dependence of the pion parton distribution function. *Phys. Rev. D* **2019**, *100*, 034505. [CrossRef]
174. Karthik, N.; Izubichi, T.; Jin, L.; Kallidonis, C.; Mukherjee, S.; Petreczky, P.; Shugert, C.; Syritsyn, S. Renormalized quasi parton distribution function of pion. *PoS* **2018**, *LATTICE2018*, 109.
175. Sufian, R.S.; Karpie, J.; Egerer, C.; Orginos, K.; Qiu, J.W.; Richards, D.G. Pion Valence Quark Distribution from Matrix Element Calculated in Lattice QCD. *Phys. Rev. D* **2019**, *99*, 074507 [CrossRef]
176. Izubuchi, T.; Jin, L.; Kallidonis, C.; Karthik, N.; Mukherjee, S.; Petreczky, P.; Shugert, C.; Syritsyn, S. Valence parton distribution function of pion from fine lattice. *Phys. Rev. D* **2019**, *100*, 034516. [CrossRef]
177. Oehm, M.; Alexandrou, C.; Constantinou, M.; Jansen, K.; Koutsou, G.; Kostrzewa, B.; Steffens, F.; Urbach, C.; Zafeiropoulos, S. $\langle x \rangle$ and $\langle x^2 \rangle$ of the pion PDF from lattice QCD with $N_f = 2+1+1$ dynamical quark flavors. *Phys. Rev. D* **2019**, *99*, 014508. [CrossRef]
178. Joó, B.; Orginos, K.; Radyushkin, A.V.; Richards, D.G.; Sufian, R.S.; Zafeiropoulos, S. Pion valence structure from Ioffe-time parton pseudodistribution functions. *Phys. Rev. D* **2019**, *100*, 114512. [CrossRef]
179. Hecht, M.B.; Roberts, C.D.; Schmidt, S.M. Valence-quark distributions in the pion. *Phys. Rev. C* **2001**, *63*, 025213. [CrossRef]
180. Chang, L.; Mezrag, C.; Moutarde, H.; Roberts, C.D.; Rodríguez-Quintero, J.; Tandy, P.C. Basic features of the pion valence-quark distribution function. *Phys. Lett. B* **2014**, *737*, 23–29. [CrossRef]
181. Ding, M.; Raya, K.; Binosi, D.; Chang, L.; Roberts, C.D.; Schmidt, S.M. Drawing insights from pion parton distributions. *Chin. Phys. C* **2020**, *44*, 031002. [CrossRef]
182. Ding, M.; Raya, K.; Binosi, D.; Chang, L.; Roberts, C.D.; Schmidt, S.M. Symmetry, symmetry breaking, and pion parton distributions. *Phys. Rev. D* **2020**, *101*, 054014. [CrossRef]

183. Raya, K.; Chang, L.; Bashir, A.; Cobos-Martinez, J.J.; Gutiérrez-Guerrero, L.X.; Roberts, C.D.; Tandy, P.C. Structure of the neutral pion and its electromagnetic transition form factor. *Phys. Rev. D* **2016**, *93*, 074017. [CrossRef]
184. Raya, K.; Ding, M.; Bashir, A.; Chang, L.; Roberts, C.D. Partonic structure of neutral pseudoscalars via two photon transition form factors. *Phys. Rev. D* **2017**, *95*, 074014. [CrossRef]
185. Barry, P.C.; Sato, N.; Melnitchouk, W.; Ji, C.R. First Monte Carlo Global QCD Analysis of Pion Parton Distributions. *Phys. Rev. Lett.* **2018**, *121*, 152001. [CrossRef]
186. Aicher, M.; Schäfer, A.; Vogelsang, W. Soft-Gluon Resummation and the Valence Parton Distribution Function of the Pion. *Phys. Rev. Lett.* **2010**, *105*, 252003. [CrossRef]
187. Westmark, D.; Owens, J.F. Enhanced threshold resummation formalism for lepton pair production and its effects in the determination of parton distribution functions. *Phys. Rev. D* **2017**, *95*, 056024. [CrossRef]
188. Novikov, I.; Abdolmaleki, H.; Britzger, D.; Cooper-Sarkar, A.; Giuli, F.; Glazov, A.; Kusina, A.; Luszczak, A.; Olness, F.; Starovoitov, P.; et al. Parton Distribution Functions of the Charged Pion Within The xFitter. *Phys. Rev. D* **2020**, *102*, 014040. [CrossRef]
189. Conway, J.S.; Adolphsen, C.E.; Alexander, J.P.; Anderson, K.J.; Heinrich, J.G.; Pilcher, J.E.; Possoz, A.; Rosenberg, E.I.; Biino, C.; Greenhalgh, J.F.; et al. Experimental study of muon pairs produced by 252-GeV pions on tungsten. *Phys. Rev. D* **1989**, *39*, 92–122. [CrossRef]
190. Barabanov, M.Y.; Bedolla, M.A.; Brooks, W.K.; Cates, G.D.; Chen, C.; Chen, Y.; Cisbani, E.; Ding, M.; Eichmann, G.; Ent, R.; et al. Diquark Correlations in Hadron Physics: Origin, Impact and Evidence. *arXiv* **2020**, arXiv:2008.07630.
191. Bhagwat, M.S.; Höll, A.; Krassnigg, A.; Roberts, C.D.; Tandy, P.C. Aspects and consequences of a dressed-quark-gluon vertex. *Phys. Rev. C* **2004**, *70*, 035205. [CrossRef]
192. Cahill, R.T.; Roberts, C.D.; Praschifka, J. Calculation of diquark masses in QCD. *Phys. Rev. D* **1987**, *36*, 2804. [CrossRef] [PubMed]
193. Maris, P. Effective masses of diquarks. *Few Body Syst.* **2002**, *32*, 41–52. [CrossRef]
194. Eichmann, G.; Cloet, I.C.; Alkofer, R.; Krassnigg, A.; Roberts, C.D. Toward unifying the description of meson and baryon properties. *Phys. Rev. C* **2009**, *79*, 012202(R). [CrossRef]
195. Segovia, J.; El-Bennich, B.; Rojas, E.; Cloet, I.C.; Roberts, C.D.; Xu, S.S.; Zong, H.S. Completing the picture of the Roper resonance. *Phys. Rev. Lett.* **2015**, *115*, 171801. [CrossRef]
196. Segovia, J.; Roberts, C.D.; Schmidt, S.M. Understanding the nucleon as a Borromean bound-state. *Phys. Lett. B* **2015**, *750*, 100–106. [CrossRef]
197. Eichmann, G.; Fischer, C.S.; Sanchis-Alepuz, H. Light baryons and their excitations. *Phys. Rev. D* **2016**, *94*, 094033. [CrossRef]
198. Lu, Y.; Chen, C.; Roberts, C.D.; Segovia, J.; Xu, S.S.; Zong, H.S. Parity partners in the baryon resonance spectrum. *Phys. Rev. C* **2017**, *96*, 015208. [CrossRef]
199. Chen, C.; El-Bennich, B.; Roberts, C.D.; Schmidt, S.M.; Segovia, J.; Wan, S. Structure of the nucleon's low-lying excitations. *Phys. Rev. D* **2018**, *97*, 034016. [CrossRef]
200. Roberts, C.D.; Holt, R.J.; Schmidt, S.M. Nucleon spin structure at very high x. *Phys. Lett. B* **2013**, *727*, 249–254. [CrossRef]
201. Segovia, J.; Chen, C.; Cloet, I.C.; Roberts, C.D.; Schmidt, S.M.; Wan, S. Elastic and transition form factors of the $\Delta(1232)$. *Few Body Syst.* **2014**, *55*, 1–33. [CrossRef]
202. Segovia, J.; Roberts, C.D. Dissecting nucleon transition electromagnetic form factors. *Phys. Rev. C* **2016**, *94*, 042201(R). [CrossRef]
203. Mezrag, C.; Segovia, J.; Chang, L.; Roberts, C.D. Parton distribution amplitudes: Revealing correlations within the proton and Roper. *Phys. Lett. B* **2018**, *783*, 263–267. [CrossRef]
204. Maris, P. Electromagnetic properties of diquarks. *Few Body Syst.* **2004**, *35*, 117–127. [CrossRef]
205. Roberts, H.L.L.; Bashir, A.; Gutiérrez-Guerrero, L.X.; Roberts, C.D.; Wilson, D.J. π- and ρ-mesons, and their diquark partners, from a contact interaction. *Phys. Rev. C* **2011**, *83*, 065206. [CrossRef]
206. Anselmino, M.; Predazzi, E.; Ekelin, S.; Fredriksson, S.; Lichtenberg, D.B. Diquarks. *Rev. Mod. Phys.* **1993**, *65*, 1199–1234. [CrossRef]
207. Aznauryan, I.; Burkert, V.; Lee, T.S.; Mokeev, V. Results from the N* program at JLab. *J. Phys. Conf. Ser.* **2011**, *299*, 012008. [CrossRef]

208. Burkert, V.D.; Lee, T.S.H. Electromagnetic meson production in the nucleon resonance region. *Int. J. Mod. Phys. E* **2004**, *13*, 1035–1112. [CrossRef]
209. Segovia, J.; Cloet, I.C.; Roberts, C.D.; Schmidt, S.M. Nucleon and Δ elastic and transition form factors. *Few Body Syst.* **2014**, *55*, 1185–1222. [CrossRef]
210. Chen, C.; Lu, Y.; Binosi, D.; Roberts, C.D.; Rodríguez-Quintero, J.; Segovia, J. Nucleon-to-Roper electromagnetic transition form factors at large Q^2. *Phys. Rev. D* **2019**, *99*, 034013. [CrossRef]
211. Chen, C.; Krein, G.I.; Roberts, C.D.; Schmidt, S.M.; Segovia, J. Spectrum and structure of octet and decuplet baryons and their positive-parity excitations. *Phys. Rev. D* **2019**, *100*, 054009. [CrossRef]
212. Lu, Y.; Chen, C.; Cui, Z.F.; Roberts, C.D.; Schmidt, S.M.; Segovia, J.; Zong, H.S. Transition form factors: $\gamma^* + p \to \Delta(1232), \Delta(1600)$. *Phys. Rev. D* **2019**, *100*, 034001. [CrossRef]
213. Cui, Z.F.; Chen, C.; Binosi, D.; de Soto, F.; Roberts, C.D.; Rodríguez-Quintero, J.; Schmidt, S.M.; Segovia, J. Nucleon elastic form factors at accessible large spacelike momenta. *Phys. Rev. D* **2020**, arXiv:2003.11655.
214. Eichmann, G.; Alkofer, R.; Krassnigg, A.; Nicmorus, D. Nucleon mass from a covariant three-quark Faddeev equation. *Phys. Rev. Lett.* **2010**, *104*, 201601. [CrossRef] [PubMed]
215. Xu, S.S.; Cui, Z.F.; Chang, L.; Papavassiliou, J.; Roberts, C.D.; Zong, H.S. New perspective on hybrid mesons. *Eur. Phys. J. A (Lett.)* **2019**, *55*, 113. [CrossRef]
216. Souza, E.V.; Ferreira, M.N.; Aguilar, A.C.; Papavassiliou, J.; Roberts, C.D.; Xu, S.S. Pseudoscalar glueball mass: A window on three-gluon interactions. *Eur. Phys. J. A (Lett.)* **2020**, *56*, 25. [CrossRef]
217. Yin, P.L.; Chen, C.; Krein, G.; Roberts, C.D.; Segovia, J.; Xu, S.S. Masses of ground-state mesons and baryons, including those with heavy quarks. *Phys. Rev. D* **2019**, *100*, 034008. [CrossRef]
218. Edwards, R.G.; Dudek, J.J.; Richards, D.G.; Wallace, S.J. Excited state baryon spectroscopy from lattice QCD. *Phys. Rev. D* **2011**, *84*, 074508. [CrossRef]
219. CLAS Collaboration. Measurement of $ep \to e'p\pi^+\pi^-$ and baryon resonance analysis. *Phys. Rev. Lett.* **2003**, *91*, 022002. [CrossRef] [PubMed]
220. Burkert, V.D. Evidence of new nucleon resonances from electromagnetic meson production. *EPJ Web Conf.* **2012**, *37*, 01017. [CrossRef]
221. Kamano, H.; Nakamura, S.X.; Lee, T.S.H.; Sato, T. Nucleon resonances within a dynamical coupled-channels model of πN and γN reactions. *Phys. Rev. C* **2013**, *88*, 035209. [CrossRef]
222. Crede, V.; Roberts, W. Progress towards understanding baryon resonances. *Rept. Prog. Phys.* **2013**, *76*, 076301. [CrossRef] [PubMed]
223. Mokeev, V.I.; Aznauryan, I.; Burkert, V.; Gothe, R. Recent results on the nucleon resonance spectrum and structure from the CLAS detector. *EPJ Web Conf.* **2016**, *113*, 01013. [CrossRef]
224. Anisovich, A.V.; Burkert, V.; Hadžimehmedović, H.; Ireland, D.G.; Klempt, E.; Nikonov, V.A.; Omerović, R.; Osmanović, H.; Sarantsev, A.V.; Švarc, A.; et al. Strong Evidence for Nucleon Resonances near 1900 MeV. *Phys. Rev. Lett.* **2017**, *119*, 062004. [CrossRef] [PubMed]
225. Braun, V.M.; Collins, S.; Gläßle, B.; Göckeler, M.; Schäfer, A.; Schiel, R.W.; Södner, W.; Sternbeck, A.; Wein, P. Light-cone Distribution Amplitudes of the Nucleon and Negative Parity Nucleon Resonances from Lattice QCD. *Phys. Rev. D* **2014**, *89*, 094511. [CrossRef]
226. Bali, G.S.; Braun, V.M.; Göckeler, M.; Gruber, M.; Hutzler, F.; Schäfer, A.; Schiel, R.W.; Simeth, J.; Södner, W.; Sternbeck, A.; et al. Light-cone distribution amplitudes of the baryon octet. *JHEP* **2016**, *02*, 070. [CrossRef]
227. Braun, V.M.; Collins, S.; Göckeler, M.; Pérez-Rubio, P.; Schäfer, A.; Schiel, R.W.; Sternbeck, A. Second Moment of the Pion Light-cone Distribution Amplitude from Lattice QCD. *Phys. Rev. D* **2015**, *92*, 014504. [CrossRef]
228. Gao, F.; Chang, L.; Liu, Y.X. Bayesian extraction of the parton distribution amplitude from the Bethe-Salpeter wave function. *Phys. Lett. B* **2017**, *770*, 551–555. [CrossRef]
229. Zhang, J.H.; Chen, J.W.; Ji, X.; Jin, L.; Lin, H.W. Pion Distribution Amplitude from Lattice QCD. *Phys. Rev. D* **2017**, *95*, 094514. [CrossRef]
230. Zhang, J.H.; Jin, L.; Lin, H.W.; Schäfer, A.; Sun, P.; Yang, Y.B.; Zhang, R.; Zhao, Y.; Chen, J.W. Kaon Distribution Amplitude from Lattice QCD and the Flavor SU(3) Symmetry. *Nucl. Phys. B* **2019**, *939*, 429–446. [CrossRef]
231. Roper, L.D. Evidence for a P-11 Pion-Nucleon Resonance at 556 MeV. *Phys. Rev. Lett.* **1964**, *12*, 340–342. [CrossRef]
232. Bareyre, P.; Bricman, C.; Valladas, G.; Villet, G.; Bizard, J.; Seguinot, J. Pion-nucleon interactions between Tlab = 300 and Tlab = 700 MeV. *Phys. Lett.* **1964**, *8*, 137–141. [CrossRef]

233. Auvil, P.; Lovelace, C.; Donnachie, A.; Lea, A. Pion-nucleon phase shifts and resonances. *Phys. Lett.* **1964**, *12*, 76–80. [CrossRef]
234. Adelman, S.L. Evidence for an N^* Resonance at 1425 MeV. *Phys. Rev. Lett.* **1964**, *13*, 555–557. [CrossRef]
235. Roper, L.D.; Wright, R.M.; Feld, B.T. Energy-Dependent Pion-Nucleon Phase-Shift Analysis. *Phys. Rev.* **1965**, *138*, B190–B210. [CrossRef]
236. Li, B.L.; Chang, L.; Gao, F.; Roberts, C.D.; Schmidt, S.M.; Zong, H.S. Distribution amplitudes of radially-excited π and K mesons. *Phys. Rev. D* **2016**, *93*, 114033. [CrossRef]
237. Li, B.L.; Chang, L.; Ding, M.; Roberts, C.D.; Zong, H.S. Leading-twist distribution amplitudes of scalar- and vector-mesons. *Phys. Rev. D* **2016**, *94*, 094014. [CrossRef]
238. Jefferson Lab Hall A Collaboration. G_{E_p}/G_{M_p} ratio by polarization transfer in $\vec{e}p \to e\vec{p}$. *Phys. Rev. Lett.* **2000**, *84*, 1398–1402. [CrossRef]
239. Cates, G.; de Jager, C.; Riordan, S.; Wojtsekhowski, B. Flavor decomposition of the elastic nucleon electromagnetic form factors. *Phys. Rev. Lett.* **2011**, *106*, 252003. [CrossRef]
240. Wilson, D.J.; Cloet, I.C.; Chang, L.; Roberts, C.D. Nucleon and Roper electromagnetic elastic and transition form factors. *Phys. Rev. C* **2012**, *85*, 025205. [CrossRef]
241. Sachs, R. High-Energy Behavior of Nucleon Electromagnetic Form Factors. *Phys. Rev.* **1962**, *126*, 2256–2260. [CrossRef]
242. Perdrisat, C.F.; Punjabi, V.; Vanderhaeghen, M. Nucleon electromagnetic form factors. *Prog. Part. Nucl. Phys.* **2007**, *59*, 694–764. [CrossRef]
243. Akhiezer, A.; Rekalo, M. Polarization effects in the scattering of leptons by hadrons. *Sov. J. Part. Nucl.* **1974**, *4*, 277.
244. Arnold, R.; Carlson, C.E.; Gross, F. Polarization Transfer in Elastic electron Scattering from Nucleons and Deuterons. *Phys. Rev. C* **1981**, *23*, 363. [CrossRef]
245. Jefferson Lab Hall A Collaboration. Measurement of G(E(p))/G(M(p)) in $\vec{e}p \to e\vec{p}$ to $Q^2 = 5.6\,\mathrm{GeV}^2$. *Phys. Rev. Lett.* **2002**, *88*, 092301. [CrossRef] [PubMed]
246. Punjabi, V.; Perdrisat, C.F.; Aniol, K.A.; Baker, F.T.; Berthot, J.; Bertin, P.Y.; Bertozzi, W.; Besson, A.; Bimbot, L.; Boeglin, W.U.; et al. Proton elastic form factor ratios to $Q^2 = 3.5\,\mathrm{GeV}^2$ by polarization transfer. *Phys. Rev. C* **2005**, *71*, 055202. [Erratum-ibid. C **71**, 069902 (2005)]. [CrossRef]
247. Puckett, A.J.R.; Brash, E.J.; Gayou, O.; Jones, M.K.; Pentchev, L.; Perdrisat, C.F.; Punjabi, V.; Aniol, V.; Averett, T.; Benmokhtar, F.; et al. Final Analysis of Proton Form Factor Ratio Data at $Q^2 = 4.0$, 4.8 and 5.6 GeV^2. *Phys. Rev. C* **2012**, *85*, 045203. [CrossRef]
248. Puckett, A.J.R.; Brash, E.J.; Jones, M.K.; Luo, W.; Meziane, M.; Pentchev, L.; Perdrisat, C.F.; Punjabi, V.; Wesselmann, F.R.; Ahmidouch, A.; et al. Recoil Polarization Measurements of the Proton Electromagnetic Form Factor Ratio to $Q^2 = 8.5\,\mathrm{GeV}^2$. *Phys. Rev. Lett.* **2010**, *104*, 242301. [CrossRef]
249. Kelly, J.J. Simple parametrization of nucleon form factors. *Phys. Rev. C* **2004**, *70*, 068202. [CrossRef]
250. Bradford, R.; Bodek, A.; Budd, H.S.; Arrington, J. A New parameterization of the nucleon elastic form-factors. *Nucl. Phys. Proc. Suppl.* **2006**, *159*, 127–132. [CrossRef]
251. Gutiérrez-Guerrero, L.X.; Bashir, A.; Cloet, I.C.; Roberts, C.D. Pion form factor from a contact interaction. *Phys. Rev. C* **2010**, *81*, 065202. [CrossRef]
252. Frank, M.; Jennings, B.; Miller, G. The Role of color neutrality in nuclear physics: Modifications of nucleonic wave functions. *Phys. Rev. C* **1996**, *54*, 920–935. [CrossRef] [PubMed]
253. Chang, L.; Liu, Y.X.; Roberts, C.D. Dressed-quark anomalous magnetic moments. *Phys. Rev. Lett.* **2011**, *106*, 072001. [CrossRef] [PubMed]
254. Kallidonis, C.; Syritsyn, S.; Engelhardt, M.; Green, J.; Meinel, S.; Negele, J.; Pochinsky, A. Nucleon electromagnetic form factors at high Q^2 from Wilson-clover fermions. *PoS* **2018**, *LATTICE2018*, 125.
255. E93-038 Collaboration. Measurements of G_E^n/G_M^n from the $^2\mathrm{H}(\vec{e}, e'\vec{n})$ reaction to $Q^2 = 1.45\,\mathrm{GeV}/c)^2$. *Phys. Rev. Lett.* **2003**, *91*, 122002. [CrossRef]
256. Riordan, S.; Abrahamyan, S.; Craver, B.; Kelleher, A.; Kolarkar, A.; Miller, J.; Cates, G.D.; Liyanage, N.; Wojtsekhowski, B.; Acha, A. et al. Measurements of the Electric Form Factor of the Neutron up to $Q^2 =$ 3.4 GeV^2 using the Reaction $^3\vec{He}(\vec{e}, e'n)pp$. *Phys. Rev. Lett.* **2010**, *105*, 262302. [CrossRef]
257. Schlessinger, L.; Schwartz, C. Analyticity as a Useful Computation Tool. *Phys. Rev. Lett.* **1966**, *16*, 1173–1174. [CrossRef]

258. Schlessinger, L. Use of Analyticity in the Calculation of Nonrelativistic Scattering Amplitudes. *Phys. Rev.* **1968**, *167*, 1411–1423. [CrossRef]
259. Tripolt, R.A.; Haritan, I.; Wambach, J.; Moiseyev, N. Threshold energies and poles for hadron physical problems by a model-independent universal algorithm. *Phys. Lett. B* **2017**, *774*, 411–416. [CrossRef]
260. Binosi, D.; Chang, L.; Ding, M.; Gao, F.; Papavassiliou, J.; Roberts, C.D. Distribution Amplitudes of Heavy-Light Mesons. *Phys. Lett. B* **2019**, *790*, 257–262. [CrossRef]
261. Xu, Y.Z.; Binosi, D.; Cui, Z.F.; Li, B.L.; Roberts, C.D.; Xu, S.S.; Zong, H.S. Elastic electromagnetic form factors of vector mesons. *Phys. Rev. D* **2019**, *100*, 114038. [CrossRef]
262. HAPPEX Collaboration. Parity-violating electron scattering from ^{4}He and the strange electric form-factor of the nucleon. *Phys. Rev. Lett.* **2006**, *96*, 022003. [CrossRef] [PubMed]
263. G0 Collaboration. Strange quark contributions to parity-violating asymmetries in the forward G0 electron-proton scattering experiment. *Phys. Rev. Lett.* **2005**, *95*, 092001. [CrossRef] [PubMed]
264. Arrington, J.; Melnitchouk, W.; Tjon, J.A. Global analysis of proton elastic form factor data with two-photon exchange corrections. *Phys. Rev. C* **2007**, *76*, 035205. [CrossRef]
265. Wojtsekhowski, B.; Cates, G.D.; Riordan, S. *Measurement of the Neutron Electromagnetic Form Factor Ratio Gen/GMn at High Q^2*; Jefferson Lab 12 GeV Experiment: E12-09-016; 2009; Approved. Available online: https://www.jlab.org/exp_prog/proposals/09/PR12-09-016.pdf (accessed on 22 August 2020).

© 2020 by the author. Licensee MDPI, Basel, Switzerland. This article is an open access article distributed under the terms and conditions of the Creative Commons Attribution (CC BY) license (http://creativecommons.org/licenses/by/4.0/).

symmetry

MDPI

Article

Two-Pole Structures in QCD: Facts, Not Fantasy!

Ulf-G. Meißner [1,2,3]

[1] Helmholtz-Institut für Strahlen- und Kernphysik and Bethe Center for Theoretical Physics, Universität Bonn, D-53115 Bonn, Germany; meissner@hiskp.uni-bonn.de
[2] Institute for Advanced Simulation, Institut für Kernphysik, and Jülich Center for Hadron Physics, Forschungszentrum Jülich, D-52425 Jülich, Germany
[3] Tbilisi State University, 0186 Tbilisi, Georgia

Received: 15 May 2020; Accepted: 4 June 2020; Published: 8 June 2020

Abstract: The two-pole structure refers to the fact that particular single states in the spectrum as listed in the PDG tables are often two states. The story began with the $\Lambda(1405)$, when in 2001, using unitarized chiral perturbation theory, it was observed that there are two poles in the complex plane, one close to the $\bar{K}p$ and the other close to the $\pi\Sigma$ threshold. This was later understood combining the SU(3) limit and group-theoretical arguments. Different unitarization approaches that all lead to the two-pole structure have been considered in the mean time, showing some spread in the pole positions. This fact is now part of the PDG book, although it is not yet listed in the summary tables. Here, I discuss the open ends and critically review approaches that cannot deal with this issue. In the meson sector, some excited charm mesons are good candidates for such a two-pole structure. Next, I consider in detail the $D_0^*(2300)$, which is another candidate for this scenario. Combining lattice QCD with chiral unitary approaches in the finite volume, the precise data of the Hadron Spectrum Collaboration for coupled-channel $D\pi$, $D\eta$, $D_s\bar{K}$ scattering in the isospin $I = 1/2$ channel indeed reveal its two-pole structure. Further states in the heavy meson sector with $I = 1/2$ exhibiting this phenomenon are predicted, especially in the beauty meson sector. I also discuss the relation of these two-pole structures and the possible molecular nature of the states under consideration.

Keywords: chiral symmetry; coupled channels; hadron spectrum; lattice QCD; chiral perturbation theory; unitarization

1. Introduction

The hadron spectrum is arguably the least understood part of Quantum Chromodynamics (QCD), the theory of the strong interactions. It is part of the successful Standard Model (SM) and, thus, we can say that structure formation, that is the emergence of hadrons and nuclei from the underlying quark and gluon degrees of freedom, is indeed the last corner of the SM that is not yet understood. For a long time, the quark model of Gell–Mann [1] and Zweig [2] (and many sophisticated extensions thereof, e.g., [3]) have been used to bring order into the particle zoo. However, already before QCD it was a puzzling fact that all observed hadrons could be described by the simplest combinations of quarks/antiquarks, namely mesons as quark–antiquark states and baryons as three quark states, while symmetries and quantum numbers would also allow for tetraquarks, pentaquarks and so on. The situation got even worse when QCD finally appeared on the scene, as it allows for the following structures (bound systems of quarks and/or gluons):

- Conventional hadrons, that is mesons and baryons as described before;
- Multiquark hadrons, such as tetraquark states (mesons from two quarks and two antiquarks), pentaquark states (baryons made from four quarks and one antiquark) and so on;
- Hadronic molecules and atomic nuclei, that is multiquark states composed of a certain number of conventional hadrons (as discussed in more detail below);

- Hybrid states, which are composed of quarks and (valence) gluons; and
- Glueballs, bound states solely made of gluons, arguably the most exotic form of matter, which has so far been elusive in all searches.

The observed hadrons are listed with their properties in the tables of the Particle Data Group (PDG) (also called "Review of Particle Physics" (RPP)) [4] within a certain rating scheme, just telling us that some states are better understood as others. What complicates matters a lot is the fact that almost all hadrons are **resonances**, that is unstable states. These decay into other hadrons and leptons, e.g., the ρ meson decays into two pions or into a pion and a photon (and other final states). Such a resonance is thus described by a complex energy, more precisely, the real part is called the mass, m_R, and the imaginary part the half-width, $\Gamma_R/2$. Consequently, all other properties are also given by complex numbers. The only model-independent way (There are very few exceptions of isolated resonances on an energy-independent background where other methods can be used, but even in such cases a unique determination of the mass and the width is not always possible, see, e.g., the discussion of the $\rho(770)$ in the RPP.) to pin down these basic resonance properties is to look for poles in the complex plane, where resonances are usually located on the second Riemann sheet at

$$z_R = (\Re z_R, \Im z_R) = (m_R, \Gamma_R/2). \tag{1}$$

The residues at these poles contain information about the possible decays of such a resonance, as discussed in more detail below.

As I discuss in what follows, the hunt for such poles in the complex plane has revealed the astonishing feature of the **two-pole structure**, namely that certain states that are listed in the RPP are indeed superpositions of two states. The most prominent example is the $\Lambda(1405)$, which is discussed in detail in Section 3. More recently, the $D_0^*(2300)$, an excited charm meson, has become another prime candidate for the two-pole scenario, paving the way for a whole set of such states in the heavy-light sector (mesons made of one light (u,d,s) and one heavy (c,b) quark) (see Section 4). Before discussing these intricate states, it is, however, necessary to review the pertinent methods underlying the theoretical analyses (see Section 2). The conclusions and outlook are given in Section 5.

2. Methods

We start with the Lagrangian of QCD for up, down and strange quarks ($N_f = 3$), which can be written as:

$$\begin{aligned} \mathcal{L}_{\rm QCD} &= \mathcal{L}^0_{\rm QCD} - \bar{q}\mathcal{M}q, \\ \mathcal{L}^0_{\rm QCD} &= -\frac{1}{2g^2}\,{\rm Tr}\left[G_{\mu\nu}G^{\mu\nu}\right] + \bar{q}i\gamma^\mu \underbrace{(\partial_\mu - iA_\mu)}_{=D_\mu} q. \end{aligned} \tag{2}$$

Here, $q = (u,d,s)^T$ is the quark triplet, A_μ is the gluon field, $G_{\mu\nu}$ is the gluon field strength tensor, g is the strong coupling constant and the color indices related to the underlying SU(3)$_c$ local gauge symmetry have not been displayed. Further, $\mathcal{M} = {\rm diag}(m_u, m_d, m_s)$ is the quark matrix and the heavy flavors charm and bottom can be added analogously (We eschew here the top quark as it does not form hadrons due to its fast decay.). In addition, gauge fixing and the CP-violating θ-term are not displayed. Remarkably, $\mathcal{L}^0_{\rm QCD}$ displays a SU(3) $\times$ SU(3) flavor symmetry (I do not discuss the additional U(1) symmetries/non-symmetries here),

$$\mathcal{L}^0_{\rm QCD}(G_{\mu\nu}, q', D_\mu q') = \mathcal{L}^0_{\rm QCD}(G_{\mu\nu}, q, D_\mu q), \tag{3}$$

in terms of left- and right-handed quark fields,

$$q' = g_R P_R q + g_L P_L q, P_{R,L} = \frac{1}{2}(1 \pm \gamma_5), g_I g_I^\dagger = \mathbb{1}, {\rm det} g_I = 1, I = L, R. \tag{4}$$

This is the **chiral symmetry** of QCD. It leads to $16 = 2 \cdot (N_f^2 - 1)$ conserved Noether currents, which can be rearranged as eight conserved vector and eight conserved axial-vector currents. However, we know that the symmetry of the ground state is not the symmetry of the QCD Hamiltonian, as, e.g., there is no parity-doubling in the spectrum. The symmetry is spontaneously broken (or hidden, as Nambu preferred to say) to its vectorial subgroup:

$$\mathrm{SU}(3)_L \times \mathrm{SU}(3)_R \to \mathrm{SU}(3)_V. \tag{5}$$

The Goldstone theorem then tells us that for each broken generator there should be a massless boson (the famous Goldstone bosons). Therefore, in the absence of quark masses, we are dealing with a theory without a mass gap, which implies that Taylor expansions are not analytic. When the quark masses are included, the pseudoscalar Goldstone bosons acquire a small mass. In fact, the lightest hadrons are the eight pseudoscalar mesons (π, K, η). All this is the basis for the formulation of an effective field theory (EFT) that allows for perturbative calculations at low energy. Similarly, for the heavy quarks c and b, one can formulate a different EFT, based on the fact that the c ad b masses are large, $m_{c,b} \gg \Lambda_{\rm QCD}$, as discussed next.

2.1. Limits of QCD

We discuss one particular limit of QCD above, namely the chiral limit of the three light flavor theory. Such a special formulation can be extended also to the heavy quark sector and to so-called heavy-light systems. The various limits of QCD are:

- **Light quarks:**
$$\mathcal{L}_{\rm QCD} = \bar{q}_L i\not{D} q_L + \bar{q}_R i\not{D} q_R + \mathcal{O}(m_f/\Lambda_{\rm QCD})[f = u, d, s]. \tag{6}$$
In this limit, left- and right-handed quarks decouple which, is the chiral symmetry. As stated, it is spontaneously broken leading to the appearance of eight pseudo-Goldstone bosons. The pertinent EFT is chiral perturbation theory (CHPT) (see Section 2.2). Note that the corrections due to the quark masses are powers in m_f.
- **Heavy quarks:**
$$\mathcal{L}_{\rm QCD} = \bar{Q}_f iv \cdot D\, Q_f + \mathcal{O}(\Lambda_{\rm QCD}/m_f)[f = c, b], \tag{7}$$
where Q denotes the field of a heavy quark. In this limit, the Lagrangian is independent of quark spin and flavor, which leads to SU(2) spin and SU(2) flavor symmetries, called HQSS and HQFS, respectively. The pertinent EFT is heavy quark effective field theory (HQEFT) (see, e.g., [5,6]). Here, the corrections due to the quark masses are powers in $1/m_Q$.
- **Heavy-light systems:** Here, heavy quarks act as matter fields coupled to light pions and one thus can combine CHPT and HQEFT, as pioneered in [7–9] (see also Section 2.3).

2.2. A Factsheet on Chiral Perturbation Theory

Chiral perturbation theory is the EFT of QCD at low energies [10,11]. For introduction and reviews, see, e.g., [12–14]. Its basic properties are:

- $\mathcal{L}$ is symmetric under some Lie group $\mathcal{G}$; here, $\mathcal{G} = \mathrm{SU}(3)_L \times \mathrm{SU}(3)_R$.
- The ground state $|0\rangle$ is asymmetric and $\mathcal{G}$ is spontaneously broken to $\mathcal{H} \subset \mathcal{G}$, leading to the the appearance of Goldstone bosons (GBs) $|\phi^i(p)\rangle$. In QCD, $\mathcal{H} = \mathrm{SU}(3)_V$ and the Goldstone bosons are the aforementioned eight pseudoscalar mesons.
- In QCD, the matrix element of the axial-vector current $\mathcal{A}^i_\mu$, $\langle 0|\mathcal{A}^i_\mu|\phi^k(p)\rangle = i\delta^{ik} p_\mu F \neq 0$ $(i = 1, \ldots, 8)$, where F is related to the pseudoscalar decay constant in the chiral limit. $F \neq 0$ is a sufficient and necessary condition for spontaneous chiral symmetry breaking.
- There are no other massless strongly interacting particles.

Universality tells us that, at low energies, any theory with these properties looks the same as long as the number of space-time dimensions is larger than two. One can readily deduce that the interactions of the GBs are weak in the low-energy regime and indeed vanish at zero energy. This allows for a systematic expansion in small momenta and energies, and the quark masses lead to finite but small GB masses, which defines a second expansion parameter. In fact, these two parameters can be merged in one. The corresponding effective Lagrangian is readily constructed; it takes the form

$$\mathcal{L}_{\text{eff}} = \mathcal{L}^{(2)} + \mathcal{L}^{(4)} + \mathcal{L}^{(6)} + \ldots, \tag{8}$$

where the superscript denotes the power of the small expansion parameter p (derivatives and/or GB mass insertions). This expansion is systematic, as an underlying power counting [11] can be derived. This shows that graphs with n loops are suppressed by powers of p^{2n} and that, at each order, we have local operators accompanied by unknown coupling constants, also called low-energy constants (LECs). These LECs must be determined from fits to experimental data or can be calculated using lattice QCD. Their specific values single out QCD from the whole universality class of theories discussed above. One important issue concerns unitarity. Leading order calculations are based on tree diagrams with insertions from $\mathcal{L}^{(2)}$, which means that such amplitudes are real. Imaginary parts are only generated at one-loop order through the loop diagrams, which means that unitarity is fulfilled perturbatively but not exactly in CHPT (for a general discussion, see [15]). We come to this issue in Section 2.4.

Matter fields such as baryons can also be included in a systematic fashion. There is one major complication, namely the matter field mass that is of the same size as the breakdown scale of the EFT, here $\Lambda_\chi \sim 1\,\text{GeV}$. Therefore, only three-momenta of the matter fields can be small and the mass must be dealt with in some manner. Various schemes such as the heavy baryon approach [16,17], infrared regularization [18] and the extended on-mass-shell scheme [19] exist to restore the power counting. For details, I refer to the reviews [20–22].

2.3. Chiral Perturbation Theory for Heavy-Light Systems

In this section, we display the effective Lagrangians that we need for the discussion of charm mesons and their interactions. Consider first Goldstone boson scattering off D-mesons. The effective Lagrangian takes the form [23–25]:

$$\begin{aligned}
\mathcal{L}_{\text{eff}} &= \mathcal{L}^{(1)} + \mathcal{L}^{(2)}, \\
\mathcal{L}^{(1)} &= \mathcal{D}_\mu D \mathcal{D}^\mu D^\dagger - M_D^2 D D^\dagger, \\
\mathcal{L}^{(2)} &= D\left[-h_0\langle\chi_+\rangle - h_1\chi_+ + h_2\langle u_\mu u^\mu\rangle - h_3 u_\mu u^\mu\right]\bar{D} + \mathcal{D}_\mu D\left[h_4\langle u^\mu u^\nu\rangle - h_5\{u^\mu, u^\nu\}\right]\mathcal{D}_\nu \bar{D}.
\end{aligned} \tag{9}$$

Here, $D = (D^0, D^+, D_s^+)$ is the D-meson triplet, M_D is the average mass of the D-mesons and we utilize the standard chiral building blocks $u_\mu \sim \partial_\mu \phi$, with ϕ a member of the GB octet and $\chi_+ \sim \text{diag}(m_u, m_d, m_s)$. The pertinent LECs can all be determined: $h_1 = 0.42$ from the D_s-D splitting, while $h_{2,3,4,5}$ are fixed from a fit to lattice data [26]. Further, h_0 can be fixed from the pion mass dependence of the D meson masses.

In what follows, we also consider $\bar{B} \to D$ transitions with the emission of two light pseudoscalars (pions). Here, chiral symmetry puts constraints on one of the two pions while the other one moves fast and does not participate in the final-state interactions. The corresponding chiral effective Lagrangian is developed in Ref. [27]:

$$\begin{aligned}
\mathcal{L}_{\text{eff}} &= \bar{B}\big[c_1\left(u_\mu t M + M t u_\mu\right) + c_2\left(u_\mu M + M u_\mu\right)t + c_3\, t\left(u_\mu M + M u_\mu\right) \\
&+ c_4\left(u_\mu\langle M t\rangle + M\langle u_\mu t\rangle\right) + c_5\, t\langle M u_\mu\rangle + c_6\langle\left(M u_\mu + u_\mu M\right)t\rangle\big]\partial^\mu D^\dagger,
\end{aligned} \tag{10}$$

in terms of the B-meson triplet $\bar{B} = (B^-, \bar{B}^0, \bar{B}^0_s)$, M is the matter field for the fast-moving pion and $t = uHu$ is a spurion field for Cabbibo-allowed decays,

$$H = \begin{pmatrix} 0 & 0 & 0 \\ 1 & 0 & 0 \\ 0 & 0 & 0 \end{pmatrix}. \tag{11}$$

In the $B \to D\pi\pi$ decays that we discuss below, only some combinations of the LECs c_i $(i = 1, \ldots, 6)$ appear (see Section 4.3).

2.4. Unitarization Schemes

As stated above, unitarity is only fulfilled perturbatively in CHPT. The hard scale in this EFT is set by the appearance of resonances, such as the $f_0(500)$ or the $\rho(770)$ in various partial waves of pion–pion scattering. CHPT is thus not the proper framework to describe resonances. One possible way to extend the energy region where this can be applied is unitarization, originally proposed in Ref. [28]. However, this comes at a price, usually crossing symmetry is violated in such type of approach and the coefficient of subleading chiral logarithms are often incorrectly given (see [15]). Here, let us just discuss a familiar approach on solving the Bethe–Salpeter equation in the on-shell approximation (see, e.g., [29,30]). To be specific, let us consider the coupled-channel process $\phi + D \to \phi + D$ (suppressing all indices). The basic unitarization method is depicted in Figure 1. It amounts to a resummation of the so-called "fundamental bubble" (the 2-point loop function). To describe resonances, e.g., the $D^*_{s0}(2317)$, one has to search for poles of the T-matrix, which is generated from the CHPT potential by unitarization.

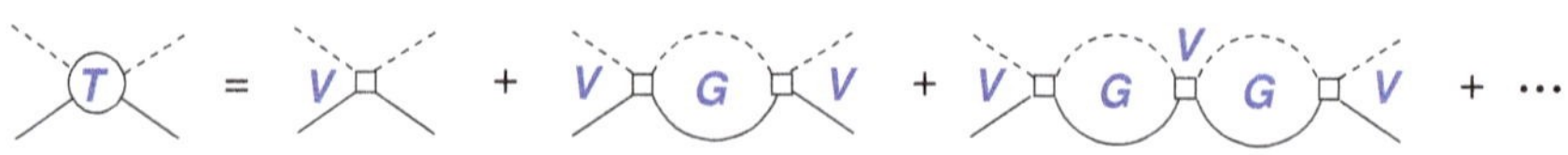

Figure 1. The T-matrix for GB (dashed lines) scattering off D-mesons (solid lines) as a bubble sum based on the effective potential V, which is obtained from the underlying chiral effective Lagrangian.

This version of unitarized CHPT is based on the fundamental equation

$$T^{-1}(s) = V^{-1}(s) - G(s), \tag{12}$$

where $V(s)$ is derived from the effective Lagrangian Equation (9) and $G(s)$ is the 2-point scalar loop function regularized by a subtraction constant $a(\mu)$,

$$\begin{aligned} G(s) = \; & \frac{1}{16\pi^2}\Big\{ a(\mu) + \ln\frac{m_2^2}{\lambda^2} + \frac{m_1^2 - m_2^2 + s}{2s}\ln\frac{m_1^2}{m_2^2} + \frac{\sigma}{2s}\Big[\ln(s - m_1^2 + m_2^2 + \sigma) \\ & - \ln(-s + m_1^2 - m_2^2 + \sigma) + \ln(s + m_1^2 - m_2^2 + \sigma) - \ln(-s - m_1^2 + m_2^2 + \sigma)\Big]\Big\}, \end{aligned} \tag{13}$$

with $\sigma = \left\{[s - (m_1 + m_2)^2][s - (m_1 - m_2)^2]\right\}^{1/2}$ and m_1 and m_2 are the masses of the two mesons in the loop, here one D-meson and one GB. μ is the scale of dimensional regularization, and a change of μ can be absorbed by a corresponding change of $a(\mu)$. Promoting $T(s)$, $V(s)$ and $G(s)$ to be matrix-valued quantities, it is easy to generalize Equation (12) to coupled channels. More details on the unitarization schemes are given in the subsequent sections.

2.5. Unitarized Chiral Perturbation Theory in a Finite Volume

To compare with lattice data, we need to formulate the unitarization scheme in a finite volume (FV). Obviously, in any FV scheme, momenta are no longer continuous but quantized,

$$\vec{q} = \frac{2\pi}{L}\vec{n}, \vec{n} \in \mathbb{Z}^3, \tag{14}$$

in a cubic volume of length L, i.e., $V = L^3$. An appropriate FV representation of the scalar 2-point function is (see [31] for details)

$$\tilde{G}(s,L) = \lim_{\Lambda\to\infty}\left[\frac{1}{L^3}\sum_{\vec{n}}^{|\vec{q}|<\Lambda} I(\vec{q}) - \int_0^\Lambda \frac{q^2 dq}{2\pi^2} I(\vec{q})\right], \tag{15}$$

with $I(\vec{q})$ the corresponding integrand. The FV energy levels of the process under consideration are then obtained from the poles of $\tilde{T}(s,L)$:

$$\tilde{T}^{-1}(s,L) = V^{-1}(s) - \tilde{G}(s,L). \tag{16}$$

Note that all volume dependence resides in $\tilde{G}(s,L)$ and the effective Lagrangian and thus the effective potential are the same as in the continuum [32]. Again, in the case of coupled channels, Equation (16) is promoted to a matrix equation.

3. The Story of the $\Lambda(1405)$

3.1. Basic Facts

In the quark model, the $\Lambda(1405)$ is a *uds* excitation with $J^P = 1/2^-$ a few hundred MeV above the ground-state $\Lambda(1116)$. The RPP gives **one** corresponding state with

$$m = 1405.1^{+1.3}_{-1.0}\,\text{MeV}\,,\Gamma = 50.5 \pm 2.0\,\text{MeV}. \tag{17}$$

In fact, the $\Lambda(1405)$ was predicted long before the quark model as a resonance in the coupled $\pi\Sigma$ and $\bar{K}N$ channels (see [33] and also [34]), and considered as a $\bar{K}N$ bound state, arguably the first "exotic" hadron ever. The analytical structure in the complex energy plane between the $\pi\Lambda$ and $\eta\Lambda$ thresholds is shown in Figure 2, together with the location of the the $\Lambda(1405)$ and the further isospin splitting of the pertinent $\pi\Sigma$ and $\bar{K}N$ thresholds. The $\Lambda(1405)$ was clearly seen in $K^-p \to \Sigma 3\pi$ reactions at 4.2 GeV at CERN [35]. The spin and parity were only recently determined directly in photoproduction reactions at Jefferson Laboratory, consistent with the theoretical expectation of $J^P = 1/2^-$ [36]. However, it is too low in mass for the quark model, but can be described in certain models such as the cloudy bag model (it is amusing to note that the two-pole structure of the $\Lambda(1405)$ was already observed in this model but little attention was paid to this work [37]) or the Skyrme model. However, these models are only loosely rooted in QCD and do not allow for controlled error estimates, an important ingredient in any theoretical prediction.

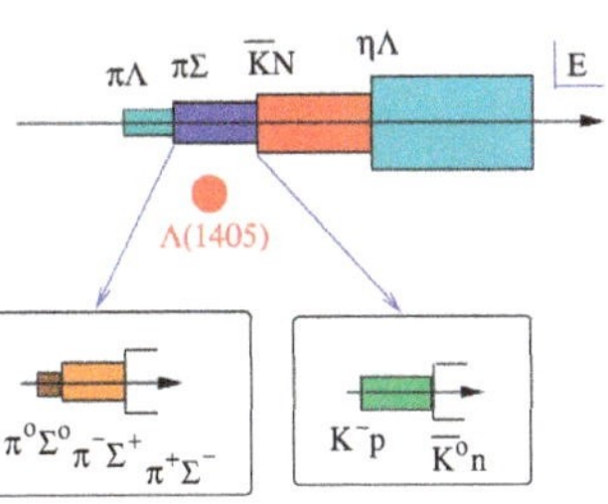

Figure 2. Complex energy plane in the vicinity of the $\Lambda(1405)$.

3.2. Enter Chiral Dynamics

An important step in the theory of the $\Lambda(1405)$ was based on the idea of combining the (leading order) chiral SU(3) meson–baryon Lagrangian with coupled-channel dynamics [38]. This study gave an excellent description of the $K^-p \to K^-p, K^0n, \pi^0\Lambda, \pi^\pm\Sigma^\mp, \pi^0\Sigma^0$ scattering data and the $\pi\Sigma$ mass distribution, and it was found that the Weinberg–Tomozawa (WT) term gave the most important contribution. In this scheme, the $\Lambda(1405)$ appears as a **dynamically generated state**, a meson–baryon molecule, connecting the pioneering coupled-channel works with the chiral dynamics of QCD. This led to a number of highly cited follow-up papers by the groups from Munich and Valencia (see, e.g., [39,40] and the early review [29]). These groundbreaking works were, however, beset by certain shortcomings. In particular, there was an unpleasant regulator dependence for the employed Yukawa-type functions or momentum cutoffs and the issue of maintaining gauge invariance in such type of regulated theories was only resolved years later [41–43].

3.3. The Two-Pole Structure

A re-analysis of coupled-channel K^-p scattering and the $\Lambda(1405)$ in the framework of unitarized CHPT was performed by Oller and Meißner [44]. This work was originally motivated by developing methods to overcome some of the shortcomings discussed before. The following technical improvements were worked out: (1) the subtracted meson–baryon loop function based on dimensional regularization, cf. Equation (13), which has become the standard regularization method; (2) a coupled-channel approach to the $\pi\Sigma$ mass distribution, which replaced the common assumption that this process is dominated by the $I = 0$ $\pi\Sigma$ system and thus can be calculated directly from the $\pi\Sigma \to \pi\Sigma$ S-wave amplitude; and (3) matching formulas to any order in chiral perturbation theory were established, which allows for a better constraining of such non-perturbative amplitudes. The most significant finding of that work was, however, the finding of the two-pole structure: "Note that the $\Lambda(1405)$ resonance is described by **two poles** on sheets II and III with rather different imaginary parts indicating a clear departure from the Breit-Wigner situation". The location of the poles are: Pole 1 at $(1379.2 - i27.6)$MeV and pole 2 at $(1433.7 - i11.0)$MeV on Sheet II, close to the $\pi\Sigma$ and K^-p thresholds, respectively. This two-pole structure was also found in follow-up works by other groups [45,46]. A better understanding of this two-pole structure was achieved in Ref. [47] using SU(3) symmetry considerations and group theory. For the case under consideration, namely the dynamical generation of resonances in Goldstone boson scattering off baryons, the following group theoretical consideration applies: The decomposition of the combination of the two octets, the Goldstone bosons and the ground-state baryons, is

$$8 \otimes 8 = \underbrace{1 \oplus 8_s \oplus 8_a}_{\text{bindingatLO}} \oplus 10 \oplus \overline{10} \oplus 27, \tag{18}$$

where using the leading order WT-term one finds poles in the singlet and in the two octets. The two octets are degenerate and the poles are located on the real axis (see Figure 3). One can now follow the developments of these poles in the complex plane. For that, one parameterizes the departure from the SU(3) limit for the GB (M_i) and baryon masses (m_i) and subtraction constants a_i (the subtraction constants can be channel-dependent but collapse to one value in the SU(3) limit) as follows:

$$M_i^2(x) = M_0^2 + x(M_i^2 - M_0^2), m_i(x) = m_0 + x(m_i - m_0), a_i(x) = a_0 + x(a_i - a_0), \tag{19}$$

with $0 \le x \le 1$, where $x = 0$ corresponds to the SU(3) limit and $x = 1$ describes the physical world. Further, $m_0 = 1151$ MeV, $M_0 = 368$ MeV and $a_0 = -2.148$. The trajectories of the various poles in the complex plane as the SU(3) breaking is gradually increased up to the physical values at $x = 1$ is shown in Figure 3. First, we observe that not all poles present in the SU(3) limit appear for $x = 1$ (using the LO WT term only).

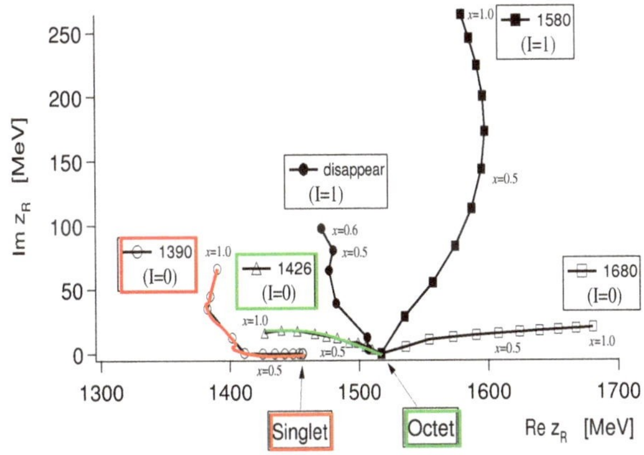

Figure 3. Trajectories of the poles in the scattering amplitudes obtained by changing the SU(3) breaking parameter x. In the SU(3) limit ($x = 0$), only two poles appear, one for the singlet and the other for the octets. The symbols correspond to the step size $\delta x = 0.1$ and the two trajectories contributing to the $\Lambda(1405)$ are highlighted.

Second, concerning the $\Lambda(1405)$, we see that the singlet pole moves towards the $\pi\Sigma$ threshold and becomes rather broad, whereas the second pole from the octet comes out close to the K^-p threshold and stays rather narrow. Thus, there are in fact two resonances. Having determined these poles, one can determine the couplings of these resonances to the physical states by studying the amplitudes in the vicinity of the poles,

$$T_{ij} = \frac{g_i g_j}{z - z_R} + \text{regularterms}, \tag{20}$$

with i, j channel indices and the couplings g_i are complex valued numbers. While the lower pole couples more strongly to the $\pi\Sigma$ channel, the higher one displays a stronger coupling to $\bar{K}N$. Consequently, it is possible to find the existence of the two resonances by performing different experiments, since in different experiments the weights by which the two resonances are excited are different (see [47] for more details). Further early support of the two-pole scenario was provided by the leading order investigation of the reaction $K^-p \to \pi^0\pi^0\Sigma^0$ [48].

3.4. Beyond Leading Order

Clearly, to achieve a better precision, one has to go beyond leading order and include the next-to-leading order (NLO) terms in the chiral SU(3) Lagrangian in the effective potential. This task was performed by three groups independently [49–51]. These investigations were also triggered by the precise measurements of the energy shift and width of kaonic hydrogen [52], which was based on the improved Deser-type formula from Ref. [53], thus resolving the long standing "kaonic hydrogen puzzle" (the discrepancy between the values of the $\bar{K}N$ scattering lengths extracted from scattering data or earlier kaonic hydrogen experiments). This allowed to pin down the subthreshold K^-p scattering amplitude to better precision and to make predictions for the K^-n scattering lengths. Most importantly, all of these works confirmed the two-pole structure as shown in Figure 4 based on the results of [51].

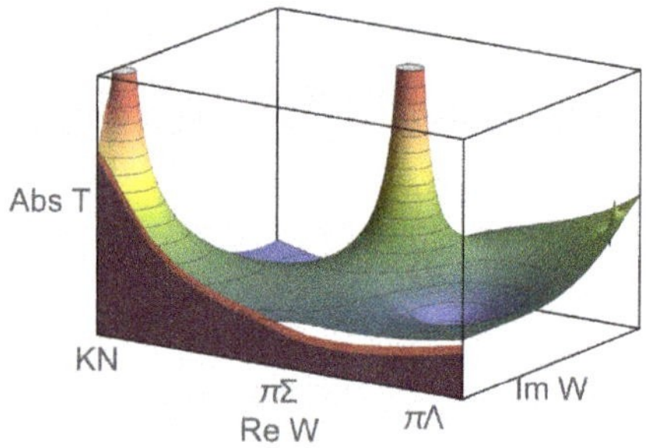

Figure 4. Three-dimensional plot of the two-pole structure of the $\Lambda(1405)$. W denotes the center-of-mass energy. Figure courtesy of Maxim Mai.

There was yet another surprise, namely by looking more closely at the scattering and kaonic hydrogen data, one can find at least eight solutions of similar quality with different pairs of poles for the $\Lambda(1405)$ (see [54]). Here, photoproduction comes to the rescue. The CLAS collaboration at Jefferson Laboratory did a superb job in measuring the $\Sigma\pi$ photoproduction line shapes near the $\Lambda(1405)$ [55]. These data were first analyzed using LO unitarized CHPT and a polynomial ansatz for the photoproduction process $\gamma p \to K^+ M_i B_i$ in Ref. [56], leading to a good fit of these data and a further check of the two-pole nature of the $\Lambda(1405)$. The same ansatz was used in Ref. [54], leaving only two of the eight solutions, as shown for one of the remaining solutions in Figure 5 for a fraction of the data (the complete fit is shown in [54]). Similarly, the spread in the two poles from the eight solutions was sizeably reduced, although the lower pole could not be pinned down as well as the higher one. This work also supplied error bands, not only for the photoproduction results but also for the underlying hadronic scattering processes. This shows that the inclusion of the photoproduction data serves as a new important constraint on the antikaon–nucleon scattering amplitude. In view of all these NLO results, the two-pole structure first appeared in the RPP in form of a mini-review [57], called "Pole structure of the $\Lambda(1405)$ region" co-authored by Hyodo and Meißner, which give also references to other works related to the two-pole scenario. However, in the RPP summary tables, the $\Lambda(1405)$ was listed (and still is) as one resonance only.

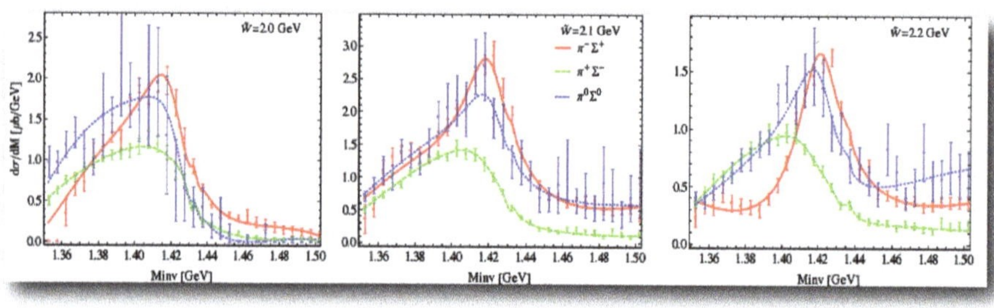

Figure 5. Result of the fits to the CLAS data in all three channels $\pi^+\Sigma^-$ (green), $\pi^-\Sigma^+$(red) and $\pi^0\Sigma^0$. Correspondingly, green (dashed), red (full) and blue (dotted) lines represent the outcome of the model for Solution 4 in the $\pi^+\Sigma^-, \pi^-\Sigma^+, \pi^0\Sigma^0$ channels, respectively. Figure courtesy of Maxim Mai.

3.5. Where Do We Stand?

A comparative analysis of the theoretical approaches that aim at description of low-energy meson–baryon interactions in the strangeness $S = -1$ sector was presented by Cieply et al. [58]. All the models discussed [49–51,59] are derived from a chiral Lagrangian that includes terms up to NLO, $\mathcal{O}(p^2)$, with the free parameters fitted to the low-energy K^-p reactions data and to the characteristics of the kaonic hydrogen as measured by the SIDDHARTA collaboration. Thus, the various models available on the market were put under a direct comparison aiming at determining the subthreshold energy dependence of the $\bar{K}N$ scattering amplitudes and on the pole content of the models related to the dynamically generated baryon resonances. The discussed approaches represent the variety of different philosophies they are built on. Most of them (the Kyoto–Munich, Murcia and Bonn ones) use dimensional regularization to tame the ultraviolet divergences in the meson–baryon loop function and treat the meson–baryon interactions on the energy shell while the Prague model introduces off-shell form factors to regularize the Green function and phenomenologically accounts for the off-shell effects. All approaches but the Bonn one are based on a potential concept, introducing an effective meson–baryon potential that matches the chiral amplitude up to a given order and is then used as a potential kernel in the Lippmann–Schwinger equation to sum a major part of the chiral perturbation series. The Bonn model differs by solving a genuine Bethe–Salpeter equation before making a projection to the S-wave and neglecting the off-shell contributions. Finally, the Kyoto–Munich and Prague models have relatively small NLO contributions (representing only moderate corrections to the LO chiral interactions) while the Murcia and Bonn models introduce sizeable NLO terms that generate inter-channel couplings very different from those obtained by only the WT interaction. Despite all these differences, the models are able to reproduce the experimental data on a qualitatively very similar level and in mutual agreement especially concerning the data available at the $\bar{K}N$ threshold. The models also tend to agree on a position of the higher energy of the two poles generated for the $\Lambda(1405)$ resonance, predicting it at the complex energy with the real part $\Re z \approx 1420\ldots1430$ MeV and the imaginary part $-\Im z \approx 10\ldots40$ MeV (see Figure 6).

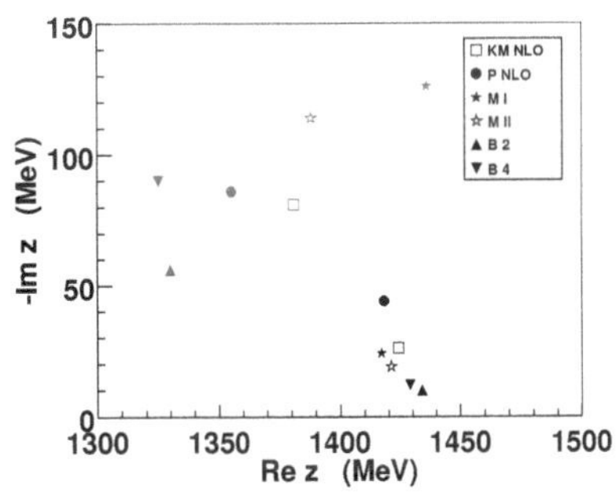

Figure 6. Pole positions for various approaches: Kyoto-Munich (KM) [49], Prag (P) [59], Murcia (M) [50] and Bonn (B) [51].

However, that is about where the agreement among the predictions ends. In particular, these different approaches lead to very different locations of the lighter pole and different predictions for the elastic K^-p and K^-n amplitudes at sub-threshold energies. This certainly has impact on the predictions one can make (or has made) for kaonic atoms and antikaon quasi-bound states (see, e.g., [60]).

Clearly, we need better data to constrain the $\pi\Sigma$ spectra in various processes besides the already mentioned photoproduction data. A step in this direction was performed by Kamiya et al. [61], who provided further constraints on the $\bar{K}N$ subthreshold interaction by analyzing $\pi\Sigma$ spectra in various processes, such as the $K^-d \to \pi\Sigma n$ reaction and the $\Lambda_c \to \pi\pi\Sigma$ decay. In addition, the yet to be measured 1S level shift of kaonic deuterium will put further constraints on the $\bar{K}N$ interaction [62,63]. For the status of such measurements, see [64]. Further progress has also been made by including the P-waves and performing a sophisticated error analysis [65], confirming again and sharpening further the two-pole scenario of the $\Lambda(1405)$ (note that the P-waves had already been included earlier with a focus on the $S = 0$ sector [66]). Despite these remaining uncertainties, it must be stated clearly that the two-pole nature of the $\Lambda(1405)$ is established!

To end this section, I briefly discuss two recent works that challenged the two-pole structure. It was claimed by Revai [67] that the two-pole nature is an artifact of the on-shell approximation used in most studies. Including off-shell effects, only one pole is generated in that study. However, as convincingly shown in Ref. [68], that approach violates constraints imposed by chiral symmetry. The origin of this violation can be traced back to the off-shell treatment of the chiral-effective vertices, in combination with the use of non-relativistic approximations and the chosen regularization scheme. Overcoming these deficiencies, the two-pole scenario reappears. This does not come as a surprise as the NLO study of Mai and Meißner [51] already went beyond the on-shell approximation. Another recent paper with only one pole is Ref. [69], based on the phenomenological Bonn–Gatchina (BnGa) approach. In this framework, a large number of scattering and photoproduction data is fitted. However, this scheme does not allow for the dynamical generation of resonances and no pole searches in the complex-energy plane are reported in Ref. [69]. In a recent update, these authors also included a second pole in their model [70]. They found that the second pole does not considerably worsen the description of the considered data but still they prefer the solution with one pole. Thus, these results are not conclusive. In addition, data that further support the two-pole nature on $\pi^- p \to K^0\pi\Sigma$ and $pp \to pK^+\pi\Sigma$ [71] as well as $K^+\Lambda(1405)$ electroproduction [72] are also not included. It can therefore safely be said that these papers do *not* challenge the two-pole scenario.

4. Meson Sector: The $D_0^*(2300)$ and Related States

Thus far, one might consider the two-pole structure a curiosity related to just one particular state. In the meson sector, a similar doubling of states was already considered in 1986 in the discussion of the mysterious decay properties of the $f_0(980)$ (then called $\delta(980)$), but that remained largely unnoticed [73]. However, let us now take a closer look at the spectrum of the excited charmed mesons, especially the $D_{s0}^*(2317)$ first observed by BaBar [74] and the $D_0^*(2300)$ first observed by Belle [75] (see also [76]). According to the recent edition of the PDG, the characteristics of these states are:

$$D_0^*(2300) \;:\; M = 2300 \pm 19\text{MeV}, \Gamma = 274 \pm 40\text{MeV}, I(J^P) = \frac{1}{2}(0^+), \tag{21}$$

$$D_{s0}^*(2317) \;:\; M = 2318.0 \pm 0.7\text{MeV}, \Gamma < 3.8\text{MeV}, I(J^P) = 0(0^+). \tag{22}$$

According to the quark model, the quark composition for these scalar mesons is $c\bar{u}$ and $c\bar{s}$, respectively. This immediately poses the question: Why is the $D_{s0}^*(2317)$ as heavy as the $D_0^*(2300)$, it should be about 100 MeV, which is the mass of the strange quark, heavier? In addition, why is the $D_{s0}^*(2317)$ about 150 MeV below the prediction of the quark model, that has been rather successful [3]? While this is an interesting question (I refer to the review [77], which has a very detailed discussion of this state), here I focus on the the non-strange charmed scalar meson, as it appears to be too heavy, but in fact gives further support to the two-pole scenario.

4.1. Two-Pole Structure

Let us consider first the fine lattice QCD work by the Hadron Spectrum Collaboration, who investigated coupled-channel $D\pi$, $D\eta$ and $D_s\bar{K}$ scattering with $J^P = 0^+$ and $I = 1/2$ in three lattice volumes, one value for the temporal and the spatial lattice spacing, respectively, at a pion mass $M_\pi = 391$ MeV and D-meson mass $M_D = 1885$ MeV [78]. They used various K-matrix type extrapolations of the type

$$K_{ij} = \left(g_i^{(0)} + g_i^{(1)}s\right)\left(g_j^{(0)} + g_j^{(1)}s\right)\frac{1}{m^2 - s} + \gamma_{ij}^{(0)} + \gamma_{ij}^{(1)}s, \tag{23}$$

to find the poles in the complex plane, by fitting the parameters g, γ to the computed energy levels, and use the T-matrix to extract the poles. They found one S-wave pole at (2275.0 ± 0.9) MeV, extremely close to the $D\pi$ threshold at 2276 MeV. This state is consistent with the $D_0^*(2300)$ of the PDG. However, the extrapolations in Equation (23) do not take into account chiral symmetry.

Therefore, this topic was revisited by Albaladejo et al. [79], who implemented the chiral Lagrangian, Equation (9), together with LECs from Ref. [26] within the finite volume formalism outlined in Section 2.5 to postdict in a parameter-free manner the energy levels measured by the Hadron Spectrum Collaboration. The stunning result of this unitarized CHPT calculation is shown in the left panel of Figure 7; a very accurate postdiction of the lattice levels is achieved (note that this is not a fit). Note further that the region above 2.7 GeV is beyond the range of applicability of this NLO calculation. The level below the $D\pi$ threshold is interpreted in Ref. [78] as a bound state associated to the $D_0^*(2300)$ as stated before. The finite-volume UCHPT calculation also finds this pole at $M = 2264^{+8}_{-14}$ MeV and half-width $\Gamma/2 = 0$ MeV, very similar to the results of Moir et al. [78]. However, there is a **second pole** at $M = 2468^{+32}_{-25}$ MeV with $\Gamma/2 = 113^{+18}_{-16}$ MeV (see also Figure 7, right). Using chiral extrapolations, one can then evaluate the spectroscopic content of the scattering amplitudes for the physical pion mass, collected in Table 1.

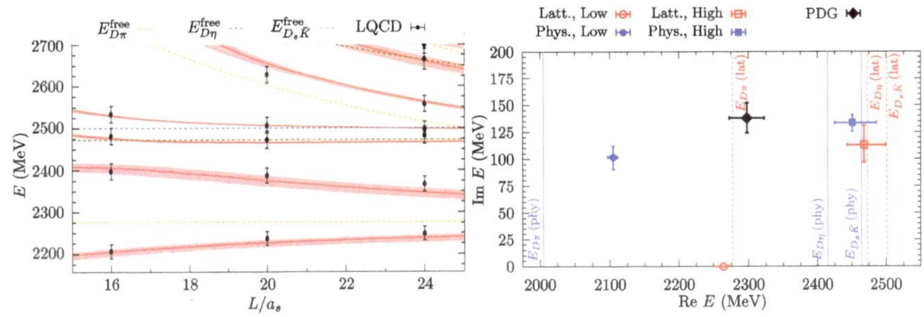

Figure 7. (**Left**) Energy levels calculated in finite-volume unitarized chiral perturbation theory with all LECs determined before (red bands representing the 1σ uncertainties) in comparison to the lattice QCD results of [78] (black circles). The dashed lines give the various free levels of the two-particle systems $D\pi$, $D\eta$ and $D_s\bar{K}$. (**Right**) Location of the two poles in the complex energy plane for the lattice masses (red symbols) and physical masses (blue symbols). The black diamond represents the PDG value. The various thresholds are indicated by the dotted lines. Figures courtesy of Feng-Kun Guo.

Table 1. Position $\sqrt{s} = M - i\Gamma/2$ (in MeV) and couplings (in GeV) of the two poles in the (0,1/2) sector using physical pion masses.

M (MeV)	$\Gamma/2$ (MeV)	$\lvert g_{D\pi}\rvert$	$\lvert g_{D\eta}\rvert$	$\lvert g_{D_s\bar{K}}\rvert$
2105^{+6}_{-8}	102^{+10}_{-12}	$9.4^{+0.2}_{-0.2}$	$1.8^{+0.7}_{-0.7}$	$4.4^{+0.5}_{-0.5}$
2451^{+36}_{-26}	134^{+7}_{-8}	$5.0^{+0.7}_{-0.4}$	$6.3^{+0.8}_{-0.5}$	$12.8^{+0.8}_{-0.6}$

The bound state below the $D\pi$ threshold evolves into a resonance above it when the physical masses are used, where the threshold is now at 2005 MeV. This behavior is typical for S-wave poles. The second pole moves very little and its couplings are rather independent of the meson masses. It is a resonance located between the $D\eta$ and $D_s\bar{K}$ thresholds on the (110) Riemann sheet, continuously connected to the physical sheet. Thus, the $D_0^*(2300)$ of the RPP is produced by two different poles, and in fact the lower pole solves the enigma discussed in the beginning of this section. Note that this two-pole structure was observed earlier by the authors of [80–82] but only explained properly in Ref. [79], as discussed next.

Consider again the SU(3) limit, where the eight Goldstone bosons take the common value M_0 and the three heavy D-mesons the common value M_{D0}, so that departures from the SU(3) limit are parameterized as

$$\begin{aligned} M_{\phi,i} &= M_{\phi,i}^{\text{phys}} + x(M_0 - M_{\phi,i}^{\text{phys}})(i = 1, \ldots, 8), \\ M_{D,j} &= M_{D,j}^{\text{phys}} + x(M_{D0} - M_{D,j}^{\text{phys}})(j = 1, \ldots, 3), \end{aligned}$$

with $x = 0$ and $x = 1$ corresponding to the physical and the SU(3) symmetric case, respectively, and $M_0 = 0.49$ GeV and $M_{D0} = 1.95$ GeV. In that study, only one subtraction constant for all channels was used and kept fixed with varying x. Note that in contrast to the work of Jido et al. [47] discussed above, here a linear extrapolation formula is used for the GB masses, which is also legitimate. As above, the two-pole nature is understood from group theory,

$$\bar{3} \otimes 8 = \underbrace{\bar{3} \oplus 6}_{\text{attractive}} \oplus \overline{15}. \tag{24}$$

This means one has attraction in the $\bar{3}$ and 6 irreducible representations (irreps) but repulsion in the $\overline{15}$ irrep at leading order in the effective potential. The most attractive irrep, the $\bar{3}$, admits a $c\bar{q}$ ($q = u, d, s$) configuration. At NLO, the potentials receive corrections, but the qualitative features remain. The evolution from the SU(3) limit to the physical case is shown in the left panel of Figure 8. Shown are the two poles corresponding to the $D_0^*(2300)$ and its strange sibling, the $D_{s0}^*(2317)$. As one moves away from the SU(3) limit, the lower pole of the $D_0^*(2300)$ moves down in the complex plane, restoring the expected ordering that the $c\bar{u}$ excitation should be lighter than its $c\bar{s}$ partner.

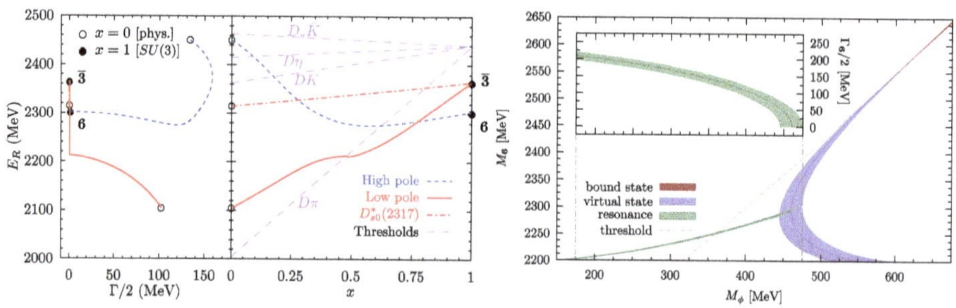

Figure 8. (**Left**) Pole paths in the complex plane when recovering the SU(3) limit (left subpanel). Mass evolution of the different poles with x. Besides the two (0,1/2) poles, denoted as high and low (blue dashed and green solid lines), the evolution of the (1,0) bound state, the $D_{s0}^*(2317)$ resonance (orange dot-dashed line), is shown on the right. Figure courtesy of Feng-Kun Guo. (**Right**) Mass of the predicted sextet state M_6 at the SU(3) symmetric point as a function of the Goldstone boson mass M_ϕ. The inset shows the half-width of the resonance for GB masses below 475 MeV.

It was already observed in Ref. [79] that the higher pole connects with a virtual state in the sextet representation due to the weaker binding. This issue was further elaborated on by Du et al. [83], who studied the SU(3) limit in more detail. In the right panel of Figure 8, the sextet pole is shown for varying GB masses. Below $M_\phi \lesssim 475$ MeV, the pole is a resonance with its imaginary part ($\Gamma_6/2$) shown in the inserted sub-figure. Above $M_\phi \simeq 475$ MeV, it evolves into a pair of virtual states, and finally it becomes a bound state at $M_\phi \simeq 600$ MeV. Such a study in the SU(3) limit $m_u = m_d = m_s$ at large GB masses could be performed on the lattice and offer another test of this scenario. Finally, let me come back to the lattice calculation in Ref. [78]. Indeed, while various of the amplitudes employed in that analysis contained a second pole, its location was strongly parameterization-dependent [84], and therefore not reported in that paper.

4.2. Other Candidates

As already noted, in the $(S, I) = (1, 0)$ sector, the same chiral Lagrangian produces a pole at 2315^{+18}_{-28} MeV which is naturally identified with the $D^*_{s0}(2317)$. It emerges from the pole in the triplet representation. The $D^*_{s0}(2317)$ is dominantly a DK molecule. Substituting the D-meson by a D^* and employing HQSS, the molecular picture naturally gives [85]

$$M_{D_{s1}(2460)} - M_{D^*_{s0}(2317)} \simeq M_{D^*} - M_D, \tag{25}$$

which is also obtained in the so-called doublet model [86–88]. Similar to the $D^*_0(2300)$, the nonstrange charmed meson D_1 also comes with two poles (see Table 2). This was noted before by Guo et al. [89]. In both cases, the single RPP pole sits in between the lower and the higher pole.

Using HQFS, one can further predict similar states in the B-meson sector, by just replacing the corresponding D-mesons with their bottom counterparts at leading order [81,85]. Using the NLO framework employed in the charm sector, one has to scale the LECs $h_{0,1,2,3} \sim m_Q$ and $h_{4,5} \sim 1/m_Q$ and the subtraction constant is adjusted, as described in Ref. [81]. This leads again to the two-pole structures also collected in Table 2.

For the lowest positive-parity heavy strange mesons, it is instructive to compare with lattice QCD results. This gives for the masses of the charm-strange mesons $M_{D^*_{s0}} = 2315^{+18}_{-28}[2348^{+7}_{-4}]$ [90], $M_{D_{s1}} = 2456^{+15}_{-21}[2459.5 \pm 0.6]$ [90], and for the strange-bottom ones $M_{B^*_{s0}} = 5720^{+16}_{-23}[5711 \pm 23]$ [91], $M_{D_{s1}} = 5772^{+15}_{-21}[5750 \pm 25]$ [91], where the first (second) number refers to the molecular (lattice QCD) prediction. The agreement is rather remarkable.

Table 2. Predicted poles corresponding to the positive-parity heavy-light nonstrange mesons given as $(M, \Gamma/2)$, with M the mass and Γ the total decay width, in units of MeV. The current RPP [4] values are listed in the last column.

	Lower Pole	Higher Pole	RPP
D^*_0	$\left(2105^{+6}_{-8}, 102^{+10}_{-11}\right)$	$\left(2451^{+35}_{-26}, 134^{+7}_{-8}\right)$	$(2300 \pm 19, 137 \pm 20)$
D_1	$\left(2247^{+5}_{-6}, 107^{+11}_{-10}\right)$	$\left(2555^{+47}_{-30}, 203^{+8}_{-9}\right)$	$(2427 \pm 26 \pm 25, 192^{+54}_{-38} \pm 37)$
B^*_0	$\left(5535^{+9}_{-11}, 113^{+15}_{-17}\right)$	$\left(5852^{+16}_{-19}, 36 \pm 5\right)$	-
B_1	$\left(5584^{+9}_{-11}, 119^{+14}_{-17}\right)$	$\left(5912^{+15}_{-18}, 42^{+5}_{-4}\right)$	-

Thus, the plot of the two-pole scenario thickens. In the absence of direct measurements of some of these states, one might ask the question: Is further experimental support for the picture just outlined?

4.3. Analysis of $B \to D\pi\pi$ Data

To further test the mechanism of the dynamical generation of the charm-light flavored mesons discussed thus far, let me turn to the high precision results of the LHCb collaboration for the decays $B \to D\pi\pi$ [92]. As shown in Ref. [83], the amplitudes with the two D^*_0 states are fully consistent with the LHCb measurements of the reaction $B^- \to D^+\pi^-\pi^-$, which are at present the best data providing access to the $D\pi$ system and thus to the nonstrange scalar charm mesons. This information is encoded in the so-called angular momenta, which are discussed in detail in the LHCb paper [92].

The theoretical framework to analyze this process is based on the unitarized chiral effective Lagrangian, Equation (10), where one pion is fast and the other participates in the $D\pi$ final-state interactions (for more details, see [83]). To be specific, consider the reaction $B^- \to D^+\pi^-\pi^-$. For sufficiently low energies in the $D\pi$ system, it suffices to include the lowest partial waves (S,P,D), so we can write the decay amplitude as

$$\mathcal{A}(B^- \to D^+\pi^-\pi^-) = \sum_{L=0}^{2} \sqrt{2L+1}\mathcal{A}_L(s)P_L(z)\,, \tag{26}$$

where $\mathcal{A}_{0,1,2}(s)$ correspond to the amplitudes with $D^+\pi^-$ in the S-, P- and D-waves, respectively, and $P_L(z)$ are the Legendre polynomials. For the P- and D-wave amplitudes, we use the same Breit–Wigner form as in the LHCb analysis [92], containing the D^* and $D^*(2680)$ mesons in the P-wave and the $D_2(2460)$ in the D-wave. For the S-wave, however, we employ

$$\begin{aligned}\mathcal{A}_0(s) &= A\Big\{E_\pi\Big[2+G_1(s)\left(\frac{5}{3}T_{11}^{1/2}(s)+\frac{1}{3}T^{3/2}(s)\right)\Big]\\ &\quad +\frac{1}{3}E_\eta G_2(s)T_{21}^{1/2}(s)+\sqrt{\frac{2}{3}}E_{\bar{K}}G_3(s)T_{31}^{1/2}(s)\Big\}+BE_\eta G_2(s)T_{21}^{1/2},\end{aligned} \tag{27}$$

where A and B are two independent couplings following from SU(3) flavor symmetry (i.e., combinations of the LECs c_i, $A=\sqrt{2}(c_1+c_4)/F_\phi$ and $B=2\sqrt{2}(c_2+c_6)/(3F_\phi)$, with F_ϕ the GB decay constant), and $E_{\pi,\eta,\bar{K}}$ are the energies of the light mesons. Further, the $T_{ij}^I(s)$ are the S-wave scattering amplitudes for the coupled-channel system with total isospin I, where i,j are channel indices with 1, 2 and 3 referring to $D\pi$, $D\eta$ and $D_s\bar{K}$, respectively, and the $G_i(s)$ are the corresponding 2-point loop functions. These scattering amplitudes are again taken from Ref. [26] where also all the other parameters are fixed. To filter out the S-wave, the following (combinations of) angular moments are used:

$$\begin{aligned}\langle P_0\rangle &\propto |\mathcal{A}_0|^2+|\mathcal{A}_1|^2+|\mathcal{A}_2|^2,\\ \langle P_2\rangle &\propto \frac{2}{5}|\mathcal{A}_1|^2+\frac{2}{7}|\mathcal{A}_2|^2+\frac{2}{\sqrt{5}}|\mathcal{A}_0||\mathcal{A}_2|\cos(\delta_2-\delta_0),\\ \langle P_{13}\rangle &= \langle P_1\rangle-\frac{14}{9}\langle P_3\rangle\propto\frac{2}{\sqrt{3}}|\mathcal{A}_0||\mathcal{A}_1|\cos(\delta_1-\delta_0),\end{aligned} \tag{28}$$

with $\delta_0, \delta_1, \delta_2$ the S-, P- and D-wave phase shift, respectively.

The best fit to the LHCb data is shown in Figure 9 together with their best fit provided by LHCb based on cubic splines (dashed lines). The bands in Figure 9 reflect the one-sigma errors of the parameters in the scattering amplitudes determined in Ref. [26]. It is worthwhile to notice that, in $\langle P_{13}\rangle$, where the $D_2(2460)$ does not play any role, the data show a significant variation between 2.4 and 2.5 GeV. Theoretically, this feature can now be understood as a signal for the opening of the $D^0\eta$ and $D_s^+K^-$ thresholds at 2.413 and 2.462 GeV, respectively, which leads to two cusps in the amplitude. This effect is amplified by the higher pole which is relatively close to the $D_s\bar{K}$ threshold on the unphysical sheet. There is some discrepancy between the chiral amplitude and the data for $\langle P_{13}\rangle$ at low energies: Does this point at a deficit of the former? Fortunately, the LHCb Collaboration provided more detailed information on their S-wave amplitude in Ref. [92]: in the analysis of the data, a series of anchor points were defined where the strength and the phase of the S-wave amplitude were extracted from the data. Then, cubic splines were used to interpolate between these anchor points.

In Figure 10, the S-amplitude fixed as described above is compared to the LHCb anchor points. Not only does this figure show very clearly that the strength of the S-wave amplitude largely determined by the fits to lattice data is fully consistent with the one extracted from the data for $B^-\to D^+\pi^-\pi^-$, but the presented amplitude also shows the importance of the $D\eta$ and $D_s\bar{K}$ cusps and thus also of the role of the higher pole in the $I=1/2$ and $S=0$ channel even more clearly than the angular moments discussed above. This clearly highlights the importance of a coupled-channel treatment for this reaction. An updated analysis of the LHC Run-2 data is called for to confirm the prominence of the two cusps.

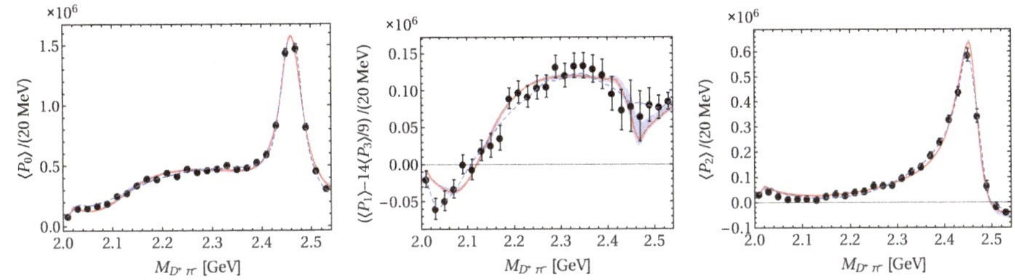

Figure 9. Fit to the LHCb data for the angular moments $\langle P_0 \rangle$, $\langle P_{13} \rangle$ and $\langle P_2 \rangle$ for the $B^- \to D^+\pi^-\pi^-$ reaction [92]. The largest error among $\langle P_1 \rangle$ and $14\langle P_3 \rangle/9$ in each bin is taken as the error of $\langle P_1 \rangle - 14\langle P_3 \rangle/9$. The solid lines show the results of Du et al. [83], with error bands corresponding to the one-sigma uncertainties propagated from the input scattering amplitudes, while the dashed lines stand for the LHCb fit using cubic splines for the S-wave [92].

LHCb presented also data on $B_s^0 \to \bar{D}^0 K^- \pi^+$, which are, however, less precise than the ones just discussed. Using the same formalism as before, with one different combination of the LECs c_i and the same resonances in the P- and D-wave as LHCb, these data can be well described by a one parameter fit (see [83] for more details). A combined analysis including also data for $B^0 \to \bar{D}^0\pi^-\pi^+$, $B^- \to D^+\pi^-K^-$ and $B_s^0 \to \bar{D}^0K^-\pi^+$ performed in Ref. [93] gives further credit to this picture.

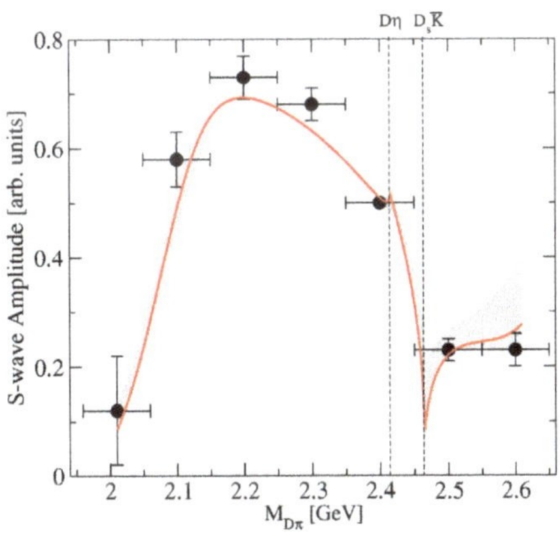

Figure 10. Comparison of the S-wave amplitude based on UCHPT to the S-wave anchor points found in the experimental analysis, shown as the data points [92]. The red line gives the best fit results and the grey band quantifies the uncertainties that emerged from the fitting procedure. The fitting range extends up to 2.55 GeV. The dashed perpendicular lines indicate the location of the $D\eta$ and $D_s\bar{K}$ threshold, respectively.

4.4. The K_1 Meson

Another state that offered support to the two-pole scenario even before the heavy-light mesons just discussed is the axial-vector meson $K_1(1270)$, which in the quark model is a kaonic excitation with angular momentum one, $I(J^P) = \frac{1}{2}(1^+)$. The two-pole nature of the $K_1(1270)$ was first noted in the study of the scattering of vector mesons off the Goldstone bosons in a chiral unitary approach at tree level [94]. This was further sharpened by Geng et al. [95]. There, the high-statistics data from the WA3 experiment on $K^-p \to K^-\pi^+\pi^-p$, analyzed by the ACCMOR collaboration [96], were reanalyzed and shown to favor a two-pole interpretation of the $K_1(1270)$. Geng et al. [95] also reanalyzed the traditional K-matrix interpretation of the WA3 data and found that the good fit of the data obtained there was due to large cancellations of terms of unclear physical interpretation. It was recently shown how this two-pole scenario can show up in D-meson decays, in particular $D^0 \to \pi^+ VP$ and $D^+ \to \nu e^+ VP$, where P and V are pseudoscalar and vector mesons, respectively [97,98].

5. Discussion and Outlook

Let me summarize briefly:

- The story with the two-pole structure started with the $\Lambda(1405)$, which can now be considered as established. However, the position of lighter pole close to the $\pi\Sigma$ threshold needs to be determined better, whereas the higher pole close to the $K^- p$ threshold is pretty well pinned down. It is comforting to note that the re-analysis of the Jülich $\bar{K}N$ meson-exchange model from the 1990s also confirmed the two-pole structure of the $\Lambda(1405)$ (see [99] and references therein). I again point out that approaches that do not allow for the dynamical generation of resonances, e.g., the BnGa model, are insufficient for describing the whole hadron spectrum.
- Further support of the two-pole scenario comes from charmed baryons. Recently, an analysis of the LHCb data on $\Lambda_b \to pD^0\pi^-$ in the near-pD^0-threshold region also revealed a two-pole structure of the $\Sigma_c(2800)^+$ when isospin-breaking is taken into account [100].
- The spectrum of excited charmed mesons, made from a heavy c quark and a light u, d, s quark, offers further support of the two-pole structure and the dynamical generation of hadron resonances. Here, a beautiful interplay of experimental results, unitarized chiral perturbation theory and lattice QCD gives very strong indications that this picture is indeed correct. Further lattice calculations and the measurement of the corresponding B-mesons will serve as further tests.
- This leads to a new paradigm in hadron physics: The hadron spectrum must not be viewed as a collection of quark model states, but rather as a manifestation of a more complex dynamics that leads to an intricate pattern of various types of states that can only be understood by an interplay of theory and experiment (cf. the light scalar mesons or the states discussed here).
- The dynamical generation of hadron states through hadron–hadron interactions ties together nuclear and particle physics, as these molecular compounds bear resemblance to the light nuclei, the deuteron, the triton and so on. Therefore, such molecular states were called "deusons" by Törnquist, one of the pioneers in the field of hadronic molecules [101].

From all this, it is rather obvious that the PDG tables published in the RPP must undergo a drastic change and finally acknowledge the two-pole structure in the main listings, not just in the review section. Time will tell how long this necessary change will take.

Funding: This research was funded in part by Deutsche Forschungsgemeinschaft (TRR 110, "Symmetries and the Emergence of Structure in QCD"), the Chinese Academy of Sciences (CAS) President's International Fellowship Initiative (PIFI) (grant No. 2018DM0034) and VolkswagenStiftung (grant No. 93562).

Acknowledgments: I am grateful to Feng-Kun Guo, Maxim Mai and Jose Oller for comments on the manuscript and FKG and MM for providing me with some of the figures shown. I also thank all my collaborators on the issues discussed here for sharing their insights. Finally, I am grateful to Dubravko Klabucar for giving me the opportunity to write up these thoughts.

Conflicts of Interest: The author declares no conflict of interest.

References

1. Gell-Mann, M. A Schematic Model of Baryons and Mesons. *Phys. Lett.* **1964**, *8*, 214–215. [CrossRef]
2. Zweig, G. *An SU(3) Model for Strong Interaction Symmetry and Its Breaking. Version 1*, CERN-TH-401; 1964. Available online: https://cds.cern.ch/record/352337 (accessed on 10 May 2020).
3. Godfrey, S.; Isgur, N. Mesons in a Relativized Quark Model with Chromodynamics. *Phys. Rev. D* **1985**, *32*, 189. [CrossRef] [PubMed]
4. Particle Data Group. Review of Particle Physics. *Phys. Rev. D* **2018**, *98*, 030001. [CrossRef]
5. Neubert, M. Heavy quark symmetry. *Phys. Rept.* **1994**, *245*, 259. [CrossRef]

6. Manohar, A.V.; Wise, M.B. *Heavy Quark Physics*, 1st ed.; Cambridge Monographs on Particle Physics, Nuclear Physics, and Cosmology; Cambridge University Press: Cambridge, UK, 2000; Volume 10.
7. Burdman, G.; Donoghue, J.F. Union of chiral and heavy quark symmetries. *Phys. Lett. B* **1992**, *280*, 287. [CrossRef]
8. Wise, M.B. Chiral perturbation theory for hadrons containing a heavy quark. *Phys. Rev. D* **1992**, *45*, R2188. [CrossRef]
9. Yan, T.M.; Cheng, H.Y.; Cheung, C.Y.; Lin, G.L.; Lin, Y.C.; Yu, H.L. Heavy quark symmetry and chiral dynamics. *Phys. Rev. D* **1992**, *46*, 1148.; Erratum in **1997**, *55*, 5851. [CrossRef]
10. Gasser, J.; Leutwyler, H. Chiral Perturbation Theory to One Loop. *Ann. Phys.* **1984**, *158*, 142–210. [CrossRef]
11. Weinberg, S. Phenomenological Lagrangians. *Physica A* **1979**, *96*, 327–340. [CrossRef]
12. Ecker, G. Chiral perturbation theory. *Prog. Part. Nucl. Phys.* **1995**, *35*, 1–80. [CrossRef]
13. Pich, A. Chiral perturbation theory. *Rep. Prog. Phys.* **1995**, *58*, 563. [CrossRef]
14. Bernard, V.; Meißner, U.-G. Chiral perturbation theory. *Ann. Rev. Nucl. Part. Sci.* **2007**, *57*, 33–60. [CrossRef]
15. Gasser, J.; Meißner, U.-G. Chiral expansion of pion form-factors beyond one loop. *Nucl. Phys. B* **1991**, *357*, 90–128. [CrossRef]
16. Jenkins, E.E.; Manohar, A.V. Baryon chiral perturbation theory using a heavy fermion Lagrangian. *Phys. Lett. B* **1991**, *255*, 558–562. [CrossRef]
17. Bernard, V.; Kaiser, N.; Kambor, J.; Meißner, U.-G. Chiral structure of the nucleon. *Nucl. Phys. B* **1992**, *388*, 315–345. [CrossRef]
18. Becher, T.; Leutwyler, H. Baryon chiral perturbation theory in manifestly Lorentz invariant form. *Eur. Phys. J. C* **1999**, *9*, 643–671. [CrossRef]
19. Fuchs, T.; Gegelia, J.; Japaridze, G.; Scherer, S. Renormalization of relativistic baryon chiral perturbation theory and power counting. *Phys. Rev. D* **2003**, *68*, 056005. [CrossRef]
20. Bernard, V.; Kaiser, N.; Meißner, U.-G. Chiral dynamics in nucleons and nuclei. *Int. J. Mod. Phys. E* **1995**, *4*, 193–344. [CrossRef]
21. Bernard, V. Chiral Perturbation Theory and Baryon Properties. *Prog. Part. Nucl. Phys.* **2008**, *60*, 82–160. [CrossRef]
22. Scherer, S.; Schindler, M.R. Chiral perturbation theory for baryons. *Lect. Notes Phys.* **2012**, *830*, 145–214. [CrossRef]
23. Cheng, H.; Cheung, C.; Lin, G.; Lin, Y.; Yan, T.; Yu, H. Corrections to chiral dynamics of heavy hadrons: SU(3) symmetry breaking. *Phys. Rev. D* **1994**, *49*, 5857–5881. [CrossRef]
24. Lutz, M.F.; Soyeur, M. Radiative and isospin-violating decays of D_s-mesons in the hadrogenesis conjecture. *Nucl. Phys. A* **2008**, *813*, 14–95. [CrossRef]
25. Guo, F.K.; Hanhart, C.; Krewald, S.; Meißner, U.-G. Subleading contributions to the width of the D*(s0)(2317). *Phys. Lett. B* **2008**, *666*, 251–255. [CrossRef]
26. Liu, L.; Orginos, K.; Guo, F.K.; Hanhart, C.; Meißner, U.-G. Interactions of charmed mesons with light pseudoscalar mesons from lattice QCD and implications on the nature of the $D_{s0}^*(2317)$. *Phys. Rev. D* **2013**, *87*, 014508. [CrossRef]
27. Savage, M.J.; Wise, M.B. SU(3) Predictions for Nonleptonic B Meson Decays. *Phys. Rev. D* **1989**, *39*, 3346.; Erratum in **1989**, *40*, 3127. [CrossRef] [PubMed]
28. Dobado, A.; Herrero, M.J.; Truong, T.N. Unitarized Chiral Perturbation Theory for Elastic Pion-Pion Scattering. *Phys. Lett. B* **1990**, *235*, 134–140. [CrossRef]
29. Oller, J.A.; Oset, E.; Ramos, A. Chiral unitary approach to meson meson and meson - baryon interactions and nuclear applications. *Prog. Part. Nucl. Phys.* **2000**, *45*, 157–242. [CrossRef]
30. Oller, J.; Coupled-channel approach in hadron-hadron scattering. *Prog. Part. Nucl. Phys.* **2020**, *110*, 103728. [CrossRef]
31. Döring, M.; Meißner, U.-G.; Oset, E.; Rusetsky, A. Unitarized Chiral Perturbation Theory in a finite volume: Scalar meson sector. *Eur. Phys. J. A* **2011**, *47*, 139. [CrossRef]
32. Gasser, J.; Leutwyler, H. Spontaneously Broken Symmetries: Effective Lagrangians at Finite Volume. *Nucl. Phys. B* **1988**, *307*, 763–778. [CrossRef]
33. Dalitz, R.H.; Tuan, S.F. A possible resonant state in pion-hyperon scattering. *Phys. Rev. Lett.* **1959**, *2*, 425. [CrossRef]

34. Kim, J.K. Low Energy K- p Interaction of the 1405 MeV Y_0^* Resonance as KbarN Bound State. *Phys. Rev. Lett.* **1965**, *14*, 29. [CrossRef]
35. Hemingway, R.J. Production of $\Lambda(1405)$ in K^-p Reactions at 4.2-GeV/c. *Nucl. Phys. B* **1985**, *253*, 742–752. [CrossRef]
36. Moriya, K.; Schumacher, R.A.; Aghasyan, M.; Amaryan, M.J.; Anderson, M.D.; Pereira, S.A.; Ball, J.; Baltzell, N.A.; Battaglieri, M.; Bellis, M.; et al. Spin and parity measurement of the $\Lambda(1405)$ baryon. *Phys. Rev. Lett.* **2014**, *112*, 082004. [CrossRef]
37. Fink, P.J., Jr.; He, G.; Landau, R.H.; Schnick, J.W. Bound States, Resonances and Poles in Low-energy $\bar{K}N$ Interaction Models. *Phys. Rev. C* **1990**, *41*, 2720. [CrossRef]
38. Kaiser, N.; Siegel, P.B.; Weise, W. Chiral dynamics and the low-energy kaon - nucleon interaction. *Nucl. Phys. A* **1995**, *594*, 325–345. [CrossRef]
39. Kaiser, N.; Waas, T.; Weise, W. SU(3) chiral dynamics with coupled channels: Eta and kaon photoproduction. *Nucl. Phys. A* **1997**, *612*, 297–320. [CrossRef]
40. Oset, E.; Ramos, A. Nonperturbative chiral approach to s wave $\bar{K}N$ interactions. *Nucl. Phys. A* **1998**, *635*, 99. [CrossRef]
41. Nacher, J.; Oset, E.; Toki, H.; Ramos, A. Radiative production of the Lambda(1405) resonance in K- collisions on protons and nuclei. *Phys. Lett. B* **1999**, *461*, 299–306. [CrossRef]
42. Borasoy, B.; Bruns, P.C.; Meißner, U.-G.; Nissler, R. Gauge invariance in two-particle scattering. *Phys. Rev. C* **2005**, *72*, 065201. [CrossRef]
43. Borasoy, B.; Bruns, P.; Meißner, U.-G.; Nissler, R. A Gauge invariant chiral unitary framework for kaon photo- and electroproduction on the proton. *Eur. Phys. J. A* **2007**, *34*, 161–183. [CrossRef]
44. Oller, J.A.; Meißner, U.-G. Chiral dynamics in the presence of bound states: Kaon nucleon interactions revisited. *Phys. Lett. B* **2001**, *500*, 263–272. [CrossRef]
45. Jido, D.; Hosaka, A.; Nacher, J.C.; Oset, E.; Ramos, A. Magnetic moments of the $\Lambda(1405)$ and $\Lambda(1670)$ resonances. *Phys. Rev. C* **2002**, *66*, 025203. [CrossRef]
46. Garcia-Recio, C.; Nieves, J.; Ruiz Arriola, E.; Vicente Vacas, M.J. S = -1 meson baryon unitarized coupled channel chiral perturbation theory and the S(01) Lambda(1405) and Lambda(1670) resonances. *Phys. Rev. D* **2003**, *67*, 076009. [CrossRef]
47. Jido, D.; Oller, J.A.; Oset, E.; Ramos, A.; Meißner, U.-G. Chiral dynamics of the two $\Lambda(1405)$ states. *Nucl. Phys. A* **2003**, *725*, 181–200. [CrossRef]
48. Magas, V.; Oset, E.; Ramos, A. Evidence for the two pole structure of the Lambda(1405) resonance. *Phys. Rev. Lett.* **2005**, *95*, 052301. [CrossRef]
49. Ikeda, Y.; Hyodo, T.; Weise, W. Chiral SU(3) theory of antikaon-nucleon interactions with improved threshold constraints. *Nucl. Phys. A* **2012**, *881*, 98–114. [CrossRef]
50. Guo, Z.H.; Oller, J.A. Meson-baryon reactions with strangeness -1 within a chiral framework. *Phys. Rev. C* **2013**, *87*, 035202. [CrossRef]
51. Mai, M.; Meißner, U.-G. New insights into antikaon-nucleon scattering and the structure of the Lambda(1405). *Nucl. Phys. A* **2013**, *900*, 51–64. [CrossRef]
52. Bazzi, M.; Beer, G.; Bombelli, L.; Bragadireanu, A.M.; Cargnelli, M.; Corradi, G.; Curceanu, C.; Fiorini, C.; Frizzi, T.; Ghio, F.; et al. A New Measurement of Kaonic Hydrogen X-rays. *Phys. Lett. B* **2011**, *704*, 113–117. [CrossRef]
53. Meißner, U.-G. Raha, U.; Rusetsky, A. Spectrum and decays of kaonic hydrogen. *Eur. Phys. J. C* **2004**, *35*, 349–357. [CrossRef]
54. Mai, M.; Meißner, U.-G. Constraints on the chiral unitary $\bar{K}N$ amplitude from $\pi\Sigma K^+$ photoproduction data. *Eur. Phys. J. A* **2015**, *51*, 30. [CrossRef]
55. Moriya, K.; Schumacher, R.A.; Adhikari, K.P.; Adikaram, D.; Aghasyan, M.; Anderson, M.D.; Pereira, S.A.; Ball, J.; Baltzell, N.A.; Battaglieri, M.; et al. Measurement of the $\Sigma\pi$ photoproduction line shapes near the $\Lambda(1405)$. *Phys. Rev. C* **2013**, *87*, 035206. [CrossRef]
56. Roca, L.; Oset, E. $\Lambda(1405)$ poles obtained from $\pi^0\Sigma^0$ photoproduction data. *Phys. Rev. C* **2013**, *87*, 055201. [CrossRef]
57. Patrignani, C.P.D.G.; Weinberg, D.H.; Woody, C.L.; Chivukula, R.S.; Buchmueller, O.; Kuyanov, Y.V.; Blucher, E.; Willocq, S.; Höcker, A.; Lippmann, C.; et al. Review of Particle Physics. *Chin. Phys. C* **2016**, *40*, 100001. [CrossRef]

58. Cieply, A.; Mai, M.; Meißner, U.-G.; Smejkal, J. On the pole content of coupled channels chiral approaches used for the $\bar{K}N$ system. *Nucl. Phys. A* **2016**, *954*, 17–40. [CrossRef]
59. Cieply, A.; Smejkal, J. Chirally motivated $\bar{K}N$ amplitudes for in-medium applications. *Nucl. Phys. A* **2012**, *881*, 115. [CrossRef]
60. Mares, J.; Barnea, N.; Cieply, A.; Friedman, E.; Gal, A.; Gazda, D. Calculations of $\bar{K}$-nuclear quasi-bound states using chiral $\bar{K}N$ amplitudes. *EPJ Web Conf.* **2014**, *66*, 09012. [CrossRef]
61. Kamiya, Y.; Miyahara, K.; Ohnishi, S.; Ikeda, Y.; Hyodo, T.; Oset, E.; Weise, W. Antikaon-nucleon interaction and $\Lambda(1405)$ in chiral SU(3) dynamics. *Nucl. Phys. A* **2016**, *954*, 41–57. [CrossRef]
62. Hoshino, T.; Ohnishi, S.; Horiuchi, W.; Hyodo, T.; Weise, W. Constraining the $\bar{K}N$ interaction from the 1S level shift of kaonic deuterium. *Phys. Rev. C* **2017**, *96*, 045204. [CrossRef]
63. Meißner, U.-G.; Raha, U.; Rusetsky, A. Kaon-nucleon scattering lengths from kaonic deuterium experiments. *Eur. Phys. J. C* **2006**, *47*, 473–480. [CrossRef]
64. Curceanu, C.; Guaraldo, C.; Sirghi, D.; Amirkhani, A.; Baniahmad, A.; Bazzi, M.; Bellotti, G.; Bosnar, D.; Bragadireanu, M.; Cargnelli, M.; et al. Kaonic Atoms to Investigate Global Symmetry Breaking. *Symmetry* **2020**, *12*, 547. [CrossRef]
65. Sadasivan, D.; Mai, M.; Döring, M. S- and p-wave structure of $S = -1$ meson-baryon scattering in the resonance region. *Phys. Lett. B* **2019**, *789*, 329–335. [CrossRef]
66. Caro Ramon, J.; Kaiser, N.; Wetzel, S.; Weise, W. Chiral SU(3) dynamics with coupled channels: Inclusion of P wave multipoles. *Nucl. Phys. A* **2000**, *672*, 249–269. [CrossRef]
67. Revai, J. Are the chiral based $\bar{K}N$ potentials really energy dependent?. *Few Body Syst.* **2018**, *59*, 49. [CrossRef]
68. Bruns, P.C.; Cieply, A. Importance of chiral constraints for the pole content of the $\bar{K}N$ scattering amplitude. *Nucl. Phys. A* **2020**, *996*, 121702. [CrossRef]
69. Anisovich, A.V.; Sarantsev, A.V.; Nikonov, V.A.; Burkert, V.; Schumacher, R.A.; Thoma, U.; Klempt, E. Hyperon I: Study of the $\Lambda(1405)$. *arXiv* **2019**, arXiv:1905.05456.
70. Anisovich, A.V.; Sarantsev, A.V.; Nikonov, V.A.; Burkert, V.; Schumacher, R.A.; Thoma, U.; Klempt, E. Hyperon III: $K^-p - \pi\Sigma$ coupled-channel dynamics in the $\Lambda(1405)$ mass region. *Eur. Phys. J. A* **2020**, *56*, 139. [CrossRef]
71. Bayar, M.; Pavao, R.; Sakai, S.; Oset, E. Role of the triangle singularity in $\Lambda(1405)$ production in the $\pi^-p \to K^0\pi\Sigma$ and $pp \to pK^+\pi\Sigma$ processes. *Phys. Rev. C* **2018**, *97*, 035203. [CrossRef]
72. Lu, H.Y.; Schumacher, R.A.; Adhikari, K.P.; Adikaram, D.; Aghasyan, M.; Amaryan, M.J.; Pereira, S.A.; Ball, J.; Battaglieri, M.; Batourine, V.; et al. First Observation of the $\Lambda(1405)$ Line Shape in Electroproduction. *Phys. Rev. C* **2013**, *88*, 045202. [CrossRef]
73. Cahn, R.; Landshoff, P. Mystery of the Delta (980). *Nucl. Phys. B* **1986**, *266*, 451–467. [CrossRef]
74. Aubert, B.; Barate, R.; Boutigny, D.; Gaillard, J.M.; Hicheur, A.; Karyotakis, Y.; Lees, J.P.; Robbe, P.; Tisserand, V.; Zghiche, A.; et al. Observation of a narrow meson decaying to $D_s^+\pi^0$ at a mass of 2.32-GeV/c^2. *Phys. Rev. Lett.* **2003**, *90*, 242001. [CrossRef] [PubMed]
75. Belle Collaboration. Study of $B^- \to D^{**0}\,\pi^-\ (D^{**0} \to D^{(*)} + \pi^-)$ decays. *Phys. Rev. D* **2004**, *69*, 112002. [CrossRef]
76. Link, J.M.; Yager, P.M.; Anjos, J.C.; Bediaga, I.; Göbel, C.; Machado, A.A.; Magnin, J.; Massafferri, A.; De Miranda, J.M.; Pepe, I.M.; et al. Measurement of masses and widths of excited charm mesons D(2)* and evidence for broad states. *Phys. Lett. B* **2004**, *586*, 11–20. [CrossRef]
77. Guo, F.; Hanhart, C.; Meißner, U.-G.; Wang, Q.; Zhao, Q.; Zou, B. Hadronic molecules. *Rev. Mod. Phys.* **2018**, *90*, 015004. [CrossRef]
78. Moir, G.; Peardon, M.; Ryan, S.M.; Thomas, C.E.; Wilson, D.J. Coupled-Channel $D\pi$, $D\eta$ and $D_s\bar{K}$ Scattering from Lattice QCD. *JHEP* **2016**, *10*, 011. [CrossRef]
79. Albaladejo, M.; Fernandez-Soler, P.; Guo, F.; Nieves, J. Two-pole structure of the $D_0^*(2400)$. *Phys. Lett. B* **2017**, *767*, 465–469. [CrossRef]
80. Kolomeitsev, E.; Lutz, M. On Heavy light meson resonances and chiral symmetry. *Phys. Lett. B* **2004**, *582*, 39–48. [CrossRef]
81. Guo, F.; Shen, P.; Chiang, H.; Ping, R.; Zou, B. Dynamically generated 0+ heavy mesons in a heavy chiral unitary approach. *Phys. Lett. B* **2006**, *641*, 278–285. [CrossRef]
82. Guo, F.; Hanhart, C.; Meißner, U.-G. Interactions between heavy mesons and Goldstone bosons from chiral dynamics. *Eur. Phys. J. A* **2009**, *40*, 171–179. [CrossRef]

83. Du, M.L.; Albaladejo, M.; Fernandez-Soler, P.; Guo, F.K.; Hanhart, C.; Meißner, U.-G.; Nieves, J.; Yao, D.L. Towards a new paradigm for heavy-light meson spectroscopy. *Phys. Rev. D* **2018**, *98*, 094018. [CrossRef]
84. David, W; Cambrige University, UK. Personal communication, 2019.
85. Cleven, M.; Guo, F.K.; Hanhart, C.; Meißner, U.-G. Light meson mass dependence of the positive parity heavy-strange mesons. *Eur. Phys. J. A* **2011**, *47*, 19. [CrossRef]
86. Bardeen, W.A.; Eichten, E.J.; Hill, C.T. Chiral multiplets of heavy-light mesons. *Phys. Rev. D* **2003**, *68*, 054024. [CrossRef]
87. Nowak, M.A.; Rho, M.; Zahed, I. Chiral doubling of heavy light hadrons: BABAR 2317-MeV/c^2 and CLEO 2463-MeV/c^2 discoveries. *Acta Phys. Pol. B* **2004**, *35*, 2377–2392.
88. Mehen, T.; Springer, R.P. Even- and odd-parity charmed meson masses in heavy hadron chiral perturbation theory. *Phys. Rev. D* **2005**, *72*, 034006. [CrossRef]
89. Guo, F.; Shen, P.; Chiang, H. Dynamically generated 1+ heavy mesons. *Phys. Lett. B* **2007**, *647*, 133–139. [CrossRef]
90. Bali, G.S.; Collins, S.; Cox, A.; Schäfer, A. Masses and decay constants of the $D_{s0}^*(2317)$ and $D_{s1}(2460)$ from $N_f = 2$ lattice QCD close to the physical point. *Phys. Rev. D* **2017**, *96*, 074501. [CrossRef]
91. Lang, C.B.; Mohler, D.; Prelovsek, S.; Woloshyn, R.M. Predicting positive parity B_s mesons from lattice QCD. *Phys. Lett. B* **2015**, *750*, 17–21. [CrossRef]
92. Aaij, R.; Adeva, B.; Adinolfi, M.; Ajaltouni, Z.; Akar, S.; Albrecht, J.; Alessio, F.; Alexander, M.; Ali, S.; Alkhazov, G.; et al. Amplitude analysis of $B^- \to D^+\pi^-\pi^-$ decays. *Phys. Rev. D* **2016**, *94*, 072001. [CrossRef]
93. Du, M.L.; Guo, F.K.; Meißner, U.-G. Implications of chiral symmetry on S-wave pionic resonances and the scalar charmed mesons. *Phys. Rev. D* **2019**, *99*, 114002. [CrossRef]
94. Roca, L.; Oset, E.; Singh, J. Low lying axial-vector mesons as dynamically generated resonances. *Phys. Rev. D* **2005**, *72*, 014002. [CrossRef]
95. Geng, L.S.; Oset, E.; Roca, L.; Oller, J.A. Clues for the existence of two $K_1(1270)$ resonances. *Phys. Rev. D* **2007**, *75*, 014017. [CrossRef]
96. Daum, C.; Hertzberger, L.; Hoogland, W.; Peters, S.; Van Deurzen, P.; Chabaud, V.; Gonzalez-Arroyo, A.; Hyams, B.; Tiecke, H.; Weilhammer, P.; et al. Diffractive Production of Strange Mesons at 63 GeV. *Nucl. Phys. B* **1981**, *187*, 1–41. [CrossRef]
97. Wang, G.Y.; Roca, L.; Oset, E. Discerning the two $K_1(1270)$ poles in $D^0 \to \pi^+ VP$ decay. *Phys. Rev. D* **2019**, *100*, 074018. [CrossRef]
98. Wang, G.Y.; Roca, L.; Wang, E.; Liang, W.H.; Oset, E. Signatures of the two $K_1(1270)$ poles in $D^+ \to \nu e^+ VP$ decay. *Eur. Phys. J. C* **2020**, *80*, 388. [CrossRef]
99. Haidenbauer, J.; Krein, G.; Meißner, U.-G.; Tolos, L. DN interaction from meson exchange. *Eur. Phys. J. A* **2011**, *47*, 18. [CrossRef]
100. Sakai, S.; Guo, F.; Kubis, B. Extraction of ND scattering lengths from the $\Lambda_b \to \pi^- pD^0$ decay and properties of the $\Sigma_c(2800)^+$. *arXiv* **2020**, arXiv:2004.09824.
101. Tornqvist, N.A. From the deuteron to deusons, an analysis of deuteron-like meson meson bound states. *Z. Phys. C* **1994**, *61*, 525–537. [CrossRef]

© 2020 by the authors. Licensee MDPI, Basel, Switzerland. This article is an open access article distributed under the terms and conditions of the Creative Commons Attribution (CC BY) license (http://creativecommons.org/licenses/by/4.0/).

symmetry MDPI

Article

Multifarious Roles of Hidden Chiral-Scale Symmetry: "Quenching" g_A in Nuclei

Mannque Rho

Institut de Physique Théorique, Université Paris-Saclay, CNRS, CEA, 91191 Gif-sur-Yvette, France; mannque.rho@ipht.fr

Abstract: I discuss how the axial current coupling constant g_A renormalized in scale symmetric chiral EFT defined at a chiral matching scale impacts on the axial current matrix elements on beta decays in nuclei with and without neutrinos. The "quenched" g_A observed in nuclear superallowed Gamow–Teller transitions, a long-standing puzzle in nuclear physics, is shown to encode the emergence of chiral-scale symmetry hidden in QCD in the vacuum. This enables one to explore how trace-anomaly-induced scale symmetry breaking enters in the renormalized g_A in nuclei applicable to certain non-unique forbidden processes involved in neutrinoless double beta decays. A parallel is made between the roles of chiral-scale symmetry in quenching g_A in highly dense medium and in hadron–quark continuity in the EoS of dense matter in massive compact stars. A systematic chiral-scale EFT, presently lacking in nuclear theory and potentially crucial for the future progress, is suggested as a challenge in the field.

Keywords: nuclear quenched g_A; Fermi-liquid fixed point; hidden scale and local symmetries in nuclear matter; genuine dilaton; anomalous dimension of gluon stress tensor in nuclei

1. Introduction

It has been recently argued [1,2] that the long-standing mystery of "quenched" g_A in nuclear super-allowed Gamow–Teller transitions can be resolved by the combination of two mechanisms: (1) strong nuclear correlations controlled by an emerging scale symmetry hidden in QCD; (2) an effect of quantum anomaly in scale symmetry inherited from QCD at the chiral scale $\Lambda_\chi \sim 4\pi f_\pi$ that defines effective field theory for nuclear dynamics. (I put the quotation marks here because what is referred to in the nuclear physics literature as "quenched g_A" is a misnomer. The coupling constant g_A appearing in the nucleonic axial current used to calculate nuclear Gamow–Teller matrix elements is in fact *not* quenched in the sense used in shell-model calculations in the literature. This is explained in what follows. I continue the discussions without the quotation marks unless otherwise noted.)

In this note, I discuss how what triggers the quenched superallowed Gamow–Teller transitions in nuclear medium encodes the emergence of chiral-scale symmetry, hidden in QCD, in strong nuclear correlations and suggest in what way the anomaly-induced breaking of scale symmetry—referred to by the acronym AISB—affects how the axial-current coupling constant g_A can indeed be "fundamentally" renormalized by the vacuum change—as opposed to the effect of nuclear correlations. This AISB is argued to have an important impact on neutrinoless double beta decay matrix elements.

2. Superallowed Gamow–Teller Transitions

To zero-in on essentials for the quenched g_A problem, first consider the superallowed Gamow–Teller decay of the doubly magic nucleus ^{100}Sn with 50 neutrons and 50 protons. I pick this case because there is what is claimed to be "accurate" data and equally importantly it offers a well-defined theoretical framework. This process allows exploiting the "extreme single-particle shell model (ESPSM)" description [3,4]. In the

ESPSM description, the GT process involves, via the spin-isospin flip, the decay of a proton in a completely filled shell $g_{9/2}$ to a neutron in an empty shell $g_{7/2}$, which can be equated nearly precisely to the GT transition of a quasi-proton to a quasi-neutron on top of the Fermi sea formulated in the Landau Fermi-liquid fixed point theory [1]. Such a feat is not usually feasible for an ESPSM description of generic (say, non-doubly-magic) nuclei.

Let us begin by defining the *quenching factor* q often used in the literature

$$M_{\rm GT} = g_A q \mathcal{M}_{\sigma\tau}. \tag{1}$$

Here, g_A is the free-space (neutron decay) axial-vector coupling constant $g_A = 1.276(4)$ and $\mathcal{M}_{\sigma\tau}$ is the proton-to-neutron single-particle GT matrix element. The quantity on the right-hand side representing nature should of course be model independent but q and $\mathcal{M}$ depend on how $\mathcal{M}$ is to be calculated, so separately model dependent. The approach adopted in [1] gives a precise meaning to what q is in the scheme and how it can be related to the experimental value. As emphasized there, given the superallowed transition with zero momentum transfer, the quantity $\mathcal{M}$ is just the spin-isospin factor for the Fermi-liquid as well as ESPSM descriptions with all the interaction effects, fundamental (i.e., AISB) as well as of pure nuclear correlations, lumped into the "quenching" factor q. In general, the two are of course intricately mixed, and it is difficult, if not impossible, to disentangle them in nuclear processes. However, if one assumes that the AISB effect is small, then the axial-current can be written with the anomaly factor $q_{\rm ssb}$ representing the scale symmetry breaking (ssb) simply multiplying the scale-invariant axial current as

$$q_{\rm ssb} g_A \bar{\psi}\tau^{\pm}\gamma_\mu\gamma_5\psi \tag{2}$$

with

$$q_{\rm ssb} = c_A + (1-c_A)\Phi^{\beta'}. \tag{3}$$

Here, c_A is a (in general) density-dependent constant and β' is the anomalous dimension of the gluon stress tensor ${\rm tr}G_{\mu\nu}^2$, both of which remain, up-to-date, unknown for QCD with the flavor number $N_f \lesssim 3$. The quantity Φ is the ratio f_π^*/f_π with the $*$ indicating density dependence of the nuclear medium. Φ has been measured by deeply bound pionic nuclei [5], so is known up to the nuclear matter density n_0. Note that, because of Φ, q_{ssb} is explicitly density-dependent. In the vacuum, $\Phi = 1$, so there is no dependence on β'. Furthermore, if c_A were equal to 1 either on symmetry grounds or by density effects in medium, there would be no dependence on β'. In both cases, the impact of scale symmetry breaking is absent in the GT transitions.

In [1], resorting to an EFT that incorporates both hidden local symmetry and hidden scale symmetry in chiral Lagrangian, the matrix element of the *superallowed* GT matrix element was calculated in the large N_c and large $\bar{N}$ approximations in the Fermi-liquid approach. In QCD, the g_A is proportional to N_c in the large N_c counting, and, in the Fermi-liquid theory, the Fermi-liquid fixed point is given by $O(1)$ term in the limit $\bar{N} = k_F/(\Lambda_{\rm FL} - k_F) \to \infty$ with $\Lambda_{\rm FL}$ the cut-off in the Fermi-liquid renormalization group decimation. In these double limits, it comes out that

$$\langle \bar{\psi}\tau^{\pm}\gamma_\mu\gamma_5\psi\rangle_{fi} = q_{\rm snc}^{\rm Landau}\langle\tau^{\pm}\sigma\rangle_{fi} \tag{4}$$

where the entire strong nuclear correlation (snc) effects are captured in

$$q_{\rm snc}^{\rm Landau} = (1-\kappa)^{-2} \tag{5}$$

with

$$\kappa = \frac{1}{3}\Phi\tilde{F}_1^{\pi}. \tag{6}$$

Here, $\tilde{F}_1^{\pi}$ is the pionic contribution to the Landau mass, governed by chiral symmetry for any density $\leq n_0 \approx 0.16$ fm^{-3}. (The pionic contribution to the Landau mass is the Fock term, which is formally of $O(1/\bar{N})$, and goes beyond the Landau Fermi-liquid fixed point approximation made in (5). However, soft theorems for both chiral symmetry (pion) and scale symmetry (dilaton) figure for the validity of the relation (5), so the pionic contribution is essential. See [2] for discussions on this matter.) Therefore, the quantity κ is almost completely controlled by low-energy theorems. In addition, due to near compensation of the density effects in the two factors, κ remains remarkably insensitive to density. It varies negligibly between n_0 and $n_0/2$. Therefore, it is applicable equally well to both light and heavy nuclei.

From (3) and (5), we get the total quenching factor

$$q = q_{\rm ssb} q_{\rm snc}^{\rm Landau}. \tag{7}$$

To see what we have, let us first ignore the AISB and set $q_{\rm ssb} = 1$. Evaluating $q_{\rm snc}$ in the Landau Fermi-liquid theory, we get at $n \approx n_0$ [1]

$$q_{\rm snc}^{\rm Landau} \approx 0.79. \tag{8}$$

Thus, what is identified as the "scale-symmetric effective g_A" given in the Landau Fermi-liquid theory is

$$g_A^{\rm ss} \equiv g_A^{\rm Landau} \approx g_A q_{\rm snc}^{\rm Landau} = 1.276 \times 0.79 \approx 1.0. \tag{9}$$

Note that $g_A^{\rm ss}$ is *not a quenched* coupling constant. The quenching is in the nuclear interactions captured in $q_{snc}^{\rm Landau}$. There is more on this crucial point below.

2.1. Fermi-Liquid Fixed Point (FLFP) ≈ Extreme Single Particle Shell Model (ESPSM)

In what sense can this result (9), which figures crucially in what follows, be taken as reliable? It relies on taking two limits. The first is the large N_c limit in QCD combined with soft theorems. The second is the large $\bar{N}$ limit in the renormalization-group approach to Landau Fermi-liquid theory in treating strongly correlated nuclear interactions. The former is akin to the Goldberger–Treiman relation which is known in the matter-free vacuum to be accurate within a few percent discrepancy. Thus, combining the two limits in nuclear processes can be considered to be comparable to applying to the Goldberger–Treiman relation in nuclear dynamics. It has been shown that the Landau Fermi-liquid fixed point approximation combined with soft theorems give very accurate predictions for such nuclear electroweak processes as the anomalous orbital gyromagnetic ratio of the proton δg_l^p and the axial charge $0^{\pm} \leftrightarrow 0^{\mp}$ $\Delta T = 1$ transitions in heavy nuclei [6,7]. To the best of my knowledge, there are no other approaches that match the precision in the predictive power for these quantities. I would consider (9) to be of the same accuracy. At present, one cannot make any more precise error assessment.

As argued above, we can equate $q_{\rm snc}^{\rm Landau}$ to the ESPSM description of the quenching factor of the doubly magic nucleus

$$q_{\rm snc}^{\rm Landau} \simeq q_{\rm snc}^{\rm ESPSM}. \tag{10}$$

Thus, the total quenching factor can be written as

$$q = q_{\rm ssb} q_{\rm snc}^{\rm Landau} \simeq q_{\rm ssb} q_{\rm snc}^{\rm ESPSM}. \tag{11}$$

2.2. Scale-Symmetric Effective $g_A^{\rm ss}$

That $g_A^{\rm ss}$ is close to 1 raises two questions: (1) What does the effective $g_A^{\rm ss}$ approaching 1 signify? (2) What does it represent in comparison with experiments?

Since the direct impact of matching to QCD at the given scale Λ_χ is factored out, q_{snc} encodes *pure* nuclear correlation effects. Thus, $g_A^{\rm ss}$ in (9) represents their effects captured in the constant $g_A q_{snc}$ multiplying the non-interacting single particle transition involving only the spin-isospin flip taking place: (a) on the Fermi surface for the LHS of (10); (b) in the ESPSM for the RHS.

To put more precisely, $g_A^{\rm ss}$ would correspond to a full-scale shell-model calculation with the *unquenched or un-renormalized* g_A that takes into account the excitation of *all* configurations connected to the parent (ground) state by nuclear forces, in particular, the nuclear tensor forces up to excitation energies of $\Delta E \lesssim (m_\Delta - m_N)$. Such a full scale microscopic calculation for a mass number of $\sim$100 would be far from feasible at present, so the relation (10) could at best be conjectural. However, highly powerful numerical techniques have been developed up-to-date that can handle the dynamics of few-nucleon systems, say, $A \lesssim 10$. Such a microscopic numerical calculation in light nuclei anchored on an effective field theory suitably defined at, say, the chiral scale would do the same physics as the mapping of the Fermi-liquid fixed point approximation to the ESPSM description for many-nucleon systems. Indeed, a recent quantum Monte-Carlo (QMC) calculation of Gamow–Teller matrix elements entering β decay and electron capture rates in $A = 3 - 10$ nuclei was shown, quite convincingly, by King et al. [8], to reproduce the experimental Gamow–Teller matrix elements at a few % level. Note that no *quenched* coupling constant is involved in this calculation. There, two-body exchange-current operators which enter, chiral-order-suppressed, at N^3LO [9] were included, but found to contribute insignificantly, say, only at the 2–3% level as predicted a long time ago [9]. As argued in [1], however, there are no theoretical justifications within the framework of chiral effective field theory to ignore next-order terms of N^rLO for $r > 3$ if the $r = 3$ terms, whether negligible or not, are taken into account. The reason for this is that there can be non-negligible cancelations between the calculated $r = 3$-order terms and the higher-order terms of comparable strength that are however not calculable due to large number of unknown counter terms. Therefore, it is argued in [1] that the many-body terms be dropped for consistency. Furthermore the possible 2–3% contribution can be of the same order as the AISB effect $q_{\rm ssb}$ contributing at density $n < n_0$, as made clear below.

One might think that $g_A^{\rm ss}$ being close to 1 may merely be an accident. Surprisingly, however, $g_A^{\rm ss} \approx 1$ is a ubiquitous and pervasive outcome of simple shell-model calculations in light nuclei (see, e.g., [10]) and has prompted many authors, starting with Wilkinson [11], to ask whether it has something fundamental intrinsic of QCD. Indeed if the Fermi-liquid theory is extended to Δ-hole configurations and the Landau–Migdal Fermi interaction g_0' is assumed to be universal, that is, $g_{NN}^0 = g_{N\Delta}^0 = g_{0\Delta\Delta}'$, the Δ-hole loop contribution can be made to screen the g_A by the same quenching factor $q \approx 0.80$ as $q_{\rm snc}^{\rm Landau}$. However, this result would make sense only if all nuclear correlations up to the energy scale of the Δ-N mass difference turned out to sum to nearly zero. However, there is absolutely no reason known why this can be so. Furthermore, even if one were to formulate, assuming it makes sense, the Fermi-liquid theory in the space of both nucleon and Δ—Landau-Migdal theory—there would be no reason the universality should hold for the g_0' channel.

In any event, I must say it is highly puzzling why the *naively calculated* Δ-N screening leads to the more or less same quenching as the purely nucleonic Landau Fermi-liquid correlations do—even if it were coincidental. This has led to a long-standing confusion in the field. This clearly points to that what is "fundamental" depends on how and at what scale the EFT is matched to QCD.

2.3. $g_A^{\rm ss}$ *and Dilaton-Limit Fixed-Point*

Let me now address what $g_A^{\rm ss} \approx 1$ could imply. As noted, the axial current $g_A \bar{\psi}\gamma_\mu\gamma_5\psi$ is scale-invariant. Thus, the nuclear interaction that leads to (10) reflects scale symmetry because it is associated with both soft-pion and soft-dilaton theorems, although the NG boson masses figure in the Landau interactions involved. Again, this is quite analogous to the Goldberger–Treiman relation which holds both with and without the pion mass as long as the mass is small compared with the QCD scale, reflecting soft theorems in chiral symmetry. That $g_A^{ss} \approx 1$ is not accidental but could perhaps reflect scale symmetry in action can be seen in what happens as baryonic matter is tweaked to the putative chiral restoration or more appropriately to the infrared (IR) fixed point in the "genuine dilaton" theory of scale symmetry [12,13]. (The idea of genuine dilaton may be controversial in technicolor approaches to the BSM, but we find it highly appropriate for nuclear dynamics, as reviewed in [2,14].) This is because, in the chiral limit, approaching the chiral restoration or the IR fixed point, the dilaton condensate and the quark condensate approach each other, $f_\chi \to f_\pi$, in arriving at what is called "dilaton-limit fixed-point (DLFP)" [15] relevant for compact-star matter [14]. In the EFT giving $g_A^{\rm Landau}$, going toward the DLFP by increasing density is found to impose the constraint on the quark axial current $g_A^Q \bar{Q}\gamma_\mu\gamma_5 Q$—where Q is the quark field—which is likewise scale-invariant

$$g_A^Q \to 1. \tag{12}$$

This corresponds to a non-interacting quasi-quark in the medium in which $f_\pi \simeq f_\chi \neq 0$ making the GT transition. It resembles—uncannily— the quasi-proton making the superallowed GT transition in the ESPSM in the doubly magic nucleus *in the absence of AISB effect*, i.e., with $q_{\rm ssb} = 1$. It is thus tempting to consider that albeit approximate scale symmetry permeates from low density—nuclei—to high density—compact-star matter. This could be considered as one of multifaceted manifestations of hadron–quark continuity discussed in [2,14]. An equally striking manifestation of continuity between hadrons and quarks, which may be related to the pervasive continuity of scale symmetry in g_A, is in the equation of state (EoS) of dense baryonic matter. Conformal symmetry, broken both by anomaly and spontaneously—and invisible in nuclear systems—emerges pervasively in the unitarity limit at low density in light nuclei [16] and in the pseudo-conformal sound speed $v_{sPC}^2/c^2 \approx 1/3$ in massive compact stars at high density, $n \gtrsim 3n_0$ [2], an observation that could be considered as a critical opalescence to the conformal speed $v_c^2/c^2 = 1/3$ expected at asymptotic density in QCD.

3. Strongly Forbidden Axial Transitions

It follows from the discussions made so far that (9) is not a *fundamental* coupling constant to be applied generically to *any* axial responses in the background defined by the modified vacuum. It encodes as completely as feasible nuclear correlations generated in the given background for the superallowed GT transitions in nuclei. It is not applicable to nuclear axial responses different from the superallowed (zero-momentum transfer) GT transitions. It is in particular *not* applicable to non-unique strongly forbidden transitions. This should be highly relevant to neutrinoless double beta decay processes [17]—important for going BSM—where the momentum transfer involved can be of order $\sim$100 MeV.

This problem was brought to light recently by Kostensalo et al. [18] in the "effective" value of g_A, denoted $\tilde{g}_A$, in the β decay spectrum-shape function involving leptonic phase-space factors and nuclear matrix elements in forbidden non-unique beta decays. The nuclear operators involved there are non-relativistic momentum-dependent impulse approximation terms. In principle, there can be n-body exchange-current corrections with $n \geq 2$. However, the corrections to the single-particle (impulse) approximation are typically of mth order with $m \gtrsim 3$ relative to the leading one-body term in the power counting in χEFT and could be suppressed so ignorable. Unlike in the superallowed GT transition however, neither soft theorems nor the Fermi-liquid renormalization-group strategy can be exploited. Hence, (9) is not relevant to the spectrum-shape function discussed in [18]. What is relevant instead is the scale-anomaly correction (3) modulo of course the reliability of the impulse-approximation operator. This means that one can identify $\tilde{g}_A$ as "fundamentally renormalized g_A" in distinction to g_A^{ss}, Equation (9).

Given that experimental data are available for the superallowed decay in ^{100}Sn [3,4], one can first analyze what one learns from it. The recent RIKEN data [4] are claimed to be a big improvement with much less error bars over what is given, among others, in [3], so let me look only at the RIKEN result. As above, I do not quote error bars in arriving at q_{ssb}. (I eschew the kind of "error analyses" done in standard chiral EFT circles which are not needed for the new ideas developed here. This is because here I am dealing with distinctively qualitative effects, not quantitative. Going beyond the accuracy commensurate with the soft theorems I relied on would require a formulation of high-order chiral-scale perturbation theory much superior in power and quality to what is available up to now.) From the RIKEN data, dividing by the ESPSM matrix element, the experimental quenching factor comes out to be

$$q_{\rm riken}^{\rm ESPSM} \approx 0.50. \tag{13}$$

Equating $q_{\rm snc}^{\rm Landau}$ to what is given in ESPSM, i.e., $q_{snc}^{\rm ESPSM}$, and dividing (13) by (8), one obtains

$$q_{\rm ssb}^{\rm RIKEN} \approx 0.63. \tag{14}$$

The deviation $\delta = (1 - q_{\rm ssb}^{\rm RIKEN})$ represents the AISB effect.

Let me suppose that the future measurements will confirm the present RIKEN data. Then, (14) can be used to deduce the axial coupling constant $\bar{g}_A = g_A q_{\rm ssb}$ in the axial current $\bar{g}_A \bar{\psi}\gamma_\mu\gamma_5\psi$ applicable to *all axial transitions in the background of given density*

$$\bar{g}_A^{\rm RIKEN} = g_A q_{\rm ssb}^{\rm RIKEN} = 1.276 \times 0.63 \approx 0.80. \tag{15}$$

In distinction to $g_A^{\rm ss}$, this is the "genuine" *quenched* axial vector constant.

Now, how does this prediction compare with the results of the analysis—referred to as "COBRA"—of the spectrum-shape factor of ^{113}Cd β decay in [18]? Listed in the increasing average χ-square are

$$g_A^{\rm COBRA} = 0.809 \pm 0.122,\ 0.893 \pm 0.054,\ 0.968 \pm 0.056. \tag{16}$$

Allowing possible uncertainty in (14), it seems reasonable to take these results [18] to be consistent with the prediction (15).

There are however two caveats to keep in mind in assessing the results given above.

The first is the accuracy of (14) extracted from the RIKEN experiment. In the previous experiments (see, e.g., [3]), various different values of $q_{\rm ssb}$ were obtained. Thus, it is highly desirable that the RIKEN measurement be reconfirmed or improved on. It is shown below that the RIKEN value for $q_{\rm ssb}$ (14) as it stands could get in tension with the theory for the AISB effect as given by Equation (3).

The second is the nuclear model dependence in the analyses leading to (16). The current operator $\bar{\psi}\gamma_\mu\gamma_5\psi$ for strongly forbidden β decay is involved. It is known in chiral perturbative approach since a long time [9]—and reconfirmed in subsequent refined calculations—that many-body (exchange current) corrections to the single-particle spatial component of the axial current are suppressed by high chiral orders, at least by three orders, so the impulse approximation—ignoring multi-body currents—in the axial channel, unless accidentally suppressed, should be sufficiently reliable in dealing with the strongly forbidden transitions involved. (The story of the unique first-forbidden transition mediated by the axial-charge operator is a totally different story. There, the two-body exchange charge operator is dominated by terms governed by soft-pion and soft-dilaton theorems as predicted by theory and confirmed by experiments a long time ago [6]. It does not however affect non-unique highly forbidden operators concerned here.) The operators involved are necessarily non-relativistic and are in principle known up to a manageable form reduced from the covariant current. However, the numerical results for $\bar{g}_A$ must sensitively depend on nuclear models that enter in the calculation of the transition matrix elements involved. It is not easy to assess the quantitative reliability of the nuclear models in giving highly forbidden non-unique transition matrix elements.

In particular, one cannot achieve for non-unique transitions the accuracy with which the q_{snc} can be computed in the superallowed kinematics of the doubly magic nucleus. This is because the mapping of the Landau Fermi-liquid fixed-point theory to the ESPSM description exploited in the superallowed GT transition is not feasible for forbidden transition kinematics.

4. Anomaly-Induced Symmetry Breaking

Let me now turn to the AISB effect (3) and how the caveats mentioned above impact on confronting theory with observations.

The two parameters that figure in the theory are quantities that are inaccessible perturbatively. They are intrinsically nonperturbative and remain un-calculated even in the matter-free vacuum [12]. Furthermore, they are inaccessible by QCD lattice simulation due to the density dependence in nuclear processes. The only thing one can say at present is that the "genuine dilaton" approach [12,13] requires that $\beta' > 0$. Thus, both are as yet unknown parameters.

Given that there are two unknown parameters, one cannot make a unique determination of $q_{\rm ssb}$ from only the RIKEN result. However, as discussed in [19], approaching dense matter in the skyrmion description in the presence of hidden local symmetry and dilaton scalar mesons—the key degrees of freedom that figure crucially in the EoS of compact stars in [14]—is found to break down disastrously unless the homogeneous Wess–Zumino (hWZ) Lagrangian $\mathcal{L}_{\rm hWZ}$—which is scale-invariant similar to the axial current—is corrected by an AISB factor of the form

$$q_{\rm hWZ} = c_{\rm hWZ} + (1 - c_{\rm hWZ})\Phi^{\beta'}. \tag{17}$$

As with c_A, $c_{\rm hWZ}$ is unknown. It turns out—fortunately—that, if one picks $1 \lesssim \beta' \lesssim 3.5$ and $0 \lesssim c_{\rm hWZ} \lesssim 0.2$, then one can not only eliminate the catastrophe but also obtain a qualitatively reasonable phase structure of the dense matter [19]. (The anomalous dimension β' has been extensively studied in nonabelian gauge theories for $N_f \gtrsim 8$ in connection with the technicolor QCD for going BSM. However, there is no information for $N_f \sim 3$ that is concerned here.) It turns out that, with $c_{\rm hWZ} \approx 0$ and $\beta' \approx (2\text{–}3)$, a satisfactory result can be obtained.

The high density-matter structure addressed in [19] is to zero-in on the density regime where the chiral condensate $\langle\bar{q}q\rangle$ tends toward zero. Here, one is dealing with $\Phi \ll 1$. The density regime involved in β decay is instead near the equilibrium density n_0 at which Φ is $\lesssim 1$. In this case, it may be unreliable to adopt the value of β' appropriate for the regime where $\Phi \ll 1$.

Just to have an idea how the expression (3) fares, let me simply take $\beta' \approx 2.5$ indicated from the dense matter case of [19]. Now, assuming that $\Phi \approx f_\pi^*/f_\pi \approx 0.79$, one can evaluate c_A's corresponding to (16). (This corresponds to density $n \approx n_0$. I am assuming that one can use the nuclear matter value although nuclei of mass number 113 are concerned.)

$$0.15 \lesssim c_A^{exp} \lesssim 0.48. \tag{18}$$

In fact, $c_A \approx 0.15$ gives $\bar{g}_A$ compatible with the superallowed RIKEN data as well as with the skyrmion crystal structure for dense matter.

It turns out however that the density dependence of Φ in nuclear β-decay processes can play an extremely important role if β' turns out to be > 1. Suppose the nuclei considered in [18] have the average density $\sim 0.6n_0$ as in the deeply bound pionic nuclei of mass number ~ 110 measured [5], then only the $q_{\rm ssb}$ that yields $\bar{g}_A \approx 1$ is acceptable. Others are ruled out. Furthermore, if $\beta' < 2$, even the RIKEN data are in serious tension with the AISB Formula (3). In this case, either the RIKEN measurement is cast in doubt or the AISB Formula (3) relying on linearized approximation is in error.

5. Remarks

While what is discussed in this note is quite interesting in that nuclear beta decay with strong sensitivity to density can give a glimpse into how hidden scale symmetry can be probed, a long-standing problem in strong-interaction physics, there are many issues to resolve to make what is obtained here truly sensible. First, there are no known reasons c_A should be directly related to c_{hWZ} except that they are of the same RG nonperturbative effects and the c coefficients for a given $\beta' > 0$ could very well be correlated. It is nonetheless exciting that nuclear processes can offer an, albeit first, hint on β' for N_f as low as 3, up-to-date a totally unknown quantity in QCD. Second, it is imperative to check that similar AISB effects do not appreciably upset the properties of normal nuclear matter as well as of dense matter that are more or less well described with the c coefficients set equal to 1, i.e., with no explicit AISB effects. Thus far, there seem to be no indications for $c \neq 1$ in the EoS in nuclear physics. If it turned out that $c_\zeta \approx 1$ for certain channels ζ and ≈ 0 for other channels, it would be absolutely necessary to understand how and why.

What transpires from the result discussed in this note is that scale symmetry combined with chiral symmetry must play a crucial role, as argued in [2], in nuclear dynamics ranging from nuclear density all the way to compact-star matter density. The work available up-to-date in the literature of incorporating scale symmetry with chiral symmetry in nuclear physics is far too involved with too many unknown parameters to be on the right track in view of the power and simplicity found in the large N_c and large $\bar{N}$ limit in resolving the intricate g_A problem. Much work needs to be done.

Funding: This research received no external funding.

Acknowledgments: I am grateful for helpful comments from Jouni Suhonen on work in progress with his collaborators and for on-going discussions/collaborations with Yong-Liang Ma.

Conflicts of Interest: The author declares no conflict of interest.

References

1. Ma, Y.L.; Rho, M. Quenched g_A in nuclei and emergent scale symmetry in baryonic matter. *Phys. Rev. Lett.* **2020**, *125*, 142501. [CrossRef] [PubMed]
2. Rho, M.; Ma, Y.L. Manifestation of hidden symmetries in baryonic matter: From finite nuclei to neutron stars. *Mod. Phys. Lett. A* **2021**, *36*, 2130012. [CrossRef]
3. Hinke, C.B.; Böhmer, M.; Boutachkov, P.; Faestermann, T.; Geissel, H.; Gerl, J. Superallowed Gamow-Teller decay of the doubly magic nucleus ^{100}Sn. *Nature* **2012**, *486*, 341. [CrossRef] [PubMed]
4. Lubos, D.; Park, J.; Faestermann, T.; Gernhäuser, R.; Krücken, R.; Lewitowicz, M.; Nishimura, S.; Sakurai, H.; Ahn, D.S.; Baba, H.; et al. Improved value for the Gamow-Teller strength of the ^{100}Sn beta decay. *Phys. Rev. Lett.* **2019**, *122*, 222502. [CrossRef] [PubMed]
5. Kienle, P.; Yamazaki, T. Pions in nuclei, a probe of chiral symmetry restoration. *Prog. Part. Nucl. Phys.* **2004**, *52*, 85. [CrossRef]
6. Kubodera, K.; Delorme, J.; Rho, M. Axial currents in nuclei. *Phys. Rev. Lett.* **1978**, *40*, 755. [CrossRef]
7. Warburton, E.K. Mesonic enhancement of the weak axial vector current evaluated from beta decay in the lead region. *Phys. Rev. Lett.* **1991**, *66*, 1823. [CrossRef] [PubMed]
8. King, G.B.; Andreoli, L.; Pastore, S.; Piarulli, M.; Schiavilla, R.; Wiringa, R.B.; Carlson, J.; Gandolfi, S. Chiral effective field theory calculations of weak transitions in light nuclei. *Phys. Rev. C* **2020**, *102*, 025501. [CrossRef]
9. Park, T.-S.; Marcucci, L.E.; Schiavilla, R.; Viviani, M.; Kievsky, A.; Rosati, S.; Kubodera, K.; Min, D.-P.; Rho, M. Parameter free effective field theory calculation for the solar proton fusion and hep processes. *Phys. Rev. C* **2003**, *67*, 055206. [CrossRef]
10. Suhonen, J.T. Value of the axial-vector coupling strength in β and $\beta\beta$ decays: A Review. *Front. Phys.* **2017**, *5*, 55. [CrossRef]
11. Wilkinson, D.H. Renormalization of the axial-vector coupling constant in nuclear beta decay. *Phys. Rev. C* **1973**, *7*, 930. [CrossRef]
12. Crewther, R.J.; Tunstall, L.C. $\Delta I = 1/2$ rule for kaon decays derived from QCD infrared fixed point. *Phys. Rev. D* **2015**, *91*, 034016. [CrossRef]
13. Crewther, R.J. Genuine dilatons in gauge theories. *Universe* **2020**, *6*, 96. [CrossRef]
14. Ma, Y.-L.; Rho, M. Towards the hadron-quark continuity via topology change in compact stars. *Prog. Part. Nucl. Phys.* **2020**, *113*, 103791. [CrossRef]
15. Beane, S.R.; van Kolck, U. The dilated chiral quark model. *Phys. Lett. B* **1994**, *328*, 137. [CrossRef]
16. König, S.; Griesshammer, H.W.; Hammer, H.W.; van Kolck, U. Nuclear physics around the unitarity limit. *Phys. Rev. Lett.* **2017**, *118*, 202501 [CrossRef] [PubMed]

7. Ejiri, H.; Suhonen, J.; Zuber, K. Neutrino–nuclear responses for astro-neutrinos, single beta decays and double beta decays. *Phys. Rep.* **2019**, *797*, 1–102. [CrossRef]
8. Kostensalo, J.; Suhonen, J.; Volkmer, J.; Zatschler, S.; Zuber, K. Confirmation of g_A quenching using the revised spectrum-shape method for the analysis of the ^{113}Cd β-decay as measured with the COBRA demonstrator. *arXiv* **2020**, arXiv:2011.11061.
9. Ma, Y.L.; Rho, M. Scale-chiral symmetry, ω meson and dense baryonic matter. *Phys. Rev. D* **2018**, *97*, 094017. [CrossRef]

symmetry

MDPI

Article

Quark Cluster Expansion Model for Interpreting Finite-T Lattice QCD Thermodynamics

David Blaschke [1,2,3,*], Kirill A. Devyatyarov [2,3] and Olaf Kaczmarek [4,5]

1 Instytut Fizyki Teoretycznej, Uniwersytet Wrocławski, 50-204 Wrocław, Poland
2 Department of Theoretical Nuclear Physics, National Research Nuclear University (MEPhI), 115409 Moscow, Russia; dka005@campus.mephi.ru
3 Bogoliubov Laboratory of Theoretical Physics, JINR Dubna, 141980 Dubna, Russia
4 Fakultät für Physik, Universität Bielefeld, D-33615 Bielefeld, Germany; okacz@physik.uni-bielefeld.de
5 Key Laboratory of Quark & Lepton Physics (MOE) and Institute of Particle Physics, Central China Normal University, Wuhan 430079, China
* Correspondence: david.blaschke@gmail.com

Abstract: In this work, we present a unified approach to the thermodynamics of hadron–quark–gluon matter at finite temperatures on the basis of a quark cluster expansion in the form of a generalized Beth–Uhlenbeck approach with a generic ansatz for the hadronic phase shifts that fulfills the Levinson theorem. The change in the composition of the system from a hadron resonance gas to a quark–gluon plasma takes place in the narrow temperature interval of 150–190 MeV, where the Mott dissociation of hadrons is triggered by the dropping quark mass as a result of the restoration of chiral symmetry. The deconfinement of quark and gluon degrees of freedom is regulated by the Polyakov loop variable that signals the breaking of the $Z(3)$ center symmetry of the color $SU(3)$ group of QCD. We suggest a Polyakov-loop quark–gluon plasma model with $\mathcal{O}(\alpha_s)$ virial correction and solve the stationarity condition of the thermodynamic potential (gap equation) for the Polyakov loop. The resulting pressure is in excellent agreement with lattice QCD simulations up to high temperatures.

Keywords: Polyakov quark–gluon plasma; hadron resonance gas; Beth–Uhlenbeck approach; lattice QCD thermodynamics

1. Introduction

Recently, continuum extrapolated lattice QCD (LQCD) thermodynamics results for physical quark masses have become available [1–4]. It has now become a major goal to construct an effective low-energy QCD model that would reproduce these results in the finite temperature and low chemical potential domain to high accuracy. Such a description could form a basis for extrapolations to the region of low temperatures and high baryochemical potentials where the sign problem still prevents LQCD obtaining benchmark solutions.

QCD Dyson–Schwinger equations [5,6] and functional renormalization group (FRG) methods [7,8] are promising tools to investigate the nonperturbative aspects of the QCD phase diagram in the vicinity of the chiral and deconfinement transitions from first principles. However, to date, they do not self-consistently account for the bound state formation. Therefore, despite a satisfactory description of the temperature dependence of the renormalized light chiral condensate [8], a question arises regarding the relationship to the chiral perturbation theory limit, where the onset of chiral condensate melting is solely due to light pseudoscalar meson excitations in the medium [9]. The account for bound states in the FRG approach to low-energy QCD is still under development [10] and not yet applicable to address the role of hadronic bound states for chiral condensate melting.

Complementary to these first-principle approaches is the Polyakov–quark–meson (PQM) model (see, e.g., [11] and references therein), which is suitable to study the interrelation of chiral and deconfinement transitions. Both transitions are closely correlated, but in comparison with LQCD, the chiral restoration temperature result is too high. It could be suspected that the inclusion of further mesonic and baryonic degrees of freedom could improve the situation. Furthermore, the dynamic quark self-energy effects due to dressing by the meson cloud should be taken into account. In the PQM model, mesons are not composites but elementary degrees of freedom.

In order to properly account for the composite nature of mesons and baryons, as well as their Mott dissociation in a hot and dense medium, a Beth–Uhlenbeck approach can be employed [12–22]. This approach has been limited to the application of low-lying mesons only and did not take into account perturbative corrections, meaning that a quantitative description of LQCD thermodynamics has not yet been possible.

To overcome these limitations, the generic behavior of the scattering phase shifts in the hadronic channels has been constructed in the spirit of a cluster expansion model which reproduces the full hadron resonance gas at low temperatures and the quark–gluon plasma (QGP) with $\mathcal{O}(\alpha_s)$ virial corrections at high temperatures [21,22]. The model embodied the main consequence of chiral symmetry restoration in the quark sector: the lowering of the thresholds for the two and three-quark scattering state continuous spectrum, which triggers the transformation of hadronic bound states to resonances in the scattering continuum. In the early version of this model, the gap equation for the Polyakov-loop was incomplete, and the model of the phase shifts was a rather complicated one.

We suggest here a Polyakov-loop quark–gluon plasma model with $\mathcal{O}(\alpha_s)$ virial correction in order to obtain a satisfactory agreement with lattice QCD simulations up to high temperatures and solve the complete stationarity condition of the thermodynamic potential (gap equation) for the Polyakov loop. The phase shift model employed in this work is simpler than in previous works and universally applicable for all hadronic species. It is in accordance with the Levinson theorem and results in the vanishing of hadronic contributions to the thermodynamics at high temperatures.

2. Cluster Virial Expansion to Quark-Hadron Matter

The main idea behind unifying the description of the quark–gluon plasma (QGP) and the hadron resonance gas (HRG) phase of low-energy QCD matter is the fact that hadrons are strong, nonperturbative correlations of quarks and gluons. In particular, mesons and baryons are bound states (clusters) of quarks and should therefore emerge in a cluster expansion of interacting quark matter as new, collective degrees of freedom.

For the total thermodynamic potential of the model, from which all other equations of state can be derived, we make the following ansatz:

$$\Omega_{\text{total}}(T;\phi) = \Omega_{\text{QGP}}(T;\phi) + \Omega_{\text{MHRG}}(T)\,, \tag{1}$$

where $\Omega_{\text{QGP}}(T;\phi) = \Omega_{\text{PNJL}}(T;\phi) + \Omega_{\text{pert}}(T;\phi)$ describes the thermodynamic potential of the quark and gluon degrees of freedom with a perturbative part $\Omega_{\text{pert}}(T;\phi)$ and a nonperturbative mean field part $\Omega_{\text{PNJL}}(T;\phi) = \Omega_{\text{Q}}(T;\phi) + \mathcal{U}(T;\phi)$ that can be decomposed into the quark quasiparticle contribution $\Omega_{\text{Q}}(\phi;T)$ and the gluon contribution that is approximated by a mean field potential $\mathcal{U}(T;\phi)$. Note that all these contributions to the QGP thermodynamic potential are intertwined by the traced Polyakov loop ϕ as the order parameter for confinement. The correlations beyond the mean field approximation which correspond to the hadronic bound states and their scattering state continuum are described by the Mott–HRG pressure $P_{\text{MHRG}}(T)$. This is an HRG pressure that takes into account the dissociation of hadrons by the Mott effect, when their masses would exceed the mass of the corresponding continuum of unbound quark states. A detailed description and numerical evaluation of these contributions is given below.

2.1. Beth–Uhlenbeck Model for HRG with Mott Dissociation

For the MHRG part of the pressure of the model, we have $P_{\rm MHRG}(T) = -\Omega_{\rm MHRG}(T)$,

$$P_{\rm MHRG}(T) = \sum_{i=M,B} P_i(T)\,, \tag{2}$$

where the sum extends over all mesonic (M) and baryonic (B) states from the particle data group (PDG), comprising an ideal mixture of hadronic bound and scattering states in the channel i that are described by a Beth–Uhlenbeck formula. Then, the partial pressure of the hadron species i is

$$P_i(T) = \mp d_i \int_0^\infty \frac{dp\, p^2}{2\pi^2} \int_0^\infty \frac{dM}{\pi}\, T \ln\left(1 \mp {\rm e}^{-\sqrt{p^2+M^2}/T}\right) \frac{d\delta_i(M;T)}{dM}\,, \tag{3}$$

where d_i is the degeneracy factor. For the phase shift of the bound states of N_i quarks in the hadron i, we adopt the simple model that is in accordance with the Levinson theorem:

$$\delta_i(M;T) = \pi[\Theta(M - M_i) - \Theta(M - M_{{\rm thr},i}(T))]\Theta(M_{{\rm thr},i}(T) - M_i). \tag{4}$$

Inserting (4) into (3) results in

$$P_i(T) = \mp d_i \int_0^\infty \frac{dp\, p^2}{2\pi^2}\, T\left[\ln\left(1 \mp {\rm e}^{-\sqrt{p^2+M_i^2}/T}\right) - \ln\left(1 \mp {\rm e}^{-\sqrt{p^2+M_{{\rm thr},i}(T)^2}/T}\right)\right] \Theta(M_{{\rm thr},i}(T) - M_i). \tag{5}$$

The temperature dependent threshold mass of the two (or three) quark continuum for mesonic (baryonic) bound state channels i is

$$M_{{\rm thr},i}(T) = \sqrt{2}[(N_i - N_s)m(T) + N_s m_s(T)]\,, \tag{6}$$

where $N_s = 0, 1, ..., N_i$ is the number of strange quarks in hadron i. The factor $\sqrt{2}$ originates from quark confinement in the following way. In the confining vacuum, the quarks are not simple plane waves with an arbitrarily long wavelength, but due to the presence of bag-like boundary conditions, their wavelength shall not exceed a certain length scale. Therefore, a minimal quark momentum applies to the quark dispersion relations $E_{q,{\rm min}}(T) = \sqrt{m_q^2(T) + p_{q,{\rm min}}^2}$, which for the choice $p_{q,{\rm min}} = m_q(T)$ results in $E_{q,{\rm min}}(T) = \sqrt{2}m_q(T)$. For details, see [23]. The chiral condensate is defined as

$$\langle\bar{\psi}\psi\rangle_{q,T} = -\frac{\partial\Omega(T)}{\partial m_q}\,,\quad q = u, d, s, \tag{7}$$

where m_l (m_s) is the current-quark mass in the light (strange) quark sector, $l = u, d$. It is an order parameter for the dynamical breaking of the chiral symmetry that is reflected in the corresponding temperature dependence of the dynamical quark masses $m_q(T)$.

In our present model, we do not treat the dynamical quark mass as an order parameter that should follow from the solution of an equation of motion (gap equation) that minimizes the thermodynamic potential, as in the case of the Polyakov-loop variable ϕ, but we use the quantity $\Delta_{l,s}(T)$ from simulations of 2 + 1 flavor lattice QCD as an input. This quantity has been introduced in [24] with the definition

$$\Delta_{l,s}(T) = \frac{\langle\bar{\psi}\psi\rangle_{l,T} - (m_l/m_s)\langle\bar{\psi}\psi\rangle_{s,T}}{\langle\bar{\psi}\psi\rangle_{l,0} - (m_l/m_s)\langle\bar{\psi}\psi\rangle_{s,0}}\,, \tag{8}$$

and was used later on, e.g., in [1,2]. Further, we assume the following for the temperature-dependent light quark mass,

$$m(T) = m(0)\Delta_{l,s}(T) + m_l\,, \tag{9}$$

where $m_l = 5.5$ MeV is the current-quark mass, and for the strange quark mass, we adopt

$$m_s(T) = m(T) + m_s - m_l = m(0)\Delta_{l,s}(T) + m_s\,, \tag{10}$$

with $m_s = 100$ MeV. The LQCD result for the temperature dependence of the chiral condensate [1,2] can be fitted by

$$\Delta_{l,s}(T) = \frac{1}{2}\left[1 - \tanh\left(\frac{T - T_c}{\delta_T}\right)\right], \tag{11}$$

where $T_c = 154$ MeV is the common pseudocritical temperature of the chiral restoration transition of both LQCD Collaborations and $\delta_T = 26$ MeV is its width for the data from [1], while $\delta_T = 22.7$ MeV for those from [2], see Figure 1. For our present applications modeling QCD thermodynamics, we use the fit of the chiral condensate (11), but with the modern value of $T_c = 156.5 \pm 1.5$ MeV [25]. We have checked that the results for the total pressure of our model are practically inert against a changing value of δ_T within the above range of variation. Inserting (9) and (10) into (6), we get

$$\begin{aligned} M_{\mathrm{thr},i}(T) &= \sqrt{2}[N_i m(T) + N_s(m_s(T) - m(T))] \\ &= \sqrt{2}[m_s N_s + m_l(N_i - N_s) + m(0)N_i\Delta_{l,s}(T)]\,, \end{aligned} \tag{12}$$

and using (9) results in

$$M_{\mathrm{thr},i}(T) = \sqrt{2}\left\{m_s N_s + m_l(N_i - N_s) + m(0)N_i\left[\frac{1}{2} - \frac{1}{2}\tanh\left(\frac{T - T_c}{\delta_T}\right)\right]\right\}. \tag{13}$$

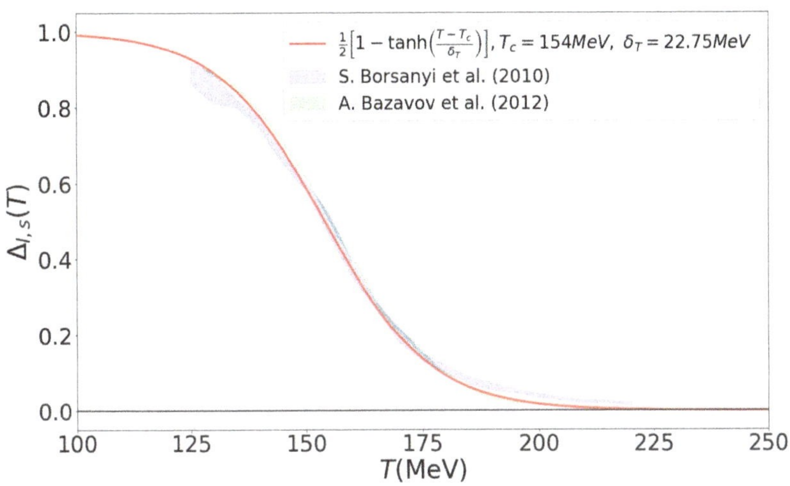

Figure 1. Comparison of the fit (11) for the temperature dependence of the chiral condensate $\Delta_{l,s}(T)$ and the LQCD data for it from the Wuppertal–Budapest Collaboration [1] and the hotQCD Collaboration [2].

In Figure 2 we show the pressure (2) as a function of temperature for the hadron resonance gas (HRG) model with stable hadrons (red line) and for the HRG with Mott dissociation of hadrons (MHRG) according to the simple phase shift model (4) employed in the present work. These results are compared to the LQCD data from the HotQCD Collaboration [4] (green band) and the Wuppertal–Budapest Collaboration [3] (blue band). We want to point out that due to the Mott dissociation effect hadrons completely vanish from the system at $T \approx 190$ MeV while the pressure of the ideal HRG model is misleadingly still in perfect agreement with LQCD data at this temperature!

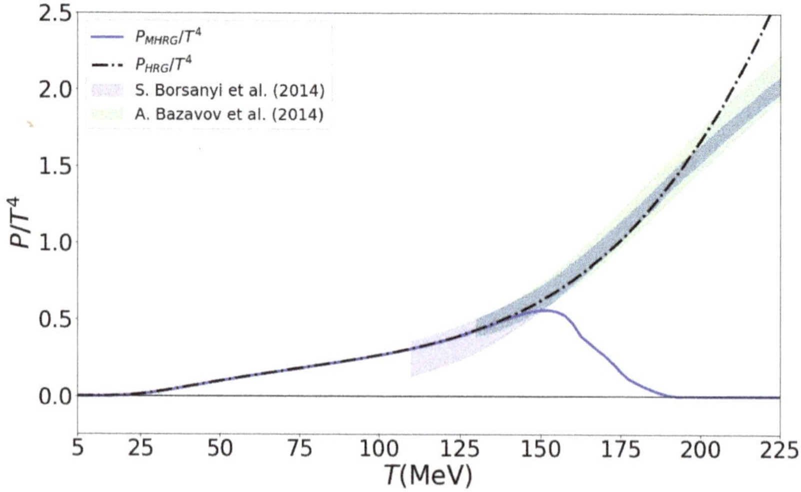

Figure 2. Pressure as a function of temperature for the HRG model with stable hadrons (red line) and for the MHRG model with Mott dissociation of hadrons according to the simple phase shift model (4) employed in the present work. These results are compared to the LQCD data from the HotQCD Collaboration [4] (green band) and the Wuppertal–Budapest Collaboration [3] (blue band).

2.2. Polyakov-Loop Improved Nambu–Jona-Lasinio (PNJL) Model

The underlying quark and gluon thermodynamics are divided into a perturbative contribution $\Omega_{\text{pert}}(T)$ which is treated as virial correction in two-loop order following [26] and a nonperturbative part described within a PNJL model in the form

$$P_{\text{PNJL}}(T;\phi) = P_Q(T;\phi) + \mathcal{U}(T;\phi)\,, \tag{14}$$

where the quark quasiparticle contribution is given by

$$P_Q(T;\phi) = 4N_c \sum_{q=u,d,s} \int \frac{dp\, p^2}{2\pi^2} \frac{T}{3} \ln\left[1 + 3\phi(1+Y_q)Y_q + Y_q^3\right],\ Y_q = e^{-\sqrt{p^2+m_q^2(T)}/T}\,, \tag{15}$$

and the Polyakov-loop potential $\mathcal{U}(T;\phi)$ takes into account the nonperturbative gluon background in a meanfield approximation using the polynomial fit of [27]

$$\mathcal{U}(T;\phi) = \frac{b_2(T)}{2}\phi^2 + \frac{b_3}{3}\phi^3 - \frac{b_4}{4}\phi^4\,, \tag{16}$$

where the temperature-dependent coefficient $b_2(T)$ is given by

$$b_2(T) = a_0 + a_1\left(\frac{T_0}{T}\right) + a_2\left(\frac{T_0}{T}\right)^2 + a_3\left(\frac{T_0}{T}\right)^3\,, \tag{17}$$

and the coefficients are given in Table 1.

Table 1. Set of values for the Polyakov-loop potential $\mathcal{U}(T;\phi)$ [27].

a_0	a_1	a_2	a_3	b_3	b_4
6.75	−1.95	2.625	−7.44	0.75	7.5

2.3. Perturbative Contribution

It is well known that the lattice QCD thermodynamics at high temperatures of $T \sim$ 1 GeV follow a Stefan–Boltzmann like behavior $\propto T^4$, but with a 15–20% reduction of the effective number of degrees of freedom. It has been observed, e.g., in [26], that this deviation can be described by the virial correction to the pressure due to the quark–gluon scattering at $\mathcal{O}(\alpha_s)$ shown in Figure 3. Here, we modify the standard expression [28] of the form

$$\Omega_{\text{pert}}(T;\phi) = -\frac{8}{\pi}\alpha_s T^4\left[I(T;\phi) + \frac{3}{\pi^2}(I(T;\phi))^2\right] \tag{18}$$

by introducing the modified integral

$$I(T;\phi) = \int_{\Lambda/T}^{\infty} \mathrm{d}x\, x f_\phi(x), \tag{19}$$

where the generalized Fermi distribution function of the PNJL model for the case of vanishing quark chemical potential considered here is defined as

$$f_\phi(x) = [\phi(1+2Y)Y + Y^3]/[1+3\phi(1+Y)Y+Y^3],\ Y = \exp(-x) \tag{20}$$

and $\Lambda = m_l(T)$ is the momentum range below which nonperturbative physics dominates and is accounted for by the dynamically generated quark mass. Here, we use a temperature-dependent, regularized running coupling [29–31]

$$\alpha_s = \frac{g^2}{4\pi} = \frac{12\pi}{11N_c - 2N_f}\left(\frac{1}{\ln(r^2/c^2)} - \frac{c^2}{r^2 - c^2}\right), \tag{21}$$

where $r = 3.2T$, $c = 350$ MeV and $N_c = N_f = 3$.

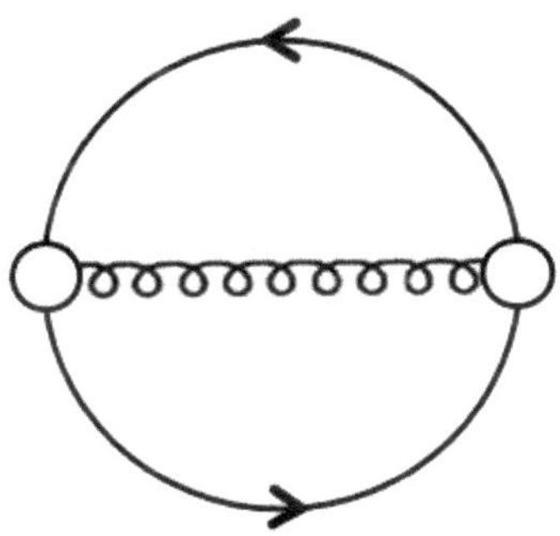

Figure 3. Two-loop diagram for the contribution of the one-gluon exchange interaction to the thermodynamic potential of quark matter.

3. Stationarity Condition for the Polyakov Loop

The pressure follows from the thermodynamic potential under the condition of stationarity with regard to variations of the order parameters. Since the chiral condensate is fixed by the fit (11) to the numerical result from lattice QCD, the Polyakov loop ϕ is the only free-order parameter in the system to be varied; this condition means

$$P_{\text{QGP}}(T) = -\min_{\phi}\{\Omega_{\text{QGP}}(T;\phi)\}\,. \tag{22}$$

This is realized by demanding

$$\frac{d\Omega_{\rm QGP}(T;\phi)}{d\phi} = \frac{dU(T;\phi)}{d\phi} + \frac{d\Omega_Q(T;\phi)}{d\phi} + \frac{d\Omega_{\rm pert}(T;\phi)}{d\phi} = 0\,, \tag{23}$$

where the separate contributions come from the variations of the Polyakov loop potential

$$\frac{dU(T;\phi)}{d\phi} = b_2(T)\phi + b_3\phi^2 - b_4\phi^3\,, \tag{24}$$

and the quark quasiparticle pressure

$$\frac{d\Omega_Q(T;\phi)}{d\phi} = 4N_c \sum_{q=u,d,s} \int \frac{dp\, p^2}{2\pi^2} \frac{(1+Y_q)Y_q}{1+3\phi(1+Y_q)Y_q + Y_q^3}\,, \tag{25}$$

with $Y_q = \exp[-\sqrt{p^2+m_q^2(T)}/T]$, and the $\mathcal{O}(\alpha_s)$ quark loop contribution

$$\frac{d\Omega_{\rm pert}(T;\phi)}{d\phi} = -\frac{8}{\pi}\alpha_s T^4 \left[\frac{dI(\phi,T)}{d\phi} + \frac{6}{\pi^2} I(\phi,T)\frac{dI(\phi,T)}{d\phi}\right]\,, \tag{26}$$

where

$$\frac{dI(T;\phi)}{d\phi} = \int_{\Lambda/T}^{\infty} {\rm d}x\, x \frac{df_\phi(x)}{d\phi}\,, \tag{27}$$

and

$$\frac{df_\phi(x)}{d\phi} = \frac{Y+2Y^2-2Y^4-Y^5}{(1+3\phi(1+Y)Y+Y^3)^2} = \frac{(1+2Y)Y-(2+Y)Y^4}{(1+3\phi(1+Y)Y+Y^3)^2}\,,\ Y=\exp(-x)\,. \tag{28}$$

The equation resulting from the stationarity condition (23) can be dubbed a "gap equation" for ϕ since it has a similar structure to the quark mass gap equation, known from Nambu–Jona-Lasinio models. In previous work [21,22], the contribution to this gap equation from the $\mathcal{O}(\alpha_s)$ quark loop diagram was omitted. Since this perturbative contribution is calculated with the Polyakov-loop generalized quark distribution functions f_ϕ, it has to be included to the generalized gap equation for the traced Polyakov loop. This has been done in the present paper for the first time.

Moreover, an infrared cutoff is placed at the loop integrals in the perturbative contribution which is set at the medium-dependent quark mass. Therefore, the full perturbative contribution in accordance with [28] is restored only at high temperatures, where $m(T) \to m_l$ and $\phi \approx 1$, while in the vicinity of the chiral and deconfinement transition, the effects of both the quark mass and Polyakov loop are taken into account. The solution of this gap equation gives the temperature dependence of the traced Polyakov loop ϕ. This is discussed in the next section.

4. Results

4.1. Polyakov Loop

We performed the numerical solution of the gap Equation (23) for the traced Polyakov loop as a function of the temperature which enters via the coefficient $b_2(T)$ of the Polyakov-loop potential (16) and the Boltzmann factors Y_q and Y of the distribution functions in the integrals (26) and (27). The result is shown in Figure 4 along with a comparison to the lattice QCD data for the renormalized Polyakov loop from the TUMQCD Collaboration [32] and the Wuppertal–Budapest Collaboration [1].

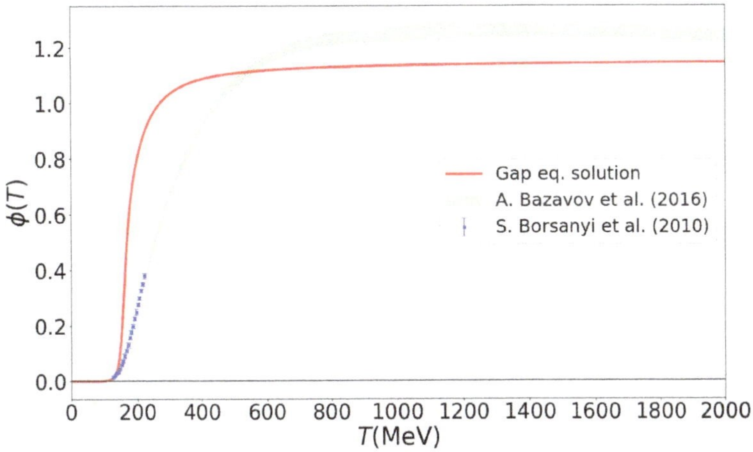

Figure 4. The traced Polyakov loop ϕ from the solution of the stationarity condition (23) on the thermodynamical potential as a function of temperature (magenta solid line) compared with the lattice results for the renormalized Polyakov loop the TU Munich QCD (TUMQCD) Collaboration [32] (green band) and the Wuppertal–Budapest Collaboration [1] (blue symbols).

In Figure 5, we compare the Polyakov-loop susceptibility $d\phi/dT$ with the chiral susceptibility $d\Delta_{l,s}/dT$ and obtain a strong synchronization effect of the chiral and Polyakov-loop crossover transitions. This is demonstrated by the almost coincident vertical lines indicating the peak positions of these transitions at $T_\chi = 156.5$ MeV and $T_\phi = 159.0$ MeV, respectively.

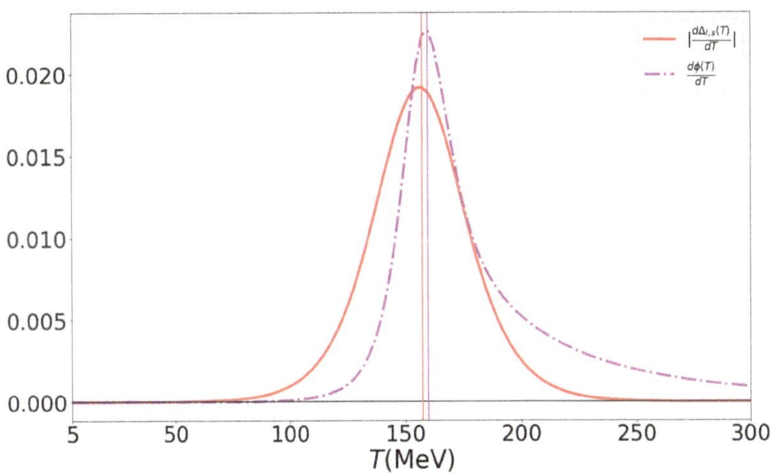

Figure 5. The temperature derivatives of the chiral condensate (chiral susceptibility $d\Delta_{l,s}/dT$, red solid line) and of the Polyakov loop (Polyakov-loop susceptibility $d\phi/dT$,) as functions of temperature. The vertical lines indicate their almost coincident peak positions at $T_\chi = 156.5$ MeV and $T_\phi = 159.0$ MeV, respectively.

4.2. Pressure

The main result of this work is a unified approach to the pressure of hadron–quark–gluon matter at finite temperatures that is in excellent agreement with lattice QCD thermodynamics (see Figure 6). The nontrivial achievement of the presented approach is that the Mott dissociation of the hadrons described by the MHRG model pressure conspires with the quark–gluon pressure described by the Polyakov-loop quark–gluon model with $\mathcal{O}(\alpha_s)$ corrections in such a way that the resulting pressure as a function of temperature yields a smooth crossover behavior. By virtue of the Polyakov-loop-improved perturbative correction, the agreement with the lattice QCD thermodynamics extends to the high temperatures of $T = 1960$ MeV reported in [33]; see Figure 7.

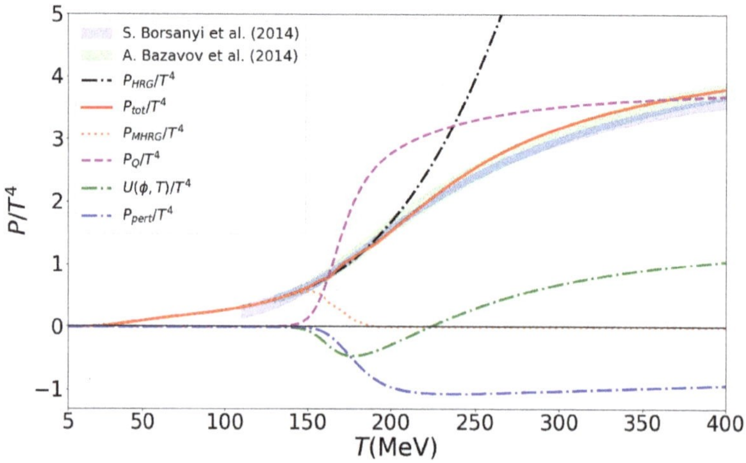

Figure 6. The temperature dependence of the total scaled pressure (red solid line) and it's constituents: MHRG (coral dotted line), quark (dashed magenta line), Polyakov-loop potential $\mathcal{U}(T;\phi)$ (dash–dotted green line) and perturbative QCD contribution (dash-dotted blue line) compared to the lattice QCD data: HotQCD Collaboration [4] (green band) and Wuppertal–Budapest Collaboration [3] (blue band).

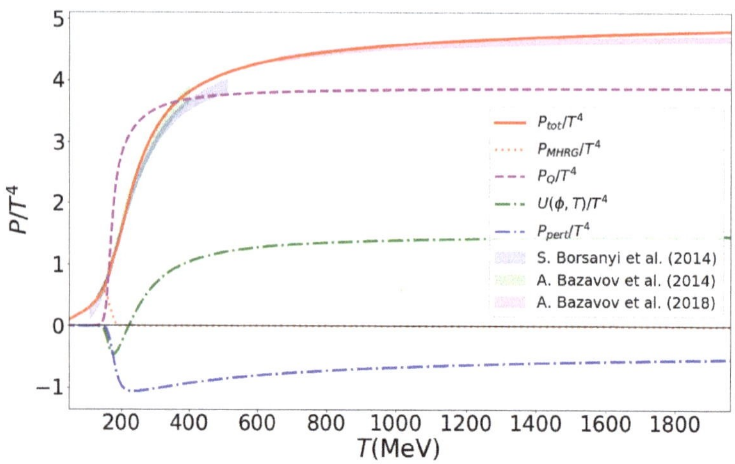

Figure 7. The temperature dependence of the total scaled pressure (red solid line) and its constituents: MHRG (coral dotted line), quark (dashed magenta line), Polyakov-loop potential $U(\phi, T)$ (dash-dotted green line) and perturbative QCD contribution (dash-dotted blue line) compared to the lattice QCD data: HotQCD Collaboration [4] (green band) and Wuppertal–Budapest Collaboration [3] (blue band), and the high-temperature result [33] (magenta band).

4.3. Quark Number Susceptibilities

In the present work, we did not consider the generalization of the approach to finite chemical potentials which would then allow us to evaluate the (generalized) susceptibilities as derivatives of pressure with respect to the corresponding chemical potential in appropriate orders. On that basis, ratios of susceptibilities could be formed as they indicate different aspects of the QCD transition between the limiting cases of a HRG and a QGP. Here, we would like to discuss, as an insight into these extensions of the approach, one of the simplest susceptibility ratios, namely the dimensionless ratio of quark number density to quark number susceptibility:

$$R_{12}(T) = \left.\frac{n_q(T)}{\mu_q\,\chi_q(T)}\right|_{\mu_q=0}, \tag{29}$$

where $n_q(T) = \partial P(T,\mu_q)/\partial\mu_q|_{\mu_q=0}$ and $\chi_q(T) = \partial^2 P(T,\mu_q)/\partial\mu_q^2|_{\mu_q=0}$. This ratio (29) has two well-known limits. At low temperatures, in the hadron resonance gas phase, it is given by

$$R_{12}^{HRG}(T) = \frac{T}{3\mu_q}\tanh\left(\frac{3\mu_q}{T}\right), \tag{30}$$

while in the QGP phase for massless quarks it approaches

$$R_{12}^{QGP}(T) = \frac{1+(1/\pi^2)(\mu_q/T)^2}{1+(3/\pi^2)(\mu_q/T)^2}. \tag{31}$$

An evaluation of (29) for the present model for the QCD pressure would require its extension to a finite μ_q, which we will perform in a subsequent work. In the present model, we used our knowledge of the composition as a function of temperature to define a proxy for (29) by interpolating between the two known limits (30) and (31) with the partial pressure of the HRG, $x_{HRG}(T) = P_{\mathrm{MHRG}}(T)/P_{\mathrm{tot}}(T)$, as

$$R_{12}(T) = x_{HRG}(T)R_{12}^{HRG}(T) + [1 - x_{HRG}(T)]R_{12}^{QGP}(T). \tag{32}$$

The result is shown in Figure 8 for two values of μ_q/T, for which lattice QCD results in the two-flavor case [34] are shown for a comparison. The fact that the present approach reproduces the transition between HRG and QGP asymptotics well in the narrow range of temperatures 150 MeV $\lesssim T \lesssim$ 190 MeV is a nontrivial result. In previous effective approaches to describe the finite-temperature lattice QCD thermodynamics results based on a spectral broadening of the HRG states, the transition to the QGP asymptotics occurred at a much higher temperature 250 MeV $\lesssim T \lesssim$ 400 MeV [35] or never [26,36,37]. In the latter case, the QGP asymptotic behavior is mimicked by an appropriate number of unaffected low-lying hadronic degrees of freedom.

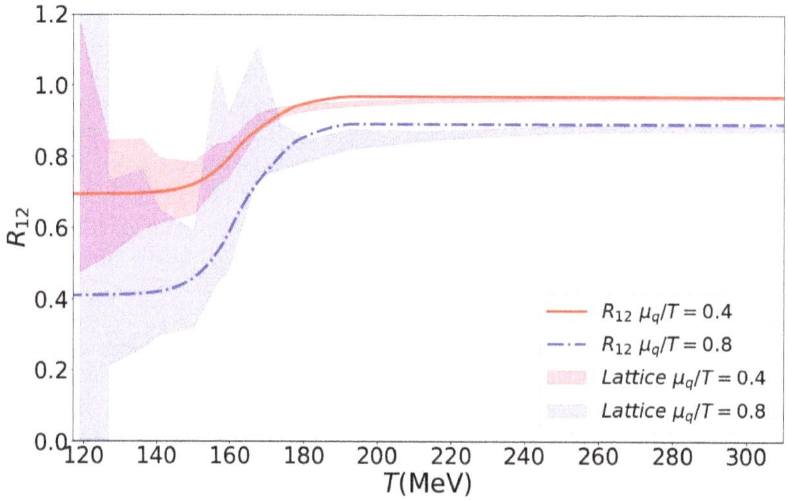

Figure 8. The dimensionless ratio of quark number density to quark number susceptibility $R_{12}(T) = n_q(T)/(\mu_q \chi_q(T))|_{\mu_q=0}$ as a function of temperature for $\mu_q/T = 0.4$ (red solid line) and $\mu_q/T = 0.8$ (blue dash-dotted line) compared to the lattice QCD data [34] $\mu_q/T = 0.4$ (red band), $\mu_q/T = 0.8$ (blue band). For details, see text.

5. Discussion and Conclusions

The main result of the present work is a unified approach to the thermodynamic potential of hadron–quark–gluon matter at finite temperatures that is in excellent agreement with lattice QCD thermodynamics on the temperature axis of the QCD phase diagram. The key aspect to this approach is the quark cluster decomposition of the thermodynamic potential within the Beth–Uhlenbeck approach [38], which allowed us to implement the effect of Mott dissociation to the hadron resonance gas phase of low-temperature/low-density QCD. The MHRG model description includes, in principle, the information about the spectral properties of all hadronic channels with their discrete and continuous part of the spectrum, encoded in the hadronic phase shifts. Instead of solving the equations of motion with a coupled hierarchy of Schwinger–Dyson equations in the one, two and many-quark channels self-consistently (a formidable task of finite-temperature quantum field theory), we applied a schematic model for the in-medium phase shifts that was in accordance with the Levinson theorem and sufficiently general to be applicable for all multiquark cluster channels. This phase shift model requires only the knowledge of the vacuum mass spectrum which can come from the particle data group tables or from relativistic quark models and the medium dependence of the multi-quark continuum threshold.

The latter requires the knowledge of the quark mass (i.e., the chiral condensate) with its medium dependence as an order parameter of the chiral symmetry breaking and restoration. Since a quark mean field model of the (P)NJL type is not sufficient as it lacks the back-reaction from the hadron resonance gas on the quark propagator properties, we employ here the chiral condensate measured in continuum-extrapolated, full lattice QCD with physical current quark masses as an input. This procedure restricts the applicability of the present model to small chemical potentials only, where lattice QCD data for the chiral condensate are available. In a further development of the model, a beyond-mean-field derivation of the quark self-energy will be given. Furthermore, at the same level of approximation, the corresponding sunset-type diagrams for the Φ functional of the 2PI approach should be derived and evaluated. This would allow us to calculate the generalized polarization-loop integrals which determine the analytic properties of the multi-quark states. These can be equivalently encoded in the corresponding medium-dependent phase shifts of the generalized Beth–Uhlenbeck approach, as has been demonstrated in particular examples for pions, diquarks [16,39] and nucleons [40] within the Polyakov-loop generalized NJL model.

Another important aspect of the present approach is that it leads to a relativistic density functional theory for QCD matter in the QCD phase diagram, with the known limits of the HRG and pQCD manifestly implemented. Such an approach allows us to predict the existence and location of critical endpoints in the QCD phase diagram, as has been demonstrated, e.g., in [41], where a dependence on a free parameter could have—besides the critical endpoint of the liquid–gas transition in the nuclear matter phase—another endpoint for the deconfinement transition or none. This "crossover all over" case of the QCD phase diagram is impossible to address with two-phase approaches that use a Maxwell construction for the phase transition. Other models that are in use for the analysis of the critical behavior of QCD (see, e.g., [42,43]) impose this by assuming a so-called "switch function" between HRG and QGP phases. They are valuable tools but do not have predictive power.

Author Contributions: Conceptualization, D.B.; Data curation, K.A.D.; Formal analysis, D.B. and K.A.D.; Funding acquisition, D.B.; Investigation, K.A.D.; Software, K.A.D.; Supervision, D.B. and O.K.; Validation, O.K.; Writing original draft, D.B.; Writing review and editing, D.B. and O.K. All authors have read and agreed to the published version of the manuscript.

Funding: The research of D.B. was supported by the Russian Fund for Basic research under grant number 18-02-40137 and from the National Research Nuclear University (MEPhI) in the framework of the Russian Academic Excellence Project under contract number 02.a03.21.0005.

Acknowledgments: D.B. acknowledges discussions with Konstantin Maslov, Krzysztof Redlich and Ludwik Turko about this work.

Conflicts of Interest: The authors declare no conflict of interest.

References

1. Borsányi, S.; Fodor, Z.; Hoelbling, C.; Katz, S.D.; Krieg, S.; Ratti, C.; Szabó, K.K.; Wuppertal-Budapest Collaboration. Is there still any T_c mystery in lattice QCD? Results with physical masses in the continuum limit III. *J. High Energy Phys.* **2010**, *1009*, 73. [CrossRef]
2. Bazavov, A.; Bhattacharya, T.; Cheng, M.; DeTar, C.; Ding, H.T.; Gottlieb, S.; Gupta, R.; Hegde, P.; Heller, U.M.; Karsch, F.; et al. The chiral and deconfinement aspects of the QCD transition. *Phys. Rev. D* **2012**, *85*, 054503. [CrossRef]
3. Borsanyi, S.; Fodor, Z.; Hoelbling, C.; Katz, S.D.; Krieg, S.; Szabo, K.K. Full result for the QCD equation of state with 2+1 flavors. *Phys. Lett. B* **2014**, *730*, 99. [CrossRef]
4. Bazavov, A.; Bhattacharya, T.; DeTar, C.; Ding, H.T.; Gottlieb, S.; Gupta, R.; Hegde, P.; Heller, U.M.; Karsch, F.; HotQCD Collaboration; et al. Equation of state in (2 + 1)-flavor QCD. *Phys. Rev. D* **2014**, *90*, 094503. [CrossRef]
5. Roberts, C.D.; Schmidt, S.M. Dyson-Schwinger equations: Density, temperature and continuum strong QCD. *Prog. Part. Nucl. Phys.* **2000**, *45*, S1–S103. [CrossRef]
6. Fischer, C.S. QCD at finite temperature and chemical potential from Dyson–Schwinger equations. *Prog. Part. Nucl. Phys.* **2019**, *105*, 1–60. [CrossRef]
7. Tripolt, R.A.; Strodthoff, N.; von Smekal, L.; Wambach, J. Spectral Functions for the Quark-Meson Model Phase Diagram from the Functional Renormalization Group. *Phys. Rev. D* **2014**, *89*, 034010. [CrossRef]
8. Gao, F.; Pawlowski, J.M. QCD phase structure from functional methods. *Phys. Rev. D* **2020**, *102*, 034027. [CrossRef]
9. Gerber, P.; Leutwyler, H. Hadrons Below the Chiral Phase Transition. *Nucl. Phys. B* **1989**, *321*, 387–429. [CrossRef]
10. Alkofer, R.; Maas, A.; Mian, W.A.; Mitter, M.; Paris-Lopez, J.; Pawlowski, J.M.; Wink, N. Bound state properties from the functional renormalization group. *Phys. Rev. D* **2019**, *99*, 054029. [CrossRef]
11. Herbst, T.K.; Mitter, M.; Pawlowski, J.M.; Schaefer, B.J.; Stiele, R. Thermodynamics of QCD at vanishing density. *Phys. Lett. B* **2014**, *731*, 248–256. [CrossRef]
12. Hüfner, J.; Klevansky, S.P.; Zhuang, P.; Voss, H. Thermodynamics of a quark plasma beyond the mean field: A generalized Beth-Uhlenbeck approach. *Ann. Phys.* **1994**, *234*, 225–244. [CrossRef]
13. Zhuang, P.; Hüfner, J.; Klevansky, S.P. Thermodynamics of a quark - meson plasma in the Nambu-Jona-Lasinio model. *Nucl. Phys. A* **1994**, *576*, 525–552. [CrossRef]
14. Wergieluk, A.; Blaschke, D.; Kalinovsky, Y.L.; Friesen, A. Pion dissociation and Levinson's theorem in hot PNJL quark matter. *Phys. Part. Nucl. Lett.* **2013**, *10*, 660–668. [CrossRef]

15. Yamazaki, K.; Matsui, T. Quark-Hadron Phase Transition in the PNJL model for interacting quarks. *Nucl. Phys. A* **2013**, *913*, 19–50. [CrossRef]
16. Blaschke, D.; Buballa, M.; Dubinin, A.; Röpke, G.; Zablocki, D. Generalized Beth–Uhlenbeck approach to mesons and diquarks in hot, dense quark matter. *Ann. Phys.* **2014**, *348*, 228. [CrossRef]
17. Yamazaki, K.; Matsui, T. Quark-hadron phase transition in a three flavor PNJL model for interacting quarks. *Nucl. Phys. A* **2014** *922*, 237–261. [CrossRef]
18. Torres-Rincon, J.M.; Aichelin, J. Equation of state of a quark-Polyakov loop-meson mixture in the PNJL model at finite temperature *arXiv* **2016**, arXiv:1601.01706.
19. Torres-Rincon, J.M.; Aichelin, J. Equation of state of a quark-meson mixture in the improved Polyakov–Nambu–Jona-Lasinio model at finite chemical potential. *Phys. Rev. C* **2017**, *96*, 045205. [CrossRef]
20. Blaschke, D.; Dubinin, A.; Radzhabov, A.; Wergieluk, A. Mott dissociation of pions and kaons in hot, dense quark matter. *Phys Rev. D* **2017**, *96*, 094008. [CrossRef]
21. Blaschke, D.; Dubinin, A.; Turko, L. Mott-hadron resonance gas and lattice QCD thermodynamics. *arXiv* **2016**, arXiv:1611.09845
22. Blaschke, D.; Dubinin, A.; Turko, L. Generalized Beth-Uhlenbeck approach to the equation of state for quark-hadron matter. *Acta Phys. Polon. Supp.* **2017**, *10*, 473–480. [CrossRef]
23. Dubinin, A.; Blaschke, D.; Kalinovsky, Y.L. Pion and sigma meson dissociation in a modified NJL model at finite temperature *Acta Phys. Polon. Supp.* **2014**, *7*, 215–223. [CrossRef]
24. Cheng, M.; Christ, N.H.; Datta, S.; Heide, J.V.; Jung, C.; Karsch, F.; Kaczmarek, O.; Laermann, E.; Mawhinney, R.D.; Miao, C.; et al The QCD equation of state with almost physical quark masses. *Phys. Rev. D* **2008**, *77*, 014511. [CrossRef]
25. Bazavov, A.; Ding, H.T.; Hegde, P.; Kaczmarek, O.; Karsch, F.; Karthik, N.; Laermann, E.; Lahiri, A.; Larsen, R.; HotQCD; et al Chiral crossover in QCD at zero and non-zero chemical potentials. *Phys. Lett. B* **2019**, *795*, 15–21. [CrossRef]
26. Turko, L.; Blaschke, D.; Prorok, D.; Berdermann, J. Mott-Hagedorn Resonance Gas and Lattice QCD Results. *Acta Phys. Polon. Supp.* **2012**, **5**, 485. [CrossRef]
27. Ratti, C.; Thaler, M.A.; Weise, W. Phases of QCD: Lattice thermodynamics and a field theoretical model. *Phys. Rev. D* **2006**, **73**, 014019. [CrossRef]
28. Kapusta, J.I. *Finite Temperature Field Theory*; Cambridge University Press: Cambridge, UK, 1989.
29. Peshier, A.; Kämpfer, B.; Soff, G. The QCD equation of state and quark star properties. *arXiv* **2003**, arXiv:hep-ph/0312080.
30. Shirkov, D.V. On the Fourier transformation of renormalization invariant coupling. *Theor. Math. Phys.* **2003**, *136*, 893. [CrossRef]
31. Blaschke, D.; Kaczmarek, O.; Laermann, E.; Yudichev, V. Heavy quark potential and quarkonia dissociation rates. *Eur. Phys. J. C* **2005**, *43*, 81. [CrossRef]
32. Bazavov, A.; Brambilla, N.; Ding, H.T.; Petreczky, P.; Schadler, H.P.; Vairo, A.; Weber, J.H. Polyakov loop in 2 + 1 flavor QCD from low to high temperatures. *Phys. Rev. D* **2016**, *93*, 114502. [CrossRef]
33. Bazavov, A.; Petreczky, P.; Weber, J.H. Equation of State in 2 + 1 Flavor QCD at High Temperatures. *Phys. Rev. D* **2018**, *97*, 014510 [CrossRef]
34. Allton, C.R.; Döring, M.; Ejiri, S.; Hands, S.J.; Kaczmarek, O.; Karsch, F.; Laermann, E.; Redlich, K. Thermodynamics of two flavor QCD to sixth order in quark chemical potential. *Phys. Rev. D* **2005**, *71*, 054508. [CrossRef]
35. Biro, T.S.; Jakovac, A. QCD above T_c: Hadrons, partons, and the continuum. *Phys. Rev. D* **2014**, *90*, 094029. [CrossRef]
36. Turko, L.; Blaschke, D.; Prorok, D.; Berdermann, J. An effective model of QCD thermodynamics. *J. Phys. Conf. Ser.* **2013**, *455*, 012056. [CrossRef]
37. Turko, L.; Blaschke, D.; Prorok, D.; Berdermann, J. Effective degrees of freedom in QCD thermodynamics. *EPJ Web Conf.* **2014**, *71*, 00134. [CrossRef]
38. Bastian, N.U.F.; Blaschke, D.; Fischer, T.; Röpke, G. Towards a Unified Quark-Hadron Matter Equation of State for Applications in Astrophysics and Heavy-Ion Collisions. *Universe* **2018**, *4*, 67. [CrossRef]
39. Blaschke, D.; Dubinin, A.; Buballa, M. Polyakov-loop suppression of colored states in a quark-meson-diquark plasma. *Phys. Rev. D* **2015**, *91*, 125040. [CrossRef]
40. Blaschke, D.; Dubinin, A.S.; Zablocki, D. NJL model approach to diquarks and baryons in quark matter. *PoS BaldinISHEPPXXII* **2015**, 083. [CrossRef]
41. Bastian, N.U.F.; Blaschke, D.B. A unified quark-nuclear matter equation of state from the cluster virial expansion within the generalized Beth–Uhlenbeck approach. *Eur. Phys. J. A* **2021**, *57*, 35. [CrossRef]
42. Plumberg, C.J.; Welle, T.; Kapusta, J.I. QCD matter with a crossover and a first-order phase transition. *PoS CORFU2018* **2018**, 157. [CrossRef]
43. Parotto, P.; Bluhm, M.; Mroczek, D.; Nahrgang, M.; Noronha-Hostler, J.; Rajagopal, K.; Ratti, C.; Schäfer, T.; Stephanov, M. QCD equation of state matched to lattice data and exhibiting a critical point singularity. *Phys. Rev. C* **2020**, *101*, 034901. [CrossRef]

symmetry

MDPI

Article

Perturbative RG Analysis of the Condensate Dependence of the Axial Anomaly in the Three-Flavor Linear Sigma Model

Gergely Fejős

Institute of Physics, Eötvös University Budapest, H-1117 Budapest, Hungary; gergely.fejos@ttk.elte.hu

Abstract: Coupling of 't Hooft's determinant term is investigated in the framework of the three-flavor linear sigma model as a function of the chiral condensate. Using perturbation theory around the minimum point of the effective action, we calculate the renormalization group flow of the first field-dependent correction to the coupling of the conventional $U_A(1)$ breaking determinant term. It is found that, at low temperatures, mesonic fluctuations make the anomaly increase when the chiral condensate decreases. As an application, we analyze the effect at the zero temperature nuclear liquid–gas transition.

Keywords: axial anomaly; nuclear liquid-gas transition; renormalization group

1. Introduction

The $U_A(1)$ subgroup of approximate $U_L(3) \times U_R(3)$ chiral symmetry is anomalously broken in quantum chromodynamics (QCD). QCD is a strongly coupled theory, and as such, most accurate results can be expected to emerge through lattice simulations. These, however, lack the ability to simulate the system at finite density due to the notorius sign problem. To tackle this issue, it is common to build effective models upon chiral symmetry, which are expected to capture essential features of QCD in the low energy regime. Even at zero density, these models are known to provide reasonable results for temperatures below that of the chiral transition [1].

In effective theories, such as the Nambu–Jona–Lasinio or linear sigma models, this is taken into account by 't Hooft's determinant term. Coefficients of operators in the Lagrangian of field theories, including the aforementioned determinant term, are considered to be (coupling) constants, without any field or environment dependence. In the quantum version of the action, however, fluctuations introduce temperature (T), baryochemical potential (μ_B) and also field dependence as they become coefficient *functions*. When talking about field dependence of a given coupling, one has in mind the resummation of higher dimensional operators that can reappear when Taylor expanding the coefficient functions in terms of the field variable(s) around a conveniently chosen expansion point.

In QCD, it is well established that the anomalous breaking of $U_A(1)$ symmetry should gradually disappear beyond the critical temperature, as at high T the instanton density causing the anomaly exponentially vanishes [2,3]. At lower temperatures, however, the situation is far from being understood in a satisfactory fashion. One has also great interest in gaining results regarding the anomaly evolution at finite μ_B due to the sign problem, as mentioned earlier.

The finite temperature and/or density behavior of the $U_A(1)$ anomaly represents an active direction of research. More conservative results usually argue that the evaporation of the anomaly should follow that of the chiral condensate and thus the $U_A(1)$ symmetry restores around the critical temperature (T_C) of the chiral transition [4–8]. There are also several arguments and results that indicate that it is visible even beyond T_C [9–13]. For example, earlier renormalization group studies indicate that when considering a field-dependent anomaly coefficient, it decreases as a function of the chiral condensate [11,14], and this profile function can also depend explicitly on the tempereture (though the former effect is more dominant). Effective restoration of the anomaly has, e.g., a consequence regarding the order of the chiral transition [15], the axion mass [16], and the fate of η' meson, whose mass if substantially drops [17–22], in a nuclear medium could lead to the formation of an η'-nucleon bound state.

The goal of this paper is to calculate the first field-dependent correction to the coupling of the 't Hooft determinant in the effective action perturbatively, i.e., we determine the first Taylor coefficient of the anomaly function. Determining this coefficient (at least qualitatively) allows for obtaining the behavior of the anomaly strength as the function of the chiral condensate. This allows for providing answer for the question whether the aforementioned results on anomaly strengthening [11,14] can be reproduced within a simple perturbative renormalizatrion group setting of the linear sigma model, or determining a full functional dependence of the effective action is necessary. Fluctuations will be included using the functional variant of the renormalization group (FRG) [23], in the so-called local potential approximation (LPA). Even though renormalizable, we think of the model as an effective field theory; therefore, an ultra violet (UV) cutoff is inherently part of the system, which we set to $\Lambda = 1\,\text{GeV}$ (we expect the linear sigma model to emerge from QCD around this scale). Our task is to integrate out all fluctuations below Λ.

The paper is organized as follows. In Section 2, we introduce the model and the corresponding method of the FRG. Section 3 is devoted for calculating the effective action and discussing the problem of the expansion point of the Taylor series. After appropriate parametrization of the model, in Section 4, as an application, we show how the anomaly strengthens at the zero temperature nuclear liquid–gas phase transition. Section 5 contains the summary.

2. Model and Method

The model we are working with in this paper is the three-flavor linear sigma model, which is defined via the following Euclidean Lagrangian:

$$\begin{aligned}\mathcal{L} &= \text{Tr}\,(\partial_i M^\dagger \partial_i M) + m^2\,\text{Tr}\,(M^\dagger M) + g_1\Big(\text{Tr}\,(M^\dagger M)\Big)^2 + g_2\,\text{Tr}\,(M^\dagger M - \text{Tr}\,(M^\dagger M))^2 \\ &\quad + a(\det M + \det M^\dagger) - (h_0 s_0 + h_8 s_8),\end{aligned} \tag{1}$$

where M contains the meson fields, $M = (s_a + i\pi_a)T_a$ [$T_a = \lambda_a/2$ are generators of the $U(3)$ group with λ_a being the Gell-Mann matrices, $a = 0, ..., 8$], m^2 is the mass parameter and g_1, g_2 refer to independent quartic couplings. As discussed in the previous section, the determinant term and the corresponding a parameter is responsible for the $U_A(1)$ anomaly. We also have explicit symmetry breaking terms containing h_0 and h_8, which represent finite quark masses.

Our main goal is to calculate the quantum effective action, Γ, built upon the theory defined via (1). As announced in the introduction, we think of (1) as an inherently effective model, which is only valid up to the scale $\Lambda = 1\,\text{GeV}$, therefore, one needs to take into account fluctuations with a cutoff Λ. The scale dependent quantum effective action, Γ_k, which includes fluctuations with momenta larger than k (i.e., they are integrated out) is defined as

$$\begin{aligned}Z_k[J] &= \int \mathcal{D}M\mathcal{D}M^\dagger \exp\Big\{-\int \mathcal{L} - \int (JM + h.c.) - \int\int M^\dagger R_k M\Big\}, \\ \Gamma_k[M] &= -\log Z_k[J] - \int (JM + h.c.) - \int\int M^\dagger R_k M,\end{aligned} \tag{2}$$

where we omitted matrix indices, J is the conjugate source variable to M, and R_k is an appropriately chosen (bilocal) regulator function freezing fluctuations with momenta smaller than k. We note that the integrals can be considered either in direct or Fourier spaces. The Γ_k functional, for homogeneous field configurations, obeys the following, so-called flow equation [23]:

$$\partial_k \Gamma_k[M] = \frac{1}{2}\,\mathrm{Tr}\int_x\int_q (\Gamma_k^{(2)} + R_k)^{-1}(q)\partial_k R_k(q), \tag{3}$$

where $\Gamma_k^{(2)}$ is the second functional derivative of Γ_k in a homogeneous background field M, thus the x integral merely gives a spacetime volume factor. We also assumed that the regulator is diagonal in momentum space.

Our aim is to calculate the scale dependent effective action, Γ_k, in an approximation that takes into account the evolution of the anomaly at the next-to-leading order, i.e., we wish to determine in Γ_k the coefficient of the operator $\mathrm{Tr}\,(M^\dagger M)\cdot(\det M + \det M^\dagger)$. Our ansatz for Γ_k is as follows:

$$\begin{aligned}\Gamma_k \;=\; & \int_x \Big[\mathrm{Tr}\,(\partial_i M^\dagger \partial_i M) + m_k^2\, \mathrm{Tr}\,(M^\dagger M) + g_{1,k}\Big(\mathrm{Tr}\,(M^\dagger M)\Big)^2 \\ & + g_{2,k}\, \mathrm{Tr}\,(M^\dagger M - \mathrm{Tr}\,(M^\dagger M))^2 - (h_0 s_0 + h_8 s_8) \\ & + a_k(\det M + \det M^\dagger) + a_{1,k}\, \mathrm{Tr}\,(M^\dagger M)\cdot(\det M + \det M^\dagger)\Big]. \end{aligned} \tag{4}$$

This is sometimes called the Local Potential Approximation (LPA), where momentum dependence is only introduced into the two point function, via the standard kinetic term in (4). Note that the LPA can be considered as the leading order of the derivative expansion, and there is substantial evidence that these kind of series do converge [23]. As seen in (4), instead of working with a completely general field-dependent potential, we are employing perturbation theory in terms of the small parameter $1/\Lambda$. That is, by gradually including higher dimensional operators, since their coefficients scale with inverse powers of the scale, the ansatz (4) in the UV can be thought of as a power series in $1/\Lambda$. We choose (4) to be compatible with (1), but all couplings come with k–dependence. The only exceptions are h_0 and h_8, as one point couplings do not flow with respect to the scale. Furthermore, notice the new term proportional to $a_{1,k}$, which is key for our purposes to determine the anomaly behavior at $k = 0$. First, our task is to calculate $\Gamma_k^{(2)}$ from (4), then plug it into (3), and identify the individual differential equations for m_k^2, $g_{1,k}$, $g_{2,k}$, a_k, and $a_{1,k}$. Finally, these equations need to be integrated from $k = 1\,\mathrm{GeV}$ to $k = 0$ to obtain $\Gamma \equiv \Gamma_{k=0}$. In the ansatz (4), obviously the actual strength of the anomaly is not described by the parameter a, but rather $a + a_1 \cdot \mathrm{Tr}\,(M^\dagger M)|_{\min}$, where we need to evaluate the chiral condensates in the minimum point of the effective action. Therefore, what we are basically after is the relative sign of a_1 to a at $k = 0$ to decide whether the anomaly strengthens or weakens as the chiral condensate gradually evaporates.

We finally note that there are various choices for the regulator function, R_k. In this paper we will stick to $R_k(q,p) \equiv R_k(q)(2\pi)^3\delta(q+p) = (k^2 - q^2)\Theta(k^2 - q^2)(2\pi)^4\delta(q+p)$, where $\Theta(x)$ is the step function. This variant has been shown to be the optimal choice for the LPA [24], maximizing the radius of convergence of an amplitude expansion.

3. Calculation of the Effective Action

The first step is to calculate $\Gamma_k^{(2)}$. In principle it is a 18×18 matrix in the $s^a - \pi^a$ space, and there is not much hope that one can invert such a complicated expression analytically. Luckily, it is not necessary at all, as in (4) we kept the field dependence up to the order of $\mathcal{O}(M^5)$. By working with a restricted background, $\Gamma_k^{(2)}$ is easily invertable and by expanding the rhs of (3) in terms of the field variables it still allows for identifying each operator that are being kept in (4).

A convenient choice is to work with $M = s_0 T_0 + s_8 T_8$. In such a background, the operators that need to be identified are as follows:

$$\rho := \mathrm{Tr}\,(M^\dagger M) = \frac{1}{2}(s_0^2 + s_8^2), \tag{5a}$$

$$\tau := \mathrm{Tr}\,(M^\dagger M - \mathrm{Tr}\,(M^\dagger M))^2 = \frac{1}{24}s_8^2(8s_0^2 - 4\sqrt{2}s_0 s_8 + s_8^2), \tag{5b}$$

$$\Delta := \det M + \det M^\dagger = \frac{1}{36}(2\sqrt{6}s_0^3 - 3\sqrt{6}s_0 s_8^2 - 2\sqrt{3}s_8^3). \tag{5c}$$

The $\Gamma_k^{(2)}$ matrix elements in the scalar sector read

$$\begin{aligned}\Gamma^{(2)}_{k,s_0s_0} &= q^2 + m_k^2 + g_{1,k}(3s_0^2 + s_8^2) + \frac{2}{3}g_{2,k}s_8^2 \\ &+\sqrt{\frac{2}{3}}a_k s_0 + a_{k,1}\Big(\frac{5}{3}\sqrt{\frac{2}{3}}s_0^3 - \frac{1}{2\sqrt{6}}s_0 s_8^2 - \frac{1}{6\sqrt{3}}s_8^3\Big),\end{aligned} \tag{6}$$

$$\begin{aligned}\Gamma^{(2)}_{k,s_0s_8} &= 2g_{1,k}s_0 s_8 + g_{2,k}\Big(\frac{4}{3}s_0 s_8 - \frac{1}{\sqrt{2}}s_8^2\Big) \\ &-\frac{a_k}{\sqrt{6}}s_8 - a_{1,k}\Big(\frac{1}{2\sqrt{6}}s_0^2 s_8 + \frac{1}{2\sqrt{3}}s_0 s_8^2 + \frac{1}{\sqrt{6}}s_8^3\Big),\end{aligned} \tag{7}$$

$$\begin{aligned}\Gamma^{(2)}_{k,s_8s_8} &= q^2 + m_k^2 + g_{1,k}(s_0^2 + 3s_8^2) + g_{2,k}\Big(\frac{2}{3}s_0^2 - \sqrt{2}s_0 s_8 + \frac{1}{2}s_8^2\Big) \\ &-a_k\Big(\frac{1}{\sqrt{6}}s_0 + \frac{1}{\sqrt{3}}s_8\Big) \\ &-a_{1,k}\Big(\frac{1}{6\sqrt{6}}s_0^3 + \frac{1}{2\sqrt{3}}s_0^2 s_8 + \sqrt{\frac{3}{2}}s_0 s_8^2 + \frac{5}{3\sqrt{3}}s_8^3\Big),\end{aligned} \tag{8}$$

$$\begin{aligned}\Gamma^{(2)}_{k,s_1s_1} &= \Gamma^{(2)}_{k,s_2s_2} = \Gamma^{(2)}_{k,s_3s_3} = q^2 + m_k^2 + g_{1,k}(s_0^2 + s_8^2) + g_{2,k}\Big(\frac{2}{3}s_0^2 + \sqrt{2}s_0 s_8 + \frac{1}{6}s_8^2\Big) \\ &+a_k\Big(-\frac{1}{\sqrt{6}}s_0 + \frac{1}{\sqrt{3}}s_8\Big) \\ &+a_{1,k}\Big(-\frac{1}{6\sqrt{6}}s_0^3 + \frac{1}{2\sqrt{3}}s_0^2 s_8 - \frac{1}{\sqrt{6}}s_0 s_8^2 + \frac{1}{3\sqrt{3}}s_8^3\Big),\end{aligned} \tag{9}$$

$$\begin{aligned}\Gamma^{(2)}_{k,s_4s_4} &= \Gamma^{(2)}_{k,s_5s_5} = \Gamma^{(2)}_{k,s_6s_6} = \Gamma^{(2)}_{k,s_7s_7} = \\ & q^2 + m_k^2 + g_{1,k}(s_0^2 + s_8^2) + g_{2,k}\Big(\frac{2}{3}s_0^2 - \frac{1}{\sqrt{2}}s_0 s_8 + \frac{1}{6}s_8^2\Big) \\ &-a_k\Big(\frac{1}{\sqrt{6}}s_0 + \frac{1}{2\sqrt{3}}s_8\Big) \\ &-a_{1,k}\Big(\frac{1}{6\sqrt{6}}s_0^3 + \frac{1}{4\sqrt{3}}s_0^2 s_8 + \frac{1}{\sqrt{6}}s_0 s_8^2 + \frac{5}{12\sqrt{3}}s_8^3\Big),\end{aligned} \tag{10}$$

while the pseudoscalar components are

$$\Gamma^{(2)}_{k,\pi_0\pi_0} = q^2 + m_k^2 + g_{1,k}(s_0^2+s_8^2) - \sqrt{\frac{2}{3}}a_k s_0 - a_{1,k}\Big(\frac{1}{3}\sqrt{\frac{2}{3}}s_0^3 + \frac{1}{2}\sqrt{\frac{3}{2}}s_0 s_8^2 + \frac{1}{6\sqrt{3}}s_8^3\Big), \tag{11}$$

$$\Gamma^{(2)}_{k,\pi_0\pi_8} = g_{2,k}\Big(\frac{2}{3}s_0 s_8 - \frac{1}{3\sqrt{2}}s_8^2\Big) + \frac{a_k}{\sqrt{6}}s_8 + a_{1,k}\Big(\frac{1}{2\sqrt{6}}s_0^2 s_8 + \frac{1}{2\sqrt{6}}s_8^3\Big), \tag{12}$$

$$\Gamma^{(2)}_{k,\pi_8\pi_8} = q^2 + m_k^2 + g_{1,k}(s_0^2+s_8^2) + g_{2,k}\Big(-\frac{\sqrt{2}}{3}s_0 s_8 + \frac{1}{6}s_8^2\Big) + a_k\Big(\frac{1}{\sqrt{6}}s_0 + \frac{1}{\sqrt{3}}s_8\Big)$$
$$+a_{1,k}\Big(\frac{5}{6}\sqrt{\frac{1}{6}}s_0^3 + \frac{1}{2\sqrt{3}}s_0^2 s_8 + \frac{1}{3\sqrt{3}}s_8^3\Big), \tag{13}$$

$$\Gamma^{(2)}_{k,\pi_1\pi_1} = \Gamma^{(2)}_{k,\pi_2\pi_2} = \Gamma^{(2)}_{k,\pi_3\pi_3} =$$
$$q^2 + m_k^2 + g_{1,k}(s_0^2+s_8^2) + g_{2,k}\Big(\frac{\sqrt{2}}{3}s_0 s_8 - \frac{1}{6}s_8^2\Big)$$
$$+a_k\Big(\frac{1}{\sqrt{6}}s_0 - \frac{1}{\sqrt{3}}s_8\Big) + a_{1,k}\Big(\frac{5}{6}\sqrt{\frac{1}{6}}s_0^3 - \frac{1}{2\sqrt{3}}s_0^2 s_8 - \frac{2}{3\sqrt{3}}s_8^3\Big), \tag{14}$$

$$\Gamma^{(2)}_{k,\pi_4\pi_4} = \Gamma^{(2)}_{k,\pi_5\pi_5} = \Gamma^{(2)}_{k,\pi_6\pi_6} = \Gamma^{(2)}_{k,\pi_7\pi_7} =$$
$$q^2 + m_k^2 + g_{1,k}(s_0^2+s_8^2) + g_{2,k}\Big(-\frac{1}{3\sqrt{2}}s_0 s_8 + \frac{5}{6}s_8^2\Big)$$
$$+a_k\Big(\frac{1}{\sqrt{6}}s_0 + \frac{1}{2\sqrt{3}}s_8\Big) + a_{1,k}\Big(\frac{5}{6}\sqrt{\frac{1}{6}}s_0^3 + \frac{1}{4\sqrt{3}}s_0^2 s_8 + \frac{1}{12\sqrt{3}}s_8^3\Big). \tag{15}$$

Using that $\partial_k R_k(q) = 2k\Theta(k^2-q^2)$ and $\Gamma_k^{(2)}(q) + R_k(q) = \Gamma_k^{(2)}(k)$ for $q<k$, from (3) we get

$$\partial_k \Gamma_k = \int_x \frac{k^5}{32\pi^2}\,\mathrm{Tr}\,\big(\Gamma_k^{(2)}(k)\big)^{-1}. \tag{16}$$

Plugging in the matrix elements calculated in the $M = s_0 T_0 + s_8 T_8$ background, we can expand the rhs of (16) in terms of s_0 and s_8. After this step, using (5) we identify the ρ, τ and Δ operators as

$$\partial_k\Gamma_k = \int_x \Big[\partial_k m_k^2\cdot\rho + \partial_k g_{1,k}\cdot\rho^2 + \partial_k g_{2,k}\cdot\tau + \partial_k a_k\cdot\Delta + \partial_k a_{1,k}\cdot\rho\Delta + \ldots\Big], \tag{17}$$

where

$$\partial_k m_k^2 = -\frac{k^5}{32\pi^2}\frac{8(15g_{1,k}+4g_{2,k})}{3(k^2+m_k^2)^2}, \quad \partial_k g_{1,k} = \frac{k^5}{32\pi^2}\frac{8(117g_{1,k}^2+48g_{1,k}g_{2,k}+16g_{2,k}^2)}{9(k^2+m_k^2)^3},$$
$$\partial_k g_{2,k} = \frac{k^5}{32\pi^2}\frac{48g_{1,k}g_{2,k}+32g_{2,k}^2}{(k^2+m_k^2)^3}, \quad \partial_k a_k = \frac{k^5}{32\pi^2}\Big(\frac{8a_k(3g_{1,k}-4g_{2,k})}{(k^2+m_k^2)^3} - \frac{24a_{1,k}}{(k^2+m_k^2)^2}\Big),$$
$$\partial_k a_{1,k} = \frac{k^5}{32\pi^2}\Big(\frac{32a_k(-9g_{1,k}^2+6g_{1,k}g_{2,k}+2g_{2,k}^2)}{(k^2+m_k^2)^4} + \frac{16a_{1,k}(33g_{1,k}-2g_{2,k})}{3(k^2+m_k^2)^3}\Big). \tag{18}$$

Note that we treated the anomaly as perturbation and dropped every term beyond $\mathcal{O}(a_k, a_{1,k})$. Introducing scale independent variables, from (18) one easily reproduces the well known 1-loop β functions of the couplings in the linear sigma model [25]. Our task now is to solve Equation (18) starting from $k = \Lambda \equiv 1\,\mathrm{GeV}$ to $k=0$.

Solving (18) all the way down to $k=0$ would require $m_k^2 > 0$ throughout the renormalization group flow. Since we wish to obtain phenomenologically reasonable results, the potential has to show spontaneous symmetry breaking. That is to say, when all fluctuations are integrated out, $m^2_{k=0}$ has to be negative. However, then there exists a critical scale $k_{\rm crit} > 0$, for which all denominators in (18) blow up and the flow equations lose their meaning. The way out is to realize is that one actually has the choice to determine the flow equations in the minimum point of the effective action, $\Gamma_k|_{s_{0,\rm min},s_{8,\rm min}}$, rather than evaluating it in a vanishing background. That is, all renormalization group flows are to be extracted at $s_{0,\rm min}$, $s_{8,\rm min}$. This way one always has a positive definite denominator and the flow equation is valid for any k.

One, therefore, repeats the calculations starting from (16), but this time expands only in terms of s_8 so that the ρ dependence of the parameters can be traced via s_0. A long but straightforward calculation leads once again to the possibility of identifying the invariants appear in (17), whose coefficients now read as

$$
\begin{aligned}
\partial_k m_k^2 &= -\frac{k^5}{32\pi^2}\Bigg[\frac{18g_{1,k}}{(k^2+m_k^2+2g_{1,k}\rho_0)^2}+\frac{6g_{1,k}}{(k^2+m_k^2+6g_{1,k}\rho_0)^2}\\
&\quad+\frac{16(3g_{1,k}+2g_{2,k})}{3(k^2+m_k^2+2g_{1,k}\rho_0+4g_{2,k}\rho_0/3)^2}+\frac{72g_{1,k}^2\rho_0}{(k^2+m_k^2+2g_{1,k}\rho_0)^3}\\
&\quad+\frac{72g_{1,k}^2\rho_0}{(k^2+m_k^2+6g_{1,k}\rho_0)^3}+\frac{64(3g_{1,k}+2g_{2,k})^2\rho_0}{9(k^2+m_k^2+2g_{1,k}\rho_0+4g_{2,k}\rho_0/3)^3}\Bigg],\\
\partial_k g_{1,k} &= \frac{k^5}{32\pi^2}\Bigg[\frac{36g_{1,k}^2}{(k^2+m_k^2+2g_{1,k}\rho_0)^3}+\frac{36g_{1,k}^2}{(k^2+m_k^2+6g_{1,k}\rho_0)^3}\\
&\quad+\frac{32(3g_{1,k}+2g_{2,k})^2}{9(k^2+m_k^2+2g_{1,k}\rho_0+4g_{2,k}\rho_0/3)^3}\Bigg],\\
\partial_k g_{2,k} &= \frac{k^5}{32\pi^2}\Bigg[\frac{6g_{2,k}^2}{(k^2+m_k^2+2g_{1,k}\rho_0)^3}-\frac{9g_{2,k}/2}{\rho_0(k^2+m_k^2+2g_{1,k}\rho_0)^2}\\
&\quad+\frac{3g_{2,k}(6g_{1,k}+g_{2,k})}{\rho_0(g_{2,k}-3g_{1,k})(k^2+m_k^2+6g_{1,k}\rho_0)^2}\\
&\quad+\frac{30g_{2,k}^2}{(k^2+m_k^2+2g_{1,k}\rho_0+4g_{2,k}\rho_0/3)^3}\\
&\quad+\frac{3g_{2,k}(g_{2,k}-21g_{1,k})/2}{\rho_0(g_{2,k}-3g_{1,k})(k^2+m_k^2+2g_{1,k}\rho_0+4g_{2,k}\rho_0/3)^2}\Bigg],\\
\partial_k a_k &= \frac{k^5}{32\pi^2}\Bigg[-\frac{36g_{1,k}(a_k+2a_{1,k}\rho_0)}{(k^2+m_k^2+2g_{1,k}\rho_0)^3}-\frac{18(a_k+a_{1,k}\rho_0)}{\rho_0(k^2+m_k^2+2g_{1,k}\rho_0)^2}\\
&\quad-\frac{12g_{1,k}(3a_k+10a_{1,k}\rho_0)}{(k^2+m_k^2+6g_{1,k}\rho_0)^3}\\
&\quad-\frac{6a_k+10a_{1,k}\rho_0}{\rho_0(k^2+m_k^2+6g_{1,k}\rho_0)^2}+\frac{16(3g_{1,k}+2g_{2,k})(3a_k+a_{1,k}\rho_0)}{3(k^2+m_k^2+2g_{1,k}\rho_0+4g_{2,k}\rho_0/3)^3}\\
&\quad+\frac{4(6a_k+a_{1,k}\rho_0)}{\rho_0(k^2+m_k^2+2g_{1,k}\rho_0+4g_{2,k}\rho_0/3)^2}\Bigg],
\end{aligned}
$$

$$\partial_k a_{1,k} = \frac{k^5}{32\pi^2}\Bigg[\frac{36g_{1,k}(a_k+2a_{1,k}\rho_0)}{\rho_0(k^2+m_k^2+2g_{1,k}\rho_0)^3}+\frac{9a_k}{\rho_0^2(k^2+m_k^2+2g_{1,k}\rho_0)^2}$$
$$+\frac{12g_{1,k}(3a_k+10a_{1,k}\rho_0)}{\rho_0(k^2+m_k^2+6g_{1,k}\rho_0)^3}-\frac{16(3g_{1,k}+2g_{2,k})(3a_k+a_{1,k}\rho_0)}{3\rho_0(k^2+m_k^2+2g_{1,k}\rho_0+4g_{2,k}\rho_0/3)^3}$$
$$-\frac{12a_k}{\rho_0^2(k^2+m_k^2+2g_{1,k}\rho_0+4g_{2,k}\rho_0/3)^2}\Bigg], \tag{19}$$

where we have denoted the expansion point by ρ_0, which is to be set to the value of ρ corresponding to the minimum point of the effective action [note that $\rho = \mathrm{Tr}\,(M^\dagger M)/2$]. As a side remark, one easily checks that choosing $\rho_0 = 0$ (19) would lead back to the earlier results, (18). Our task is to integrate the system of equations (19) from $k = \Lambda \equiv 1\,\mathrm{GeV}$ down to $k = 0$ with the boundary conditions $m_\Lambda^2 = m^2$, $g_{1,\Lambda} = g_1$, $g_{2,\Lambda} = g_2$, $a_\Lambda = a$, $a_{1,\Lambda} = 0$, where m^2, g_1, g_2, a are such constants that reproduce as accurately as possible the mesonic spectrum in the infrared. Here we used that at the UV scale the coefficient of the operator $\rho\Delta$ can be set to zero due to perturbative renormalizability. This might be questionable if the UV scale was not high enough, as being a dimension 5 operator, dimensional analyis suggests that its coefficient, a_1, is of $\mathcal{O}(1/\Lambda)$. Obviously, if the linear sigma model was not an effective theory, and Λ could be sent to infinity, the term in question would not be present. But, in principle the a_1 coupling could be included already in the UV action. Investigation of such a scenario is beyond the scope of this paper.

Before solving the coupled system of equations (19), we need to fix the explicit symmetry breaking terms, i.e., the values for h_0, h_8. Instead of h_0 and h_8, we will work in the nonstrange–strange basis, i.e., $h_{\mathrm{ns}} = \sqrt{\frac{2}{3}}h_0 + \frac{1}{\sqrt{3}}h_8$, $h_s = \frac{1}{\sqrt{3}}h_0 - \sqrt{\frac{2}{3}}h_8$. The partially conserved axialvector current (PCAC) relations give

$$m_\pi^2 f_\pi = h_{\mathrm{ns}}, \quad m_K^2 f_K = \frac{h_{\mathrm{ns}}}{2} + \frac{h_s}{\sqrt{2}}, \tag{20}$$

where $m_\pi^2 = \delta^2\Gamma/\delta\pi_i^2(q=0)$ $[i = 1,2,3]$ and $m_K^2 = \delta^2\Gamma/\delta\pi_j^2(q=0)$ $[j = 4,5,6,7]$. Using physical pion and kaon masses, $\sim 140\,\mathrm{MeV}$, $\sim 494\,\mathrm{MeV}$, respectively, and decay constants, $f_\pi = 93\,\mathrm{MeV}$, $f_K = 113\,\mathrm{MeV}$, one gets

$$h_{\mathrm{ns}} = m_\pi^2 f_\pi \approx (122\,\mathrm{MeV})^3, \quad h_{\mathrm{s}} = \frac{1}{\sqrt{2}}(2m_K^2 f_K - m_\pi^2 f_\pi) \approx (335\,\mathrm{MeV})^3, \tag{21}$$

which is equivalent to

$$h_0 = \sqrt{\frac{2}{3}}(m_\pi^2 f_\pi/2 + m_K^2 f_K) \approx (285\,\mathrm{MeV})^3, \quad h_8 = \frac{2}{\sqrt{3}}(m_\pi^2 f_\pi - m_K^2 f_K) \approx -(310\,\mathrm{MeV})^3. \tag{22}$$

Now we use that Ward identities of chiral symmetry lead to

$$\frac{\delta\Gamma}{\delta s_{\mathrm{ns}}}(q=0) = m_\pi^2 s_{\mathrm{ns}} - h_{\mathrm{ns}}, \quad \frac{\delta\Gamma}{\delta s_{\mathrm{s}}}(q=0) = \frac{m_K^2 - m_\pi^2}{\sqrt{2}} s_{\mathrm{ns}} + m_K^2 s_{\mathrm{s}} - h_{\mathrm{s}}. \tag{23}$$

Combined with (21), this shows that no matter how we choose the remaining parameters m^2, g_1, g_2, a, in the minimum point of the effective action

$$s_{\mathrm{ns,min}} = f_\pi, \quad s_{\mathrm{s,min}} = \sqrt{2}(f_K - f_\pi/2). \tag{24}$$

That leads to $\rho_0 = (s_{\mathrm{ns,min}}^2 + s_{\mathrm{s,min}}^2)/2$, and thus we are ready to fix the aforementioned parameters. Solving (19), the values $\{m^2, g_1, g_2, a\} \approx \{0.6835\,\mathrm{GeV}^2, 29.7, 91.5, -4.4\,\mathrm{GeV}\}$ lead to the masses of the pion, kaon, η, η' as $m_\pi \approx 133\,\mathrm{MeV}$, $m_K \approx 494\,\mathrm{MeV}$, $m_\eta \approx 537\,\mathrm{MeV}$, $m_{\eta'} \approx 957\,\mathrm{MeV}$, respectively. Note that the strength of the axial anomaly, a, is negative, and it remains so throughout the renormalization group flow. However, after solving (19), one concludes that at $k = 0$ the coefficient $a_{1,k=0}$ is positive. That is to say, since the actual strength of the determinant term is $A := a_{k=0} + a_{1,k=0} \cdot \rho_0$, when the chiral condensate evaporates, the absolute value of A becomes larger. This shows that at low temperatures T, where the T dependence of the anomaly parameters is negligible, the anomaly is actually strengthening as chiral symmetry gradually restores. That is one of the main results of the paper.

In what follows, we provide a rough estimate how the anomaly behaves at the zero temperature nuclear liquid–gas transition.

4. Anomaly Strengthening at The Nuclear Liquid–Gas Transition

In this section we apply our results to the zero temperature nuclear liquid–gas transition. We assume that the nucleon field couples to the mesons via Yukawa interaction, $\mathcal{L}_{\text{int}} = g\bar{\psi}M_5\psi$, $\psi^T = (p, n)$, $M_5 = \sum_{a=\text{ns},1,2,3}(s^a + i\pi^a\gamma_5)T^a$, where the nonstrange generator is $T^{\text{ns}} = \sqrt{2/3}T^0 + 1/\sqrt{3}T^8$, while γ_5 is the fifth Dirac matrix. In principle, one would also need to include the dynamics of an ω vector particle into the system [26,27] that models the repulsive interaction between nucleons, but as we will see in a moment, for our purposes it plays no role.

First, we exploit some of the zero temperature properties of nuclear matter. Note that, in the current model, the nucleon mass entirely originates from the spontaneous breaking of chiral symmetry,

$$m_N(s_{\text{ns}}) = g_Y s_{\text{ns}}/2, \tag{25}$$

and since $m_N(f_\pi) \approx 939\,\text{MeV}$ in the vacuum, we arrive at $g_Y \approx 20.19$. Normal nuclear density, $n_N \approx 0.17\,\text{fm}^{-3} \approx (109.131\,\text{MeV})^3$ leads to the Fermi momentum, p_F, of the nucleons, since at the mean field level, for $T = 0$ we have

$$n_N = 4\int_p n_F\left(\sqrt{p^2 + m_N^2} - p_F\right)\Big|_{T=0} \equiv \frac{2}{3\pi^2}p_F^3, \tag{26}$$

therefore, $p_F \approx 267.9\,\text{MeV} \approx 1.36\,\text{fm}^{-1}$. This leads to the nonstrange condensate in the liquid phase, $s_{\text{ns,liq}}$, because the Landau mass, which is defined as

$$M_L = \sqrt{p_F^2 + m_N^2(s_{\text{ns,liq}})} \tag{27}$$

is known to be $M_L \approx 0.8m_N(f_\pi) \approx 751.2\,\text{MeV}$, and thus $s_{\text{ns,liq}} \approx 69.52\,\text{MeV}$ [26,27]. This shows that as we increase the chemical potential, the nonstrange chiral condensate, s_{ns}, jumps: $f_\pi \to s_{\text{ns,liq}}$. This will definitely be accompanied by a jump in the strange condensate, but it has been shown to be significantly smaller [14]. Neglecting the change in s_s, the ρ chiral invariant jumps as $(f_\pi^2 + s_{\text{s,min}}^2)/2 \to (s_{\text{ns,liq}}^2 + s_{\text{s,min}}^2)/2$. As discussed in the previous section, the anomaly strength is $A = a_{k=0} + a_{1,k=0} \cdot \rho$, which also jumps accordingly, and the change in A becomes

$$\Delta A = a_{1,k=0} \cdot \Delta\rho, \tag{28}$$

where $\Delta\rho = (s_{\text{ns,liq}}^2 - f_\pi^2)/2$. Solving (19) one gets $a_{k=0} \approx -9.05\,\text{GeV}$ and $a_{1,k=0} \approx 494.5\,\text{GeV}^{-1}$, therefore, the relative change in the anomaly at the liquid–gas transition is

$$\frac{\Delta A}{A} = \frac{a_{1,k=0} \cdot \Delta\rho}{a_{k=0} + a_{1,k=0} \cdot \rho_0} \approx 0.2 = 20\%, \tag{29}$$

which is in the ballpark of the result of [14]. One can now check how robust this result is with respect to changing the cutoff Λ. A thorough investigation reveals that in a cutoff interval of 0.8–1.5 GeV, while the mass spectrum can be maintained within a few percent error after reparametrization, the $\Delta A/A$ ratio is less stable. One finds that the latter is a monotonically decreasing function of the cutoff and varies roughly between 15–40% in the above interval. Results show that when going below 1 GeV the cutoff dependence gets stronger, which is understandable, since non-renormalizable operators are absent at the UV scale. That is, if the latter is chosen to be too small, the model cannot provide robust results (more operators would be needed). Going beyond 1.5–2 GeV, in turn, would be physically inappropriate as at those scales quark degrees of freedom would definitely play a crucial role. From these findings it is safe to say is that the relative change of the anomaly strength is of $\mathcal{O}(10\%)$ at the transition point.

At this point, we once again wish to emphasize that we have neglected the drop in the strange condensate, and also, the present analysis is based on perturbation theory. In principle higher order operators that break the $U_A(1)$ subgroup should also be resummed, e.g., terms such as $\sim \left(\text{Tr}\,(M^\dagger M)\right)^n (\det M + \det M^\dagger)$ could be of huge importance. The lesson we wish to point out here is that the present, rather simple perturbative calculation can also capture the phenomenon of strengthening anomaly as the chiral condensate (partially) evaporates.

5. Conclusions

In this paper, we investigated how the $U_A(1)$ anomaly behaves as a function of the chiral condensate. We worked with the three-flavor linear sigma model, and calculated the leading correction in a $1/\Lambda$ expansion to the conventional anomaly term caused by quantum fluctuations. We have found that the coefficient of the aforementioned operator, $\sim \text{Tr}\,(M^\dagger M) \cdot (\det M + \det M^\dagger)$, causes the actual strength of the anomaly to become larger once the chiral condensate evaporates. For the sake of an example, we demonstrated that at the zero temperature nuclear liquid–gas transition, where (on top of a jump in the nuclear density) the chiral condensate partially restores, the actual strength of the anomaly increases. This could also happen toward the full restoration of chiral symmetry, where quark dynamics also play a significant role. Note that our findings are based solely on calculating mesonic fluctuations, and no instanton effects have been taken into account.

The linear sigma model, being an effective field theory, cannot accommodate instantons as the fundamental model of QCD. Still, there are at least two directions worth exploring in the effective model framework. Recently it has been shown [28] that $3Q$-point interactions are generated by instantons with Q topological charge, which can be embedded into the linear sigma model via $\sim [(\det M^\dagger)^Q + (\det M)^Q]$ operators. Another important direction could be to assign environment dependence even to the bare anomaly coefficient(s) from QCD data and see how these compete against thermal effects caused by mesonic fluctuations.

Finally, we wish to point out that our study calls for an extension via a non-perturbative treatment, where fluctuations are taken into account beyond the $\mathcal{O}(a)$ order, and the coefficient function of the determinant term is obtained in a functional fashion, rather than at the lowest order of its Taylor series. The aforementioned directions are under progress and will be reported in a separate study.

Funding: This work was supported by the National Research, Development and Innovation Fund under Project No. PD127982, by the János Bolyai Research Scholarship of the Hungarian Academy of Sciences, and by the ÚNKP-20-5 New National Excellence Program of the Ministry for Innovation and Technology from the source of the National Research, Development and Innovation Fund.

Acknowledgments: The author thanks A. Patkos and Zs. Szep for discussions on related topics.

Conflicts of Interest: The funders had no role in the design of the study; in the collection, analyses, or interpretation of data; in the writing of the manuscript, or in the decision to publish the results.

References

1. Fukushima, K.; Hatsuda, T. The phase diagram of dense QCD. *Rept. Prog. Phys.* **2011**, *74*, 014001. [CrossRef]
2. Hooft. G. Computation of the quantum effects due to a four-dimensional pseudoparticle. *Phys. Rev. D* **1976**, *14*, 3432.
3. Gross, D.J.; Pisarski, R.D.; Yaffe, L.G. QCD and instantons at finite temperature. *Rev. Mod. Phys.* **1981**, *53*. [CrossRef]
4. Mitter, M.; Schaefer, B.-J. Fluctuations and the axial anomaly with three quark flavors. *Phys. Rev. D* **2014**, *89*, 054027. [CrossRef]

5. Cossu, G.; Fukaya, H.; Hashimoto, S.; Noaki, J.; Tomiya, A. On the Axial $U(1)$ Symmetry at Finite Temperature. Available online: http://arxiv.org/abs/1511.05691 (accessed on 12 March 2021).
6. Ishii, M.; Yonemura, K.; Takahashi, J.; Kouno, J.; Yahiro, M. Determination of $U(1)_A$ restoration from pion and a0-meson screening masses: Toward the chiral regime. *Phys. Rev. D* **2016**, *93*, 016002. [CrossRef]
7. Nicola, A.G.; de Elvira, J.R. Chiral and $U(1)_A$ restoration for the scalar and pseudoscalar meson nonets. *Phys. Rev. D* **2018**, *98*, 014020. [CrossRef]
8. Bottaro, S.; Meggiolaro, E. QCD axion and topological susceptibility in chiral effective Lagrangian models at finite temperature. *Phys. Rev. D* **2020**, *102*, 014048. [CrossRef]
9. Dick, V.; Karsch, F.; Laermann, E.; Mukherjee, S.; Sharma, S. Microscopic origin of $U_A(1)$ symmetry violation in the high temperature phase of QCD. *Phys. Rev. D* **2015**, *91*, 094504. [CrossRef]
10. Sharma, S.; Dick, V.; Karsch, F.; Laermann, E.; Mukherjee, S. The topological structures in strongly coupled QGP with chiral fermions on the lattice. *Nucl. Phys. A* **2016**, *956*, 793. [CrossRef]
11. Fejos, G.; Hosaka, A. Thermal properties and evolution of the $U_A(1)$ factor for 2+1 flavors, *Phys. Rev. D* **2016**, *94*, 036005. [CrossRef]
12. Li, X.; Fu, W.; Liu, Y. New insight about the effective restoration of $U_A(1)$ symmetry. *Phys. Rev. D* **2020**, *101*, 054034. [CrossRef]
13. Braun, J.; Leonhardt, M.; Pawlowski, J.M.; Rosenblüh, D. Chiral and Effective $U(1)_A$ Symmetry Restoration in QCD. Available online: http://arxiv.org/abs/2012.06231 (accessed on 12 March 2021).
14. Fejos, G.; Hosaka, A. Axial anomaly and hadronic properties in a nuclear medium. *Phys. Rev. D* **2018**, *98*, 036009. [CrossRef]
15. Pelissetto, A.; Vicari, E. Relevance of the axial anomaly at the finite-temperature chiral transition in QCD. *Phys. Rev. D* **2013**, *88*, 105018. [CrossRef]
16. Horvatic, D.; Kekez, D.; Klabucar, D. Temperature dependence of the axion mass in a scenario where the restoration of chiral symmetry drives the restoration of the $U_A(1)$ Symmetry. *Universe* **2019**, *10*, 208. [CrossRef]
17. Nagahiro, H.; Takizawa, M.; Hirenzaki, S. eta- and eta'-mesic nuclei and $U_A(1)$ anomaly at finite density. *Phys. Rev. C* **2006**, *74*, 045203. [CrossRef]
18. Sakai, S.; Jido, D. In medium eta' mass and eta' N interaction based on chiral effective theory. *Phys. Rev. C* **2013**, *88*, 064906. [CrossRef]
19. Sakai, S.; Hosaka, A.; Nagahiro, H. Effect of the final state interaction of eta'N on the eta' photoproduction off nucleon. *Phys. Rev. C* **2017**, *95*, 045206. [CrossRef]
20. Jido, D.; Masutani, H.; Hirenzaki, S. Structure of eta' mesonic nuclei in a relativistic mean field theory. *Prog. Theor. Exp. Phys.* **2019**, *2019*, 053D02. [CrossRef]
21. Horvatic, D.; Kekez, D.; Klabucar, D. Eta' and eta mesons at high T when the $U_A(1)$ and chiral symmetry breaking are tied. *Phys. Rev. D* **2019**, *99*, 014007. [CrossRef]
22. Rai, S.K.; Tiwari, V.K. Exploring axial $U(1)$ restoration in a modified 2+1 flavor Polyakov quark meson model. *Eur. Phys. J. Plus* **2020**, *135*, 844. [CrossRef]
23. Kopietz, P.; Bartosch, L.; Schütz, F. *Introduction to the Functional Renormalization Group*; Springer: New York, NY, USA, 2010.
24. Litim, D. Optimisation of the exact renormalisation group. *Phys. Lett. B* **2000**, *486*, 92–99. [CrossRef]
25. Pisarski, R.D.; Wilczek, F. Remarks on the chiral phase transition in chromodynamics. *Phys. Rev. D* **1984**, *29*, 338. [CrossRef]
26. Floerchinger, S.; Wetterich, C. Chemical freeze-out in heavy ion collisions at large baryon densities. *Nucl. Phys. A* **2012**, *890*, 891. [CrossRef]
27. Drews, M.; Weise, W. Functional renormalization group studies of nuclear and neutron matter. *Prog. Part. Nucl. Phys.* **2017**, *93*, 69. [CrossRef]
28. Pisarski, R.D.; Rennecke, F. Multi-instanton contributions to anomalous quark interactions. *Phys. Rev. D* **2020**, *101*, 114019. [CrossRef]

symmetry

MDPI

Review

Dichotomy of Baryons as Quantum Hall Droplets and Skyrmions: Topological Structure of Dense Matter

Yong-Liang Ma [1,2,*] and Mannque Rho [3,*]

1 School of Fundamental Physics and Mathematical Sciences, Hangzhou Institute for Advanced Study, UCAS, Hangzhou 310024, China
2 International Center for Theoretical Physics Asia-Pacific (ICTP-AP) (Beijing/Hangzhou), University of Chinese Academy of Sciences, Beijing 100190, China
3 Institut de Physique Théorique, Université Paris-Saclay, CNRS, CEA, 91191 Gif-sur-Yvette, France
* Correspondence: ylma@ucas.ac.cn (Y.-L.M.); mannque.rho@ipht.fr (M.R.)

Abstract: We review a new development on the possible direct connection between the topological structure of the $N_f = 1$ baryon as a FQH droplet and that of the $N_f \geq 2$ baryons (such as nucleons and hyperons) as skyrmions. This development suggests a possible "domain-wall (DW)" structure of compressed baryonic matter at high density expected to be found in the core of massive compact stars. Our theoretical framework is anchored on an effective nuclear effective field theory that incorporates two symmetries either hidden in the vacuum in QCD or emergent from strong nuclear correlations. It presents a basically different, hitherto undiscovered structure of nuclear matter at low as well as high densities. Hidden "genuine dilaton (GD)" symmetry and hidden local symmetry (HLS) gauge-equivalent at low density to nonlinear sigma model capturing chiral symmetry, put together in nuclear effective field theory, are seen to play an increasingly important role in providing hadron–quark duality in baryonic matter. It is argued that the FQH droplets could actually figure essentially in the properties of the vector mesons endowed with HLS near chiral restoration. This strongly motivates incorporating both symmetries in formulating "first-principles" approaches to nuclear dynamics encompassing from the nuclear matter density to the highest density stable in the Universe.

Keywords: topological structure; compressed baryonic matter; hadron-quark duality

1. Introduction

In this review note, we recount what we have carried out in the past few years to uncover a totally novel structure, not found in the literature, of dense nuclear matter relevant to massive compact stars. Our approach is anchored on two symmetries hidden in dilute hadronic systems, i.e., chiral symmetry and scale symmetry, that could play a crucial role as density increases high. In the accompanying article written by one of us (MR), how the very symmetries involved in dense compact-star matter also manifest at nuclear matter density in one of the outstanding problems in nuclear physics for more than four decades, namely, the "quenching" of the axial-vector coupling constant g_A in nuclear Gamow–Teller transitions.

The structure of highly dense matter found in massive compact stars is a totally uncharted domain. Unlike at high temperature, at high density it can be accessed neither by lattice QCD nor by terrestrial experiments. While, as comprehensively reviewed recently [1,2], finite nuclei as well as infinite nuclear matter can be fairly accurately accessed by nuclear effective field theories, pionless or pionful, referred to herein as "standard nuclear effective field theory (SnEFT)" anchored on relevant symmetries and invariances along the line of Weinberg's Folk Theorem [3], SnEFTs, as befits their premise, are expected to break down at some high density (and low temperature) relevant to, say, the interior of massive stars.

In $SnEFT$, the power counting in density is $O(k_F^q)$ where k_F is the Fermi momentum and increasing density involves going to higher q. For the "normal" nuclear matter with density $n_0 \approx 0.16\ \mathrm{fm}^{-3}$, the expansion is required to go up to $q \sim 5$ [2]. The breakdown must occur as k_F goes beyond the nuclear matter density, say, at $n \sim (2-4)n_0$. On the other hand, if one organizes effective field theories in renormalization-group (RG) approach built on Fermi surface, the power counting in k_F comes out to be $O((1/\bar{N})^\kappa)$ where $\bar{N} = k_F/(\tilde{\Lambda} - k_F)$ with $\tilde{\Lambda}$ being the cutoff on top of the Fermi sea measured relative to the center of the Fermi sphere. The expansion in $\kappa \geq 0$ leads to Fermi-liquid fixed-point theory [4,5]. The equilibrium nuclear matter is given at the Fermi-liquid fixed point with $\bar{N} \to \infty$. Approaching Fermi-liquid theory starting from $SnEFT$ for nuclear (or neutron) matter valid up to roughly $\sim n_0$ has been formulated [2,6]. It has also been formulated using the V_{lowk} RG approach applicable to both finite nuclei and infinite matter, taking into account $1/\bar{N}$ corrections [7].

Given that the k_F expansion must inevitably break down—and hence $SnEFT$ becomes no longer valid—at some high density above n_0, a potentially promising and justifiable approach is to go over to the Fermi-liquid structure starting from the normal nuclear matter density at which the Fermi-liquid structure is fairly well established to hold. Our strategy is to build a model, that we shall refer to as "$GnEFT$," that while capturing fully what $SnEFT$ successfully does up to n_0, can be extrapolated up beyond the density at which $SnEFT$ is presumed to break down. Such an approach extensively developed in [8] is anchored on a Lagrangian that incorporates, in addition to the pions and nucleons, the lowest-lying vector mesons ρ and ω and the scalar meson χ standing for $f_0(500)$. We treat the vector mesons $V = (\rho, \omega)$ as "dynamically generated" fields possessing hidden local symmetry (HLS) [9]—equivalently "composite gauge fields" [10]—and the scalar χ as a "genuine dilaton (GD)" [11], a (pseudo-)Nambu–Goldstone (NG) boson of hidden scale symmetry [12]. We cannot say whether these symmetries are intrinsic in QCD. If they are intrinsic then they must be hidden in the vacuum since they are not visible. What must happen in our systems in nuclear medium is that they become "unhidden" by strong nuclear correlations as nuclear matter is highly compressed. Or they can emerge from highly correlated hadronic interactions such as those in condensed matter physics. One of our basic assumptions is that the HLS is consistent with the Suzuki theorem [10], and the scale symmetry with genuine dilaton has an infrared (IR) fixed point with both the chiral and scale symmetries realized in the NG mode [11] at some high density. How these symmetries, invisible in free space, could appear in dense medium has been the subject of the past efforts [8] in nuclear astrophysics and motivates us to go beyond what has been explored so far. We approach this issue by analyzing the structure of cold dense baryonic matter with density $n > (2-3)n_0$ in terms of a "baryon-quark duality" n QCD. It should be noted that the emphasis, as is to be clarified, is on "duality," not on "continuity", discussed in connection with confinement–deconfinement issue.

In this paper, we find that the combined hidden scale symmetry (HSS) and HLS, suitably formulated so as to access high density compact-star matter [8], not necessarily intrinsic in QCD but interpreted as "emergent" from strong nuclear correlations, reveals a dichotomy in the structure of baryons treated in terms of topology in the large N_c approximation and discuss how it could affect the equation of state (EoS) at high density relevant to the cores of massive compact stars. The merit of this work is that it exploits in strongly interacting baryonic matter a certain ubiquitous topological structure of highly correlated fermions, similar to electrons in condensed matter, thereby bringing in a possible paradigm change in nuclear dynamics.

2. The Problem: Dichotomy

Consider the baryons made up of the quarks with nearly zero current quark masses. We will be dealing primarily with two flavors, u(p) and d(own). However, for the role of scale symmetry, it is essential to think in terms of three flavors [11] as we will explain below. For three flavors, all octet baryons $B^{(\alpha)}$, $\alpha = 1, ..., 8$, can be obtained as solitons, i.e., skyrmions [13], from the octet mesons. This has to do with the homotopy group $\pi_3(SU(3)) \simeq \mathcal{Z}$ and is justified in QCD at the large N_c limit. However there has been one annoying puzzle in this matter: There is no skyrmion associated with the singlet meson η'. This is because $\pi_3(U(1)) = 0$. The resolution to this conundrum was suggested in 2018 by Komargodski [14]: The baryon for η', while not a skyrmion soliton, turns out to be also a topological object at the large N_c limit but more appropriately a fractional quantum Hall (FQH) droplet, somewhat similar to a pancake (or perhaps pita [15,16]).

One way of seeing how this FQH droplet comes about in QCD, the approach we adopt in this paper, is in terms of the "Cheshire Cat phenomenon" (CCP) formulated a long time ago [17].

In the CCP, the trade-in of topology for hadron–quark continuity for low energy/long wavelength nuclear processes involving the u quark and d quark is via the "infinite hotel mechanism" where N_c quarks disappear into the vacuum, with the baryon charges taken up by the triplet pions, i.e., skyrmions. The $N_f = 1$ quarks, on the other hand, forbidden to fall into the infinite hotel and become skyrmions, go instead into a 2-dimensional quantum Hall (QH) droplet described by Chern–Simons (CS) theory. In [18], this was described as the anomaly cancelation, known as the "anomaly inflow" from the bulk of a system to its boundary [19]. Here, what is involved is that the quarks propagating in one direction flow to one higher dimension making a sheet in (x, y) with the anomaly caused by the "confinement" boundary condition, with the resulting system given by abelian CS theory for the FQH droplet. One can think of this as a topological object of η' with a topology different from that of the skyrmions of π's. We denote this baryon $B^{(0)}$.

There is another way of interpreting the CCP construction of the fractional quantum Hall droplet that could be more appropriate for resolving the "dichotomy problem" mentioned below. That is to formulate it in terms of a domain-wall (DW) structure. Consider the confinement wall at $x_3 = 0$. The vacua at $x_3 > 0$ and $x_3 < 0$ are clearly different. Then, the confinement wall makes the DW at $x_3 = 0$. As an example, think of the region $x_3 < 0$ as the region in which quarks are confined in the sense of the MIT bag. Then, calculating the spectral asymmetry in the limit of the thin wall with the chiral bag boundary condition set at $x_3 = 0$, one reproduces the same baryon charge obtained in the anomaly-inflow mechanism [20].

One important consequence of this observation is that the QH droplet has the spin $J = N_c/2$, namely 3/2 for $N_c = 3$, corresponding to the baryon resonance $\Delta(3/2, 3/2)$. The mass of the $B^{(0)}$ in the large N_c limit is of course $\sim O(N_c)$ but it can also receive $O(N_c^0)$ contribution [14]. In the skyrmion system, there is also a baryon of the same quantum numbers $(3/2, 3/2)$ but there is no correction coming at $O(N_c^0)$ that distinguishes spin-1/2 and spin-3/2. There is the Casimir contribution to the skyrmion mass that comes at $O(N_c^0)$ but that is common to the skyrmions of both spins 1/2 and 3/2. The first correction to the skyrmion mass in the large N_c limit comes at $O(1/N_c)$ arising from the rotational quantization. This presents a "dichotomy problem" (dubbed DP for short).

One can see this dichotomy if one applies the same argument made for the CCP for the QH droplet for $B^{(0)}$ to $N_f = 2$ systems, namely, the nucleon. Instead of dropping into the "infinite hotel" in the CC mechanism for the $N_f = 2$ skyrmion when the bag is shrunk to zero [17,21], there seems to be nothing that would prevent the quarks from undergoing the anomaly inflow into fractional quantum Hall droplets making the CS theory nonabelian [18]. Why not form a sheet-structured matter arranged, say, in the lasagne arrays seen at high density in crystal lattice simulations of compact-star matter (to be mentioned below)?

The pertinent question then is: What dictates the $N_f = 2$ quarks to (A) drop in the ∞-hotel skyrmions or (B) instead to flow to nonabelian FQH droplets? Or could it be (A) and (B) in some combination? This sharpens and generalizes the dichotomy problem raised above. A solution to this dichotomy problem has recently been addressed by Karasik [15] in terms of a "generalized" current that unifies the $N_f = 1$ baryon, QH droplet, and the $N \geq 2$ baryon, skyrmion. Here, we explore whether and how one can go from one to the other for the $N_f = 2$ systems in terms of the EoS for dense baryonic matter. We do this by "dialing" baryon density. The hope is that this will unravel the putative hadron–quark duality possibly involved in the physics of massive compact stars. The strategy here is to extract the conceptual insights gained in the phenomenological development discussed in [8] for the physics of massive compact stars, the only system currently available in nature for high density $n \gg n_0$ at low temperature and translate them into a scheme that could address, at least qualitatively, the dichotomy problem.

3. GnEFT Lagrangian

We begin by writing the effective Lagrangian involved, in as simple a form as possible, which allows us to capture the basic idea developed. The details look rather involved, but the basic idea is in fact quite simple. We will first deal with the mesonic sector with baryons generated as solitons and later explicitly incorporate baryons. In developing the basic idea, we will frequently switch back and forth between the former description and the latter.

3.1. Scale-Invariant Hidden Local Symmetric (sHLS) Lagrangian

To address the dichotomy problem (DP) in highly compressed baryonic matter, we incorporate the η' field, $\eta' \in U_A(1)$, in addition to the pseudo-scalar NG bosons $\pi \in SU(2)$ and the vectors $\rho_\mu \in SU(2)$ and $\omega \in U(1)$. The reason for this, as we will argue, is that although η' is massive compared with the mesons we will take into account, it goes massless in the limit that $N_c \to \infty$ and plays a crucial role for bringing in Chern–Simons topological field theory structure in (possibly) dense baryonic systems. In the modern development, it is being suggested that the role of η' in the guise of fractional quantum Hall "pancake" for the flavor singlet baryon $B^{(0)}$ plays an indispensable role for QCD phase structure at high density. We will comment on this matter later although it is not well understood at present.

The Crewther's GD approach to scale symmetry [11] adopted in this paper—which will turn out to play a crucial role in our theory—necessitates the kaons on par with the dilaton. For our purpose, however, we can ignore the strange quark—given the presence of the η' meson—and focus on the two light flavors. For a reason that will become clear later, unless otherwise specified, the ρ and ω fields will be treated in $U(2)$-nonsymmetric way. In the Seiberg-type dual approach to HLS, the ω meson is not pure $U(1)$ of $U(N_f)$ that contains ρ but a mixture of $U(1)$s. This feature will appear later in Section 5 where baryonic matter with η' present is treated.

We write the chiral field U as (Unless otherwise noted, we shall use the convention of [22])

$$U = \xi^2 = e^{i\eta'} e^{i\tau_a \pi_a / f} \tag{1}$$

and the HLS fields as

$$V_\mu^\rho = \frac{1}{2} g_\rho \rho_\mu^a \tau^a, \quad V_\mu^\omega = \frac{1}{2} g_\omega \omega_\mu. \tag{2}$$

Expressed in terms of the Maurer-Cartan 1-forms

$$\hat{\alpha}^\mu_{\parallel,\perp} = \frac{1}{2i}\left(D^\mu \xi \cdot \xi^\dagger \pm D^\mu \xi^\dagger \cdot \xi\right) \tag{3}$$

where $D_\mu \xi = (\partial_\mu - iV_\mu^\rho - iV_\mu^\omega)\xi$, the HLS Lagrangian we are concerned with is of the same form as the HLS Lagrangian for three flavors [22] with the parity-anomalous homogeneous Wess–Zumino (hWZ) Lagrangian composed of three terms (in the absence of external fields). For the $SU(2) \times U(1)$ case we are dealing with, there is no 2-dimensional Wess-Zumino (WZ) term.

To implement scale symmetry à la GD [11], we are to use as explained below the conformal compensator field $\chi = f_\chi e^{\chi/f_\chi}$ which has both mass dimension and scale dimension 1. The structure of the Lagrangian that we use in this discussion was noted previously in our reviews [8] and we shall note it again here with η' incorporated. In order to justify its extension in the density domain where FQH droplets could play a role, an additional ingredient to what figures in [8] is needed. It will be given below.

The scale-symmetrized Lagrangian that we denote as sHLS is of the form

$$\begin{aligned}\mathcal{L}_{sHLS} =& f^2\Phi^2\mathrm{Tr}\left(\hat{\alpha}_\perp^\mu\hat{\alpha}_{\perp\mu}\right) + f_{\sigma\rho}^2\Phi^2\mathrm{Tr}\left(\hat{\alpha}_\parallel^\mu\hat{\alpha}_{\parallel\mu}\right)\\ &+ f_0^2\Phi^2\mathrm{Tr}\left(\hat{\alpha}_\parallel^\mu\right)\mathrm{Tr}\left(\hat{\alpha}_{\parallel\mu}\right) + \mathcal{L}_{hWZ}\\ &- \frac{1}{2g_\rho^2}\mathrm{Tr}(V_{\mu\nu}V^{\mu\nu}) - \frac{1}{2g_0^2}\mathrm{Tr}(V_{\mu\nu})\mathrm{Tr}(V^{\mu\nu})\\ &+ \frac{1}{2}\partial_\mu\chi\partial^\mu\chi + V(\chi)\end{aligned} \tag{4}$$

where $\mathcal{L}_{hWZ}$ is the hWZ term that conserves parity and charge conjugation but violates intrinsic parity, Φ is defined as

$$\Phi = \chi/f_\chi \tag{5}$$

and V_χ is the dilaton potential, the explicit form of which is not needed for our purpose. $V^{\mu\nu}$ is the usual field tensor

$$V^{\mu\nu} = \partial^\mu V^\nu - \partial^\nu V^\mu - i[V^\mu, V^\nu] \tag{6}$$

with $V^\mu = V_\rho^\mu + V_\omega^\mu$. In (4)

$$f_0^2 = \frac{f_{\sigma\omega}^2 - f_{\sigma\rho}^2}{2}, \qquad \frac{1}{g_0^2} = \frac{1}{2}\left(\frac{1}{g_\omega^2} - \frac{1}{g_\rho^2}\right), \tag{7}$$

where $f_{\sigma V}$ figures in the mass formula $m_V^2 = f_{\sigma V}^2 g_V^2$. The $U(2)$ symmetry is recovered at the classical level by setting $g_\omega = g_\rho$ and $f_0 = 1/g_0 = 0$ in (4). The hWZ Lagrangian that will be found to unify the $N_f = 1$ and $N_f \geq 2$ baryons consists of four terms with an external potential J_μ^{ext} included:

$$\mathcal{L}_{hWZ} = \frac{N_c}{16\pi^2}\sum_{i=1}^{4} c_i\mathcal{L}_i. \tag{8}$$

We have not written down the explicit expressions here because they are not very illuminating and are not needed for what follows (see [22] for the detailed forms.) They are constructed of the Maurer–Cartan 1-forms with the covariant derivatives in the vector field V_μ plus the J_μ^{ext} which is either $U(1)$ baryonic potential or EM potential.

We should make two remarks on the Lagrangian (4) that need to be kept in mind in what follows. First, it is $O(p^2)$ in power counting [22] except for the hWZ term which while $O(p^4)$ is indispensable for unifying FQH droplets and skyrmions [15]. Second, it is made scale-invariant by the conformal compensator (CC) except for the dilaton potential $V(\chi)$ which could contain scale-symmetry explicit breaking, e.g., quark mass terms. The hWZ terms are scale-invariant, so are free of the CC fields. The rationale for this strategy for scale symmetry which resorts to what is known as "quantum scale invariance" [23] is explained below.

3.2. "Genuine Dilaton" Scenario (GDS)

In accordance with the GD scheme [11] with the IR fixed point specified above, even if explicit symmetry breaking is ignored, scale symmetry can be spontaneously broken by dilaton condensate generating masses to the hadrons. The scheme follows roughly the line of ideas based on "hidden quantum scale invariance" [23]. The underlying reason is that sHLS that we are exploiting is connected with strong–weak dualities à la Seiberg, typically associated with supersymmetric gauge theories [24–27]. This connection allows us to exploit the possible duality of HLS to the gluons that will figure in the problem. In addition, the applicability of duality in non-supersymmetric case as ours is made feasible if scale symmetry is broken by the dilaton in terms of the conformal compensator [26]. Thus, the theory at the tree level contains renormalization with the regularizer that brings scale elevated to a field, the CC field χ.

From the point of view of our bottom-up approach, it is important to note that the HLS we are dealing with is dynamically generated [9]. This means that the coefficients c_is in the hWZ term that play a key role in the unification of solitonic description of both $N_f = 1$ and $N_f \geq 2$ baryons are constants that cannot be fixed by the theory. For low density, therefore, they are to be determined by experiments. They could be fixed by holographic QCD, but there are no known holographic QCD models that possess possible "orange" [15]—not to mention ultraviolet—completion that would allow for approaching the density above n_0. However, as we will see later, in approaching QH droplet baryons bottom-up in density, the coefficients will be "quantized" by topology [15,16]. This takes place because there can be a phase transition from a Higgs mode to a topological phase in which the HLS fields are supposed to be (Seiberg-)dual to the gluons of QCD [24].

Unhiding Hidden Scale Symmetry in Nuclei

Underlying the idea of duality in hadronic matter is that scale symmetry is hidden. The VeV of the dilaton χ breaks scale-symmetry spontaneously in nuclear medium and slides with the density of the matter. This spontaneous scale-symmetry breaking in nuclear medium via the dilaton condensate dependent on density can have a highly subtle impact on certain nuclear properties. One of the celebrated cases highlighting the possible restoration of scale symmetry by nuclear renormalization is the g_A quenching in nuclear Gamow–Teller transitions [28,29]. It has been shown that the hidden scale invariance emerges precociously in nuclear medium via strong nuclear correlations to lead to an effective g_A, say $g_A^{ss} \to 1$, approaching what is referred to as "dilaton limit fixed point." The Lagrangian as given in (4) is in the leading order in scale-chiral expansion [12,30]. In the sector where η' plays no role—or a negligible role—and the dilaton field is ignored, the HLS Lagrangian is gauge-equivalent to nonlinear sigma model; therefore, one can do a systematic chiral-perturbation calculation similar to the standard χPT [22]. The treatment of many-body systems resorted to in [28] is in Landau Femi-liquid fixed point theory (FLFP) with the Lagrangian (10) (given below) that figures in [8] amounts to doing nuclear higher-loop renormalization calculations in Wilsonian-RG approach on Fermi surface and what leads to $g_A^{\rm eff} \to 1$ aptly captures the restoring of scale symmetry hidden at the tree level to pseudo-conformal symmetry at the quantum level.

Now, going to high density beyond the normal nuclear matter, an important issue of the EoS of massive compact stars is the role of the dilaton and indispensably the scale symmetry in QCD. The story of scale symmetry in gauge theories has a long history dating from 1960s and it remains still a highly controversial issue in particle physics going beyond the Standard Model (BSM). Here, we will confine ourselves to QCD for $N_f \leq 3$ for which we adhere to the notion that $f_0(500)$ is a "genuine" dilaton being associated with hidden scale symmetry [11,12]. The distinctively characteristic feature of the "genuine dilaton" scenario (GDS for short) is the presence of the IR fixed point signaling the scale invariance at which both scale and chiral symmetries are in the NG mode admitting massive particles, such as nucleons, vector mesons etc. It may be that this notion of the GDS is not widely accepted in the particle physics community working on BSM [31]. It seems, however, consistent with the notion of hidden quantum scale invariance [23]. In our approach to dense matter, it turns out, as recalled below, that the GDS is consistent with the general structure of scale symmetry that manifests as an emerging symmetry from nuclear correlations at what we call "dialton-limit fixed point (DLFP)" in dense matter.

4. Baryonic Matter without η'

We first consider baryonic matter where $U_A(1)$ anomaly does not figure. In (4), we set η' equal to zero or properly integrated out given its massiveness in nature. The property of dense matter described by the theory, G*n*EFT, is analyzed in some detail in [8]. How to address many-nucleon systems directly from the Lagrangian that contains meson fields only, that is in the class of skyrmion approach, has not been worked out in a way suitable for dense matter physics. Therefore, a direct exploitation of the Lagrangian (4) treated entirely in terms of skyrmions is not feasible at present for studying the properties of dense baryonic matter. However, an astute way is to map what are established to be "robust" topological properties of skyrmions obtained with (4) to a density functional-type theory—referred frequently to as "DFT" in nuclear physics circles—by introducing explicitly baryon fields, and suitably coupling them, to (4). The strategy is to capture as fully as feasible non-perturbative properties associated with the topological structure involved. One possible way of how this can be achieved is discussed in detail in [8]. Here, we summarize what one finds in the mean-field approximation of G*n*EFT which corresponds to Landau fixed-point theory. Going beyond the approximation can be formulated in what is known as "V_{lowk}" RG approach and applied to compact stars in [8].

4.1. Dilaton Limit Fixed Point (DLFP)

To exploit the mapping of topological inputs of the sHLS Lagrangian into a mean-field approximation with G*n*EFT, we add the nucleon coupling to the sHLS fields implementing both HLS and scale symmetry as

$$\begin{aligned}\mathcal{L}_N =&\bar{N}(i\not{D} - \Phi m_N)N + g_A\bar{N}\not{\hat{\alpha}}_\perp\gamma_5 N \\ &+ \bar{N}\left(g_{V\rho}\not{\hat{\alpha}}_\parallel + g_{V0}\mathrm{Tr}[\not{\hat{\alpha}}_\parallel]\right)N + \cdots,\end{aligned} \tag{9}$$

with the covariant derivative $D_\mu = \partial_\mu - iV_\mu^\rho - iV_\mu^\omega$ and dimensionless parameters g_A, $g_{V\rho}$ and $g_{V0} \equiv \frac{1}{2}(g_{V\omega} - g_{V\rho})$. The ellipsis stands for higher derivative terms that will not be taken into account in what follows. The Lagrangian concerned that we shall refer to as *bs*HLS is

$$\mathcal{L}_{bsHLS} = \mathcal{L}_{sHLS} + \mathcal{L}_N. \tag{10}$$

With the explicit presence of the baryon field, the role of the hWZ terms is relegated to the baryon-field coupling to the vector and scalar fields that takes over the ω repulsion in dense baryonic matter.

We consider what happens when the density goes up and approaches the DLFP first considered by Beane and van Kolck [32] and apply it to the model we are considering in [8]. To do this we assume that approaching the DLFP at high density amounts to going toward the IR fixed point à la CT [11,12] described above where both chiral symmetry and scale symmetry are realized in the NG mode with the dilaton mass and pion mass going to zero in the chiral limit.

Starting from the vacuum where chiral symmetry is realized nonlinearly, as density increases, one would like to arrive, at some point near n_0, at the linear realization of chiral symmetry, say, in the form of the Gell-Mann–Lévy (linear) sigma model [33] which qualitatively captures nuclear dynamics as the Walecka mean-field model does. This means transforming the nonlinear structure of sHLS that is the habitat of the skyrmion structure to a form more adapted to dense matter, namely, the half-skyrmion structure developed for the EoS of massive compact stars in [8]. This feature of transformation is encoded in the hidden scale symmetry as pointed out by Yamawaki [34]. This point will be further elaborated on in Section 7.

To see how the *bs*HLS Lagrangian behaves as density is increased, we follow Beane and van Kolck [32] and transform *bs*HLS to a linear basis, $\Sigma = \frac{f_\pi}{f_\chi} U\chi \propto \sigma' + i\vec{\tau}\cdot\vec{\pi}'$. We interpret taking the limit $\mathcal{S} \equiv \mathrm{Tr}(\Sigma^\dagger\Sigma) \to 0$ as approaching the DLFP. Now how to take the dilaton limit requires a special interpretation. In mapping the key information of topological structure of baryonic matter to G*n*EFT as explained in [8], it is essential to interpret the limiting $\mathcal{S} \to 0$ in the same sense as in going from the skyrmion phase to the half-skyrmion phase at a density above that of normal nuclear matter. In going from skyrmions to half-skyrmions in skyrmion crystal simulation, the quark condensate $\langle\bar{q}q\rangle$ is found to globally go to zero at a density denoted $n_{1/2} > n_0$. To give an idea, $n_{1/2}$ in massive compact stars comes in [8] at $\sim 3n_0$. The condensate, however, is non-zero locally, thereby supporting a chiral density wave in skyrmion crystal [35]. This seems to be the case in general as observed in various models [36]. As a consequence, the pion decay constant is non-zero, hence the state is in the NG mode. The same is true for the dilaton condensate with inhomogeneity in consistency with the GDS. This feature resembles the "pseudogap" structure in condensed matter physics, as there the issue is subtle and highly controversial (see [37] for a comprehensive discussion on this matter). In what follows, we interpret the limiting $\mathcal{S} \to 0$ in this sense. The order parameters for the symmetries involved in medium up to the possible IR fixed point of [11] are more complicated, involving higher dimensional field operators [38]. Approaching the DLFP, the quantities involved will be denoted by asterisk $*$ as $\langle\chi\rangle^* \propto f_\pi^*$ and $\langle\chi\rangle^* \propto f_\chi^*$ with $f_\pi^* \sim f_\chi^* \neq 0$. The matter in the half-skyrmion phase going toward the DLFP then has a resemblance to the pseudo-gap phase with fractional skyrmions present in SYK superconductivity [39].

One finds that in the limit $\mathcal{S} \to 0$, there develop singularities in the thermodynamic potential. Imposing that there be no singular terms in that limit gives what we identify as DLFP "constraints" [8]

$$f_\pi \to f_\chi \neq 0,\ g_A \to g_{V\rho} \to 1. \tag{11}$$

Furthermore, since the ρ-meson coupling to the nucleon is given by

$$g_{\rho NN} = g_\rho(g_{V\rho} - 1), \tag{12}$$

one sees that the ρ meson decouples—independently of the "vector manifestation (VM)" with $g_\rho \to 0$ [22]—from the nucleons as the DLFP is approached. On the other hand, the ω-NN coupling $g_{\omega NN} = g_\omega(g_{V\omega} - 1)$ remains nonzero for $g_\omega \neq 0$ because $g_{V\omega} - 1 \not\to 0$ in the DLFP limit. This has been verified at one-loop order in the renormalization-group equation. In fact, the EoS at high density relevant to massive compact stars requires this for the stability of the matter. It is well known that there is a delicate balance between the dilaton condensate which enters in the dilaton mass m_χ^* and the ω-NN coupling $g_{\omega NN}$. In fact this balance is the well-known story of the roles of the scalar attraction and vector repulsion in nuclear physics at normal nuclear matter density. It becomes more acute at higher densities.

The broad phase structure involved is depicted in Figure 1. Apart from the nuclear matter equilibrium density n_0 and the topology change density $n_{1/2}$, other densities are not precisely pinned down. What is given in the review [8] does not represent precise values, hence Figure 1 should be taken at best schematic.

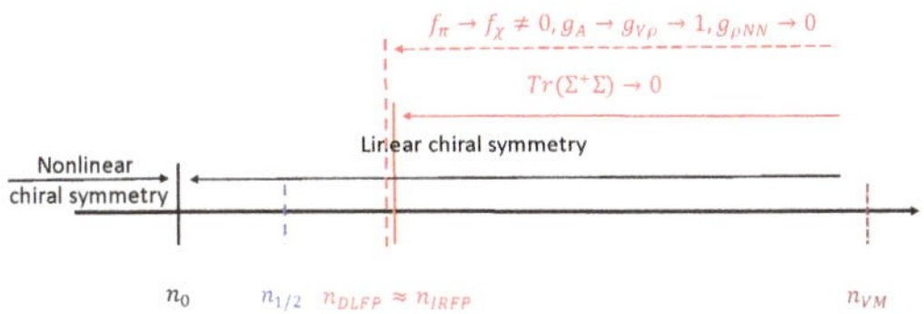

Figure 1. The proposed schematic phase structure for density regimes: n_0 stands for equilibrium nuclear matter density, $n_{1/2}$ for onset density of half-skyrmions, $n_{\rm DLFP}$ for dilaton limit fixed point, $n_{\rm IRFP}$ for IR fixed point and $n_{\rm VM}$ for vector manifestation fixed point.

4.2. Interplay between $g_{\omega NN}$ and $\langle\chi\rangle^*$

The nucleon in-medium mass is connected to the ω-nucleon coupling by the equations of motion for χ and ω and the in-medium property of the χ condensate, $\langle\chi\rangle^*$, or more appropriately the in-medium dilaton decay constant f_χ^* which controls the in-medium mass of the dilaton, hence the nucleon mass, at high density. This means that up to the DLFP, the effective nucleon mass will remain constant proportional to the dilaton condensate $\langle\chi\rangle^*$. This is seen in Figure 2. This $\langle\chi\rangle^*$ comes out to be equal to the scale-chiral invariant mass of the nucleon m_0 that figures in the parity-doubling model for the nucleon [40]. This then suggests that m_0 can show up, signaling the presence at a higher density of the DLFP through strong nuclear correlations even if it is not explicit in the QCD Lagrangian.

We claim that this is in accord with the GDS ("genuine dilaton" scenario) with the nucleon mass remaining massive together with the non-zero dilaton decay constant.

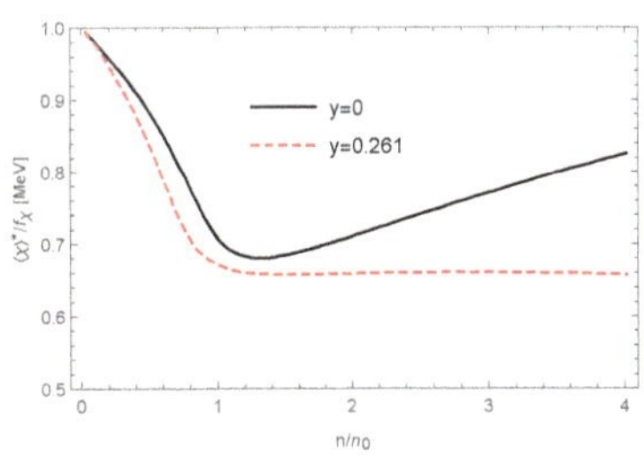

Figure 2. The ratio $\langle\chi\rangle^*/\langle\chi\rangle_0$ where $\langle\chi\rangle^* \propto f_\chi^*$ as a function of density n for varying "induced density dependence" (DD$_{induced}$)—distinct from IDD (intrinsic density dependence) inherited from QCD—of $g_{V\omega}^*$ which is parameterized as $g_{V\omega}^* - 1 = (g_{V\omega} - 1)(1 + y\frac{n}{n_0})^{-1}$. The density at which the ratio $\langle\chi\rangle^*/\langle\chi\rangle_0$ becomes constant is not given by the theory but comes out to be $\sim 3n_0$ in compact-star phenomenology. This density can be identified with $n_{1/2}$, the density at which skyrmion matter transitions to half-skyrmion matter.

The remarkable interplay between the dilaton condensate $\langle\chi\rangle^*$ and the ω-NN coupling has an important impact on the EoS for density $n \gtrsim n_{1/2}$ at which $\langle\chi\rangle^*$ flattens in density (The flattening to a density-independent constant of $\langle\chi\rangle^*/f_\chi$ at $n_{1/2}$ arising from an intricate interplay between $g_{\omega NN}$ and $\langle\chi\rangle^*$ in Figure 2 is related to that of $\langle\bar{q}q\rangle^*/f_\pi$ in a skyrmion-crystal simulation of HLS [41]. It is not obvious how to correctly implement the dilaton field in the crystal simulation, so the relation between the dilaton and quark condensates does not seem to come out correctly on skyrmion crystals. However, in the genuine dilaton scenario incorporated in GnEFT, we believe they should be tightly related—as we argued—as density approaches the IR fixed point density.). As noted above, the induced density dependence for the ρ-NN coupling $\propto (g_{V\rho} - 1)$ drops rapidly such that the ρ decouples from nucleons at the DLFP whereas $(g_{V\omega} - 1)$ remains non-zero. How this impacts on nuclear tensor forces and consequently on the symmetry energy E_{sym} deserves to be investigated in nuclear structure. Furthermore, the vector manifestation leads to the gauge coupling $g_\rho \to 0$ [22] whereas that for ω coupling g_ω drops only slightly. The delicate balance between the attraction due to the scalar (dilaton) exchange and the repulsion due to the ω exchange plays a crucial role for the EoS for $n \gtrsim n_{1/2}$ of massive neutron stars [8].

4.3. The Trace Anomaly and Pseudo-Conformal Symmetry

A striking consequence of the interplay between the $g_{\omega NN}$ coupling and the condensate $\langle\chi\rangle^*$ at $n \gtrsim n_{1/2}$ in the GnEFT framework, not shared by other models in the literature, is the precocious emergence of hidden scale symmetry in nuclear interactions. The details are involved but the phenomenon can be clearly seen in the mean-field approximation with the *bs*HLS Lagrangian (10).

The vacuum expectation value of the trace of the energy-momentum tensor θ_μ^μ is given by

$$\langle\theta_\mu^\mu\rangle = 4V(\langle\chi\rangle) - \langle\chi\rangle\left(\frac{\partial V(\chi)}{\partial\chi}\right)\bigg|_{\chi=\langle\chi\rangle^*} + \cdots \quad (13)$$

where the ellipsis stands for chiral symmetry breaking (quark mass) terms. Now, if one ignores the quark mass terms, then given that the $\langle\chi\rangle^*$ which should be identified with the dilaton decay constant is independent of density as illustrated in Figure 2, we have

$$\frac{\partial\langle\theta_\mu^\mu(n)\rangle}{\partial n} = \frac{\partial\epsilon(n)}{\partial n}(1 - 3v_s^2/c^2) = 0. \quad (14)$$

One expects that $\frac{\partial\epsilon(n)}{\partial n} \neq 0$ and hence, within the range of density where (14) holds, say, $\sim (3-7)n_0$, we arrive at what is commonly associated with the "conformal sound speed"

$$v_s^2/c^2 = 1/3. \quad (15)$$

Since, however, the trace of the energy-momentum tensor is not zero at the density involved which is far from asymptotic, it should be more appropriately called "pseudo-conformal" velocity.

This prediction made in the mean field for a neutron star of mass $M \simeq 2M_\odot$ has been confirmed—modulo of course quark-mass terms—in going beyond the mean-field approximation using the V_{lowk}RG approach [8]. Needless to say, the quark mass terms could affect this result, bringing in possible deviation from (15), but it seems reasonable to assume that the corrections cannot be significant.

5. Baryonic Matter with η'

To date, baryonic matter without the η' degree of freedom is treated as density increases toward the DLFP. The baryons involved there are skyrmions for $N_f = 2$. It has been assumed that the $U_A(1)$ anomaly plays no role at high density for compact-star physics.

However, there are at least three reasons why the η' degree of freedom cannot be ignored in nuclear dynamics. First, it is known that the $U_A(1)$ anomaly plays a crucial role for the color-charge conservation in the CCP [42] and consequently for the flavor-singlet axial-vector coupling constant of the proton $g_A^0 \ll 1$ [43]. Second, the η', though massive at low density, may become lighter and become relevant at high density. Third, it has been suggested that the FQH droplet structure of $N_f = 1$ baryon [14] can be unified in scale-symmetric HLS theory with the skyrmion structure of $N_f \geq 2$ baryons [15]. The third, while giving a possible solution to the dichotomy problem raised in Section 2, could influence the EoS at high density.

We now discuss how to approach quantum Hall droplets from skyrmions.

5.1. From sHLS to the η' Ring

In following Karasik's arguments [15,16], we take the sHLS Lagrangian (4) and focus on the terms that involve the η' field in that Lagrangian. In doing this manipulation, the role of the conformal compensator present to provide (Seiberg-)duality plays the crucial role in contrast to what was conducted in [15,16] where the the role of dilaton effects is performing double trace, which is ignored. With the baryons generated as solitons in sHLS, the parameters of the Lagrangian contain the "intrinsic density dependence" (IDD) inherited from QCD at the matching between EFT and QCD. First we ask what one should "dial" in the parameters of GnEFT—in the spirit of the strategy used—to access compact stars so that the system approaches the η' sheet. Next, we ask whether high baryon density supplied by gravity makes the η' ring "visible".

Suppose we increase density beyond $n_{1/2}$. Recalling from what we have learned in the mean-field result with (10) (with the nucleon explicitly included), it seems reasonable to assume the ρ decouples first at some density above $n_{1/2}$ before reaching the DLFP. Since the gauge coupling g_ρ goes to zero in approaching the vector manifestation fixed point n_{VM} (say, $n \gtrsim 25n_0$) [8]), the mass $m_\rho \sim f_\pi g_\rho$ goes to zero independently of whether or not f_π goes to zero and the ρ decouples from the pions. The Lagrangian $\mathcal{L}_{sHLS}$ (4) will then reduce to what was noted by Karasik [15,16]

$$\begin{aligned}\mathcal{L}_{sHLS} =& \frac{1}{2}(\partial_\mu\chi)^2 + V(\chi) + \frac{1}{2}\Phi^{*2}(\partial_\mu\eta')^2 - \frac{1}{4}(\omega_{\mu\nu})^2 + \frac{1}{2}m_\omega^2\Phi^{*2}\omega_\mu\omega^\mu \\ &- c_3\frac{N_c}{8\pi^2}\epsilon^{\mu\nu\alpha\beta}\omega_\mu\partial_\nu\omega_\alpha\partial_\beta\eta' + \cdots,\end{aligned} \tag{16}$$

where we have written $f = f_\eta$ and $f_{\sigma\rho}^2 = af_\pi^2$ and $\Phi^* \equiv f_\chi^*/f_\chi \approx f_\pi^*/f_\pi$ with the $*$ standing for density dependence in baryonic matter. In this operation, we assumed that the limit $\mathcal{S} \to 0$ is equivalent to having, as in the crystal simulation, both the dilaton and chiral condensates—space averaged globally—go to zero while they locally support density waves with their decay constants remaining nonzero. As density increases further beyond the DLFP, the condensate will vanish locally, hence $\Phi^* \to 0$ and the kinetic energy term of η' field, ill defined due to singularity, is killed and the ω mass $m_\omega \propto \langle\chi\rangle$ goes to zero. Indeed, it is explicitly shown in [41] in the crystal approach that at a density $n \gg n_{1/2}$ the phase becomes homogeneous—without density waves—so that $f_\pi \propto f_\chi \propto \langle\chi\rangle = 0$. We interpreted this phase as deconfined since both chiral and scale symmetries are restored. Note that the density involved must be considerably higher than the core density of massive stars. We will be left with the massless χ and ω fields and the $\omega\eta'$ coupling coming from $\mathcal{L}_3$ in the hWZ term, (8).

It is here that our approach to scale symmetry à la GD [11,23] brings the role of the sliding dilaton condensate at high density into contact via the Seiberg-type duality with the η' ring structure.

The last term of (16) can be written as

$$\mathcal{L}_{CS\eta'} = -c_3\frac{N_c}{4\pi}J_{\mu\nu\alpha}\omega^\mu\partial^\nu\omega^\alpha. \tag{17}$$

with the topological $U(1)$ 2-form symmetry current

$$J_{\mu\nu\alpha} = \frac{1}{2\pi}\epsilon_{\mu\nu\alpha\beta}\partial^\beta\eta'. \tag{18}$$

Now, a pertinent observation is that the charged objects under these symmetries are metamorphosed to infinitely extended sheets that interpolate from $\eta' = 0$ on one side to $\eta' = 2\pi$ on the other [14,15], involving a sheet $\eta' = \pi$. The current is conserved because η' in the space of η' configuration is a circle and $\pi_1(S^1) = \mathcal{Z}$. The Lagrangian (17) corresponds to the CS field identified with the ω field coupling to the baryon charge. The CS field is a topologically non-trivial gauge field and hence gauge invariance requires that the c_3 be quantized $c_3 = 1$. This argument holds if we assume that the matter is in the topological phase where HLS is Seiberg-dual to QCD [24]. Karasik identifies the hWZ term with constrained coefficients as "hidden WZ" term in contrast to the "homogenous WZ term" [22] relevant at lower density. Note that going from "homogeneous hWZ" to "hidden hWZ" by density would require the unhiding of hidden scale invariance [11,23]. Now, going beyond the DLFP as the system is brought toward the putative density at which $\langle\chi\rangle \to 0$ and $\langle\bar{q}q\rangle \to 0$, $f_\pi \sim f_\chi \to 0$ and $m_\chi \to 0$, $m_\omega \to 0$ etc., one is left with a quantum Hall baryon with $B = 1$ and $J = N_c/2$ [14]. This essentially rephrases Karasik's argument in terms of the GnEFT to arrive at the $N_f = 1$ baryon from the $N_f = 2$ baryons.

In what is described above, we have assumed that the ρ field decouples first before reaching the DLFP as indicated in Section 4.1. This is what seems to take place in compact-star matter studied in [8]. Instead of $U(1)$ CS field theory, however, one can generalize the discussion to nonabelian CS field theory from the sHLS Lagrangian (4). The Lagrangian (17) is modified to [24] (This matter will be discussed elsewhere. We merely note that this Lagrangian figures as the boundary term in the chiral bag model to account for the color anomaly in the chiral bag model [42]. It illustrates that η' could enter non-trivially in baryon structure in the background of the η' singular ring invisible at low density (and low temperature) far away from the chiral restoration regime that will be addressed later.).

$$\mathcal{L}'_{CS\eta'} = -c_3\frac{N_c}{4\pi}J_{\mu\nu\alpha}\mathrm{Tr}\left(V^\mu\partial^\nu V^\alpha + \frac{2}{3}V^\mu V^\nu V^\alpha\right) \tag{19}$$

where $V_\mu = \frac{1}{2}(\tau\cdot\rho_\mu + \omega_\mu)$, assuming $U(2)$ symmetry, is restored at the high density concerned. We now identify the source of the baryon number as $B = (N_c/N_f)\mathrm{Tr}V$ and differentiating the action $\int \mathcal{L}'_{CS\eta'}$ with respect to B, we obtain the baryon density

$$\rho_B = c_3\frac{1}{4\pi^2}\epsilon^{ijk}\mathrm{Tr}(\partial_i V_j)\partial_k\eta' + \cdots. \tag{20}$$

With the configuration $\eta' = 0$ at $x_3 = -\infty$ and $\eta' = 2\pi$ at $x_3 = \infty$, the baryon number is obtained by integrating over x_3. That requires again

$$c_3 = 1. \tag{21}$$

This exposes the η' ring in the $N_f = 2$ setting. This observation is relevant to the possible decay of the η' ring to a pionic sheet described in the next subsection.

Above, we have seen that in some density regime, one arrives at a CS theory coupled to a baryon-charge one object that could be identified with the η' ring. This is achieved, we suggest, by what amounts to going beyond the DLFP in the GnEFT Lagrangian. To do this more realistically, it may be necessary to include higher-lying vectors and scalars as in holographic models [24]. This is beyond our scheme so we will not go into the matter further. What is interesting is to view the process in terms of the $N_f = 2$ skyrmion given by the sHLS Lagrangian, namely, how the hedgehog ansatz (For the $SU(2)$ flavor chiral field $U = e^{i\tau\cdot\pi/f}$, which in [15] is written as $U = \sigma + i\tau\cdot\pi$, the hedgehog ansatz is $U = e^{i\theta(r)\vec{\tau}\cdot\hat{r}}$. $\theta(r)$ is sometimes called chiral angle.) in the background of the η' field is "deformed" as density goes up. It seems plausible, as suggested in [15], that high density first impacts on the EoS such that

$$\pi_1 = \pi_2 = 0,\ V_\mu^1 = V_\mu^2 = 0, V_\mu^3 = \omega_\mu \tag{22}$$

and then distorts the hedgehog configuration to

$$(\sigma + i\pi_3)/\sqrt{\sigma^2 + \pi_3^2} = e^{i\eta'}. \tag{23}$$

This suggests that while at low density η' in $U = e^{i\eta'}e^{i\tau\cdot\pi/f}$ present in the η' ring plays no significant role, except perhaps, giving an $O(N_c^0)$ correction to the $\Delta - N$ mass difference which could not be significant, the η' ring becomes important as density increases.

5.2. Going from the η' Ring to the Pionic Sheet

We consider the density regime where the ρ mesons are decoupled from the nucleons and the η' ring is unstable, and so decays to skyrmions. Noting that the η' ring, i.e., $\mathcal{L}_{CS\eta'}$, is embedded in the full hWZ term, we should look at the hWZ term (8). Following [15], we write (in the unitary gauge $\xi_R = \xi_L^\dagger = \xi$)

$$\begin{aligned}\mathcal{L}_{hWZ} =& \frac{N_c}{24\pi^2}\epsilon^{\mu\nu\rho\sigma} g_\omega \omega_\mu \\ &\times \mathrm{Tr}\left[\left(\frac{3}{8}\kappa_1\right)2\partial_\nu\xi\xi^\dagger\partial_\rho\xi\xi^\dagger\partial_\sigma\xi\xi^\dagger\right. \\ &\left.+\left(\frac{1}{2}\kappa_2\right)3iV_\nu(\partial_\rho\xi\partial_\sigma\xi^\dagger - \partial_\rho\xi^\dagger\partial_\sigma\xi) + \left(\frac{1}{2}\kappa_3\right)3i\partial_\nu V_\rho(\partial_\sigma\xi\xi^\dagger - \partial_\sigma\xi^\dagger\xi)\right]\end{aligned} \tag{24}$$

where only the terms contributing to the $N_f = 2$ completion of the topological term $\mathcal{L}_{CS\eta'}$ are retained. The coefficients κ_is can be identified with c_is of the hWZ term (8)

$$\kappa_1 = c_1 - c_2,\ \kappa_2 = c_1 + c_2, \kappa_3 = c_3. \tag{25}$$

Under gauge transformation $\omega_\mu \to \omega_\mu - \frac{1}{g_\omega}\partial_\mu\lambda$, one has

$$\begin{aligned}\delta S =& \frac{N_c}{12\pi^2}\epsilon^{\mu\nu\rho\sigma}\partial_\mu\lambda\mathrm{Tr}\left[\left(\frac{3}{8}\kappa_1\right)\partial_\nu\xi\xi^\dagger\partial_\rho\xi\xi^\dagger\partial_\sigma\xi\xi^\dagger\right] \\ &+\frac{iN_c}{8\pi^2}\epsilon^{\mu\nu\rho\sigma}\partial_\mu\lambda\partial_\nu\mathrm{Tr}\left[\left(\frac{1}{2}\kappa_2\right)V_\rho(\partial_\sigma\xi\xi^\dagger - \partial_\sigma\xi^\dagger\xi)\right].\end{aligned} \tag{26}$$

Then the gauge invariance yields the constraints

$$\int d\phi\left[\left(\frac{1}{2}\kappa_2\right)\left(\omega_\phi + V_\phi^3\right) + \left(\frac{3}{8}\kappa_1\right)4\pi_1\partial_\phi\pi_2\right] = const. \tag{27}$$

So that, on the world-sheet for the $N_f = 1$ baryon, one has [15]

$$\frac{1}{2}\kappa_2\int d\phi(\omega_\phi + V_\phi^3) = 2\pi,\ \pi_{1,2} = 0. \tag{28}$$

Thus, the η' is "seen". Note that the mixing of two $U(1)$s $(\omega_\phi + V_\phi)$ in this formula (and also the equality $V_3 = \omega$ in (22)) follows from the assumed duality of HLS [27].

Now, suppose the η' sheet structure, a background buried in the system of $N_f = 2$ skyrmions, is unstable and could subsequently decay into skyrmions in a different sheet structure containing the isovector degrees of freedom

$$\omega_\phi + V_\phi^3 = 0,\ \frac{3}{8}\kappa_1\int d\phi\pi_1\partial_\phi\pi_2 = \frac{\pi}{2}. \tag{29}$$

The question is: What is the structure of the matter encoded in the condition $\frac{3}{8}\kappa_1\int d\phi\pi_1\partial_\phi\pi_2 = \pi/2$ to which the η' decays? Could this be a sort of droplets that can be described in a topological field theory, involving isovector degrees of freedom, e.g., the $\pi^\pm$, the ρ vectors etc. as in the form of a nonabelian CS Lagrangian that seems to arise in the Cheshire Cat for

$N_f = 2$ baryons? We have no answer to this question. Clearly, isovector mesons must figure. This has to do with the quantization of other coefficients than the one giving the η' ring.

6. Ubiquitous Sheet Structure of Baryonic Matter

While it is not clear how the background of the η' ring, perhaps insignificant in the dynamics of strongly interacting many-nucleon matter at low density, affects the process of going toward the DLFP—and beyond—to "expose" the η' ring structure, we first inquire how the QH droplets structure could manifest in the sheet structure of dense matter as seen in the EoS of massive compact stars described with fair success in [8].

6.1. Crystal Skyrmions

We return to the skyrmion crystal simulation on which the GnEFT for massive compact stars is anchored [8]. As detailed there, the topological structure of the skyrmions simulated on crystal is translated into the parameters of the GnEFT Lagrangian, which is then treated in an RG-approach to many-nucleon interactions. The key role played in this procedure is the topological feature encoded in the skyrmion structure of hidden scale symmetry and local symmetry of sHLS. Notable there are the cusp in the symmetry energy of dense matter due to the "heavy" degrees of freedom, the parity doubling in the baryon spectra, and a "pseudo-gap" structure of the half-skyrmion phase. These properties encapsulated in the RG-approach with GnEFT led to the prediction of possible precocious emergence of scale symmetry in massive-star matter with the pseudo-conformal sound velocity of star $v_{pcs}^2/c^2 \simeq 1/3$ at a density $n \gtrsim 3n_0$.

Let us explore what this skyrmion crystal structure suggests for a possible sheet structure of dense matter.

It is observed in molecular dynamics simulation of nuclear matter expected in neutron-star crust and core-collapse supernova at a density a packing fraction of $\sim 5/16$ of nuclear saturation density $n_0 \sim 2 \times 10^{14}$ g/cm^3 that a system of "sheets" of lasagne, among a variety of complex shapes of so-called "nuclear pasta", could be formed and play a significant role in the EoS in low-density regime of compact star matter [44]. Involved here are standard nuclear interactions between neutrons and protons in addition to electromagnetic interactions.

At higher densities, say, at densities $\sim (2-7)n_0$, it is seen in skyrmion crystal simulations that a stack of lasagne sheets [45] or of tubes or spaghettis [46] is energetically favored over the homogenous structure. Involved here are fractionalized skyrmions, 1/2-baryon-charged for the former and $1/q$-charged for the latter with q odd integer. Those fractionalized skyrmions can be considered "dual" to (constituent) quarks in the sense of baryon-quark duality in QCD. There is an indication that the sheet structure of the stack of lasagnes could give a consistent density profile of even finite nuclei [45]. In fact, there seems little doubt that an inhomogeneity is favored in dense matter at non-asymptotic densities [35,36]. Thus, it could be considered robust.

The two phenomena at low and high densities involve basically different aspects of strong interactions, but there is a tantalizing hint that something universal is in action in both cases. We are tempted to consider that topology is involved there. This is particularly plausible at high density given that the "pasta" structure, be that lasagne or tubes (or spaghetti), is found to be strikingly robust. To date, the analysis has been made with an ansatz for the pion field, i.e., the Atiyah–Manton ansatz for the lasagne sheet and a special ansatz allowing analytical treatment for the tubes. The robustness must have to do with the fact that what is crucially involved is the topology and it is the pion field that carries the topology. What is striking is the resulting structure does not seem to depend on the presence of other massive degrees of freedom such as the vector mesons or scalar [41,47]. There, adding an infinite number of higher derivative terms to the Skyrme Lagrangian is found not to modify the ansatz for the tubes. It is, therefore, highly likely that the same structure would arise from the presence of the hidden scale-local degrees of freedom manifested in different forms of sHLS.

6.2. Density Functional Theory (DFT)

Our sHLS Lagrangian could contain the unified descriptions of both $N_f = 1$ droplet—η' ring—and $N_f = 2$ skyrmions in an EoS, but we have not been able to capture both in a unified way. That is, how the infinite-hotel and the FQH structures combine in the EoS, and whether and how the latter structure figures in compact-star physics.

A significant recent development, relevant conceptually to this matter, is the work treating the fractional quantum Hall phenomenon in the Kohn's functional density approach à la Kohn–Sham [48]. The key ingredient in this approach is the weakly interacting composite fermions (CF) formed as bound states of electrons and (even number of) quantum vortices. Treated in Kohn–Sham density functional theory, one arrives at the FQH state that captures certain strongly correlated electron interactions. The merit of this approach is that it *maps* the Kohn–Sham density functional, a microscopic description, to the CS Lagrangian, a coarse-grained macroscopic description, for the fractional quantum Hall effect.

Now, the possible relevance of this development to our problem is as follows. First, Kohn–Sham theory more or less underlies practically all nuclear EFTs employed with success in nuclear physics, as for instance, energy density functional approaches to nuclear structure. Second, our GnEFT approach belongs to this class of density-functional theories in the strong-interaction regime. Third, the successful working of the GnEFT model backed by robust topology and implemented with intrinsic density dependence inherited from QCD could very well be attributed to the power of the (Kohn–Sham-type) density functional in baryonic matter at high density $n \gtrsim 3n_0$. These three observations combined suggest that we approach the dichotomy problem in a way related to what was performed for FQHE.

The first indication that GnEFT anchored on the topology change could be capturing the weak CF structure of [48] in FQHE is seen in the nearly non-interacting quasiparticle behavior in the chiral field configuration U in the half-skyrmion phase (see Figure 8 in [8]). This feature may be understood as follows. Due to hidden $U(1)$ gauge symmetry in the hedgehog configuration, the half-skyrmion carries a magnetic monopole associated with the dual ω [49,50]. The energy of the "bare" monopole in the half-skyrmion diverges when separated, but the divergence is tamed by interactions in the skyrmion as a bound state of two half-skyrmions where the divergence is absent. In a way analogous to what happens in the Kohn–Sham theory of FQHE [48], there could intervene the gauge interactions between the skyrmions pierced by a pair of monopoles in sHLS—as composite fermions—possibly induced by the Berry phases due to the magnetic vortices. Thus, it is possible that the topological structure of the FQH is buried in the bound half-skyrmion structure at high density. A possible avenue to the problem is to formulate the EoS in terms of a stack of ordered coupled sheets of CS droplets. We hope to elucidate this in future work.

6.3. Hadron—Quark Continuity a.k.a. Duality

In the dilaton limit where the constraints (11) set in, there are NG excitations and the nucleon mass is $O(m_0)$ with $m_0 \simeq y m_N$, $y \sim (0.6 - 0.9)$. In compact stars treated in [8], the core of the massive stars with $M \sim 2M_\odot$ has density $\sim (6-7)n_0$. A natural question one raises is whether the core of the star contains "deconfined" quarks either co-existing with or without baryons. In the framework of [8], the constituents of the core are fractional-baryon-charged quasiparticles. They are neither baryons nor quarks. The fractional-charged phase arises without order-parameter change and hence considered evolving continuously from baryonic phase with a certain topology change. In certain models having domain walls, those fractional-charged objects can be "deconfined" on the domain wall [51]. If the sheets in the skyrmion matter discussed above are domain walls, then it is possible that the fractional-charged objects are "deconfined" on the sheets in the sense discussed in [51].

There are two significant issues raised here.

One is the possible observation of an evidence for "quarks" in the core of massive neutron stars [52]. Very recently, combining the astrophysical observations and theoretical ab initio calculations, Annal et al. concluded that inside the maximally massive stars there could very well be a quark core consisting of "deconfined quarks" [52]. Their analysis is based on the observation that in the core of the massive stars, the sound velocity approaches the conformal limit $v_s/c \to 1/\sqrt{3}$ and the polytropic index takes the value $\gamma < 1.75$, 1.75 being the value close to the minimal one obtained in the hadronic model. It has been found [53] that in the pseudo-conformal structure of our GnEFT, the sound velocity becomes *nearly* conformal $v_{pcs}/c \sim 1/\sqrt{3}$ and the γ goes near 1 at $n \gtrsim 3n_0$. Thus, at least at the maximum density relevant for $\sim 2M_\odot$ stars, what could be interpreted as "deconfined quarks" can be more appropriately fractionally charged quasipartlcles. Are these "deconfined" objects on domain walls as in [51] or confined two half-skyrmions as mentioned in Section 6.1?

The other is what is referred to as "baryon-quark continuity" in [8] in the domain of density relevant to compact-star phenomenology. This is not in the domain of density relevant to the color-flavor locking which is to take place at asymptotic density [54]. It seems more appropriate to say that the gauge degrees of freedom we are dealing with should be considered as "dual" to the gluons in QCD [24].

6.4. Hadron—Quark Continuity or Deconfinement

It has recently been argued that the hadron–quark continuity in the sense of [54] is ruled out on the basis of the existence of a nonlocal order parameter involving a (color-)vortex holonomy [55]. However, such a "theorem", perhaps holding at asymptotic density, could very well be irrelevant even at the maximum possible density observable in nature, whatever the maximum mass of the star stable against gravitational collapse might be. The argument of [55] cannot rule out the baryon-quark duality argument given in [8] and in this paper which is far below asymptotic density. The presence of the scale-chiral-invariant nucleon mass m_0 testifies for this assertion.

6.5. Emergence of Hidden Scale Symmetry in Nuclear Matter

The hadron–quark duality for which the hidden scale symmetry—together with the HLS—figures crucially in our discussion of resolving the dichotomy, we argue, leaves a trail of other observables where its effect has impacts on. One prominent case is the long-standing mystery described above of the "quenched" axial-vector coupling constant g_A in nuclear medium reviewed in the note that accompanies this note [29].

We recall as the density approaches the DLFP density, the constraints (11) require that the effective $g_A \to 1$. Surprisingly, as has been recently shown [28], the effective g_A^{ss} in Gamow–Teller transitions (most accurately measurable in doubly-magic nuclei)—which is denoted as $g_A^{\rm Landau}$ in [28]—is predicted to be "quenched" in the presence of emerging scale invariance from $g_A = 1.27$ in free space to $g_A^{ss} \approx 1$ at a much lower density, say, in finite nuclei.

Now, here is an intriguing possibility. While the precocious onset of the conformal sound speed $v_s^2/c^2 = 1/3$ in massive stars at density $n \gtrsim 3n_0$ is a signal for an emergence of an albeit approximate scale symmetry, there has been, to date, no evident indication of the pseudo-conformal structure at low density $n \lesssim 3n_0$. As suggested in [56], it is an appealing possibility that $g_A^{\rm eff} \approx 1$ in nuclei (modulo possible, as yet unknown, β' corrections) and $g_A = 1$ at the DLFP is the continuity of the "emergent" scale symmetry reflecting hidden quantum scale invariance between low and high densities.

Regardless of whether the hidden nature of scale symmetry is appropriate for the "genuine dilaton" of Crewther [11] or quantum scale invariance [23] or the dilaton in the conformal window [31], scale symmetry is intrinsically hidden. This point has been clearly illustrated in Yamawaki's argument [34]. Yamawaki starts with the $SU(2)_L \times SU(2)_R$ linear sigma model with two parameters which corresponds to the Standard Model Higgs Lagrangian, makes a series of field re-parameterizations and writes the SM Higgs model in two terms, a scale invariant term and a potential term which breaks scale symmetry which depends on one dimensionless parameter λ. By dialing $\lambda \to \infty$, he obtains the nonlinear sigma model with the scale symmetry breaking shoved into the NG boson field kinetic energy term, and by dialing $\lambda \to 0$ he obtains scale-invariant theory going toward the conformal window. This suggests that one can think of what is happening in baryonic matter as dialing the parameter λ in terms of nuclear dynamics. For compact-star physics, it is the density that does the dialing.

7. Comments and Further Remarks

The principal proposition of this article is that the effective low-energy Lagrangian sHLS that incorporates hidden scale and local symmetries containing, in addition to the (octet) pions, the η' degree of freedom could contain both $N_f = 1$ baryons and $N_f = 2$ baryons through hidden scale and local symmetries dual to the gluons in agreement with Karasik. What is highly non-trivial is that the GnEFT with η' included in the chiral field U could contain both the topological structure of quantum Hall baryons and that of skyrmion baryons. How to write the ansatz for the former as one does for the latter is unclear, but it should be feasible to do so and would allow one to see how the former comes into the latter to resolve the dichotomy problem.

We are uncovering an interesting role that could be played by the scale symmetry with its dilaton and the hidden local symmetry with the vector mesons dual to the gluons in "unifying" the two different topological sheet structures. The analysis made in the GnEFT framework based on sHLS indicates that in the density regime relevant to massive compact stars, the chiral condensate and dilaton condensate go proportional to each other in the NG mode. In going beyond the regime of massive compact stars, we find the DLFP approaching, if not coinciding with, the IR fixed point with $f_\pi = f_\chi$. How and where the density regime for the IR fixed point is approached cannot at present be fixed.

It should be re-stressed that the scale symmetry involved in dense matter is basically different from the presently favored scenario invoking the "conformal window" in the domain of BSM with $N_f \sim 8$ typically characterized by the ratio [31] $F_\pi^2/F_d^2 \sim 0.1$. This is in stark contrast with $f_\pi^2/f_\chi^2 \sim 1$ which seems to be encoded in the pseudo-conformal structure in dense nuclear systems. Another notable observation is that zeroing-in on the deep IR regime associated with the η' singularity involved in the domain-wall topological structure of baryons as argued in [15,16] uncovers the ω (a.k.a. Chern–Simons) mass going to zero as the fermion ("quark") mass $m \to \infty$. This contrasts with how the η' ring structure could be "exposed" in dense nuclear processes at high density as the ω mass is to go to zero with the dilaton mass $m_\chi \propto \langle\chi\rangle$ going to zero. This could be explainable in terms of a (Seiberg-type) duality between the gluons (in the topological phase) and the HLS mesons (in the Higgs phase). See a related discussion in [24]. In the former, the vector dominance is found to play a crucial role for the $N_f = 1$ baryon structure [16] whereas in the latter, the VD—unless a high tower of vector mesons is taken into account [57]—famously fails to work for the $N_f = 2$ (i.e., nucleon) EM form factors. How to correlate or reconcile these two processes appears highly challenging.

Finally, there is an intriguing, so far un-explored, role that the η' singular ring (a.k.a. FQH droplet) could play in the properties of HLS mesons near chiral restoration. This could be particularly relevant to dilepton production in relativistic heavy-ion collisions in search of dropping vector meson mass in extreme conditions, in particular at high temperature $T_c \sim 150$ MeV near chiral restoration. What our work suggested is that near the chiral transition where the dialton decay constant goes to zero, there would be exposed FQH droplets. Kitano and Matsudo have argued [58], based on their analysis of vector mesons on domain wall and on what is conjectured in [24], that when the chiral phase transition is approached, the FQH pancake configurations are absolutely essential for the description of the behavior of the vector mesons near the phase transition. To date, the theoretical analyses on the vector mesons in the dilepton experiments did not take into account, in addition to the what was argued in a similar vein in [59], the possibly crucial role of the FQH pancakes; therefore, the failure of the dilepton experiments to see the dropping-mass effect could not be concluded as "ruling out" the vector manifestation fixed point as was performed in the heavy-ion community [60].

In contrast, given that the density of the core of massive neutron stars is predicted to be far below the putative chiral restoration density, the phase with the FQH-pancakes with conformal invariance may not have a direct impact on the EoS for neutron stars. Nevertheless, it may not be ignorable in that at densities much lower than that relevant to the FQH pancake configurations it could have an "un-nuclear matter" structure in the star properties—such as "unnuclei" [61] along the notion of "unparticles" [62], a possible example being the pseudo-conformal sound speed predicted in the *Gn*EFT approach [8].

Author Contributions: All authors contributed equally to the review. All authors have read and agreed to the published version of the manuscript

Funding: The work of YLM was supported in part by the National Science Foundation of China (NSFC) under Grant No. 11875147 and 11475071.

Institutional Review Board Statement: Not applicable.

Informed Consent Statement: Not applicable.

Data Availability Statement: Not applicable.

Conflicts of Interest: The authors declare no conflict of interest.

References

1. Hammer, H.-W.; König, S.; van Kolck, U. Nuclear effective field theory: status and perspectives. *Rev. Mod. Phys.* **2020**, *92*, 025004 [CrossRef]
2. Holt, J.W.; Rho, M.; Weise, W. Chiral symmetry and effective field theories for hadronic, nuclear and stellar matter. *Phys. Rept.* **2016**, *621*, 2. [CrossRef]
3. Weinberg, S. What is quantum field theory, and what did we think it is? *arXiv* **1997**, arXiv:hep-th/9702027.
4. Shankar, R. Renormalization group approach to interacting fermions. *Rev. Mod. Phys.* **1994**, *66*, 129. [CrossRef]
5. Polchinski, J. Effective field theory and the Fermi surface. *arXiv* **1992**, arXiv:hep-th/9210046.
6. Holt, J.W.; Kaiser, N.; Weise, W. Chiral Fermi liquid approach to neutron matter. *Phys. Rev. C* **2013**, *87*, 014338. [CrossRef]
7. Holt, J.W.; Brown, G.E.; Holt, J.D.; Kuo, T.T.S. Nuclear matter with Brown-Rho-scaled Fermi liquid interactions. *Nucl. Phys. A* **2007**, *785*, 322. [CrossRef]
8. Ma, Y.L.; Rho, M. Towards the hadron-quark continuity via a topology change in compact stars. *Prog. Part. Nucl. Phys.* **2020**, *113*, 103791. [CrossRef]
9. Bando, M.; Kugo, T.; Uehara, S.; Yamawaki, K.; Yanagida, T. Is ρ meson a dynamical gauge boson of hidden local symmetry? *Phys. Rev. Lett.* **1985**, *54*, 1215. [CrossRef]
10. Suzuki, M. Inevitable emergence of composite gauge bosons. *Phys. Rev. D* **2017**, *96*, 065010. [CrossRef]
11. Crewther, R.J. Genuine dilatons in gauge theories. *Universe* **2020**, *6*, 96. [CrossRef]
12. Crewther, R.J.; Tunstall, L.C. $\Delta I = 1/2$ rule for kaon decays derived from QCD infrared fixed point. *Phys. Rev. D* **2015**, *91*, 034016 [CrossRef]
13. Skyrme, T.H.R. A unified field theory of mesons and baryons. *Nucl. Phys.* **1962**, *315*, 556. [CrossRef]
14. Komargodski, Z. Baryons as quantum Hall droplets. *arXiv* **2018**, arXiv:1812.09253.
15. Karasik, A. Skyrmions, quantum Hall droplets, and one current to rule them all. *SciPost Phys.* **2020**, *9*, 8. [CrossRef]
16. Karasik, A. Vector dominance, one flavored baryons, and QCD domain walls from the "hidden" Wess-Zumino term. *SciPost Phys.* **2021**, *10*, 138. [CrossRef]
17. Nadkarni, S.; Nielsen, H.B.; Zahed, I. Bosonization relations as bag boundary conditions. *Nucl. Phys. B* **1985**, *253*, 308. [CrossRef]
18. Ma, Y.L.; Nowak, M.A.; Rho, M.; Zahed, I. Baryon as a quantum Hall droplet and the hadron-quark duality. *Phys. Rev. Lett.* **2019**, *123*, 172301. [CrossRef]
19. Callan, C.G., Jr.; Harvey, J.A. Anomalies and fermion zero modes on strings and domain Walls. *Nucl. Phys. B* **1985**, *250*, 427. [CrossRef]
20. Ma, Y.L.; Rho, M. Mapping topology to nuclear dilaton-HLS effective field theory for dense baryonic matter. *arXiv* **2021**, arXiv:2103.01860.
21. Goldstone, J.; Jaffe, R.L. The baryon number in chiral bag models. *Phys. Rev. Lett.* **1983**, *51*, 1518. [CrossRef]
22. Harada, M.; Yamawaki, K. Hidden local symmetry at loop: A New perspective of composite gauge boson and chiral phase transition. *Phys. Rep.* **2003**, *381*, 1–233. [CrossRef]
23. Shaposhnikov, M.; Zenhausern, D. Quantum scale invariance, cosmological constant and hierarchy problem. *Phys. Lett. B* **2009**, *671*, 162–166. [CrossRef]
24. Kan, N.; Kitano, R.; Yankielowicz, S.; Yokokura, R. From 3d dualities to hadron physics. *Phys. Rev. D* **2020**, *102*, 125034. [CrossRef]

25. Komargodski, Z. Vector mesons and an interpretation of Seiberg duality. *J. High Energy Phys.* **2011**, *2011*, 1–21. [CrossRef]
26. Abel, S.; Barnard, J. Seiberg duality versus hidden local symmetry. *J. High Energy Phys.* **2012**, *2012*, 1–42. [CrossRef]
27. Kitano, R.; Nakamura, M.; Yokoi, N. Making confining strings out of mesons. *Phys. Rev. D* **2012**, *86*, 014510. [CrossRef]
28. Ma, Y.L.; Rho, M. Quenched g_A in nuclei and emergent scale symmetry in baryonic matter. *Phys. Rev. Lett.* **2020**, *125*, 142501. [CrossRef]
29. Rho, M. Multifarious roles of hidden chiral-scale symmetry: "Quenching" g_A in nuclei. *Symmetry* **2021**, *13*, 1388. [CrossRef]
30. Li, Y.L.; Ma, Y.L.; Rho, M. Chiral-scale effective theory including a dilatonic meson. *Phys. Rev. D* **2017**, *95*, 114011. [CrossRef]
31. Appelquist, T.; Ingoldby, J.; Piai, M. Nearly conformal composite Higgs model. *Phys. Rev. Lett.* **2021**, *126*, 191804. [CrossRef]
32. Beane, S.R.; van Kolck, U. The dilated chiral quark model. *Phys. Lett. B* **1994**, *328*, 137. [CrossRef]
33. Gell-Mann, M.; Lévy, M. The axial vector current in beta decay. *Nuovo Cim.* **1960**, *16*, 705. [CrossRef]
34. Yamawaki, K. Hidden local symmetry and beyond. *Int. J. Mod. Phys. E* **2017**, *26*, 1740032. [CrossRef]
35. Harada, M.; Lee, H.K.; Ma, Y.L.; Rho, M. Inhomogeneous quark condensate in compressed Skyrmion matter. *Phys. Rev. D* **2015**, *91*, 096011. [CrossRef]
36. Buballa, M.; Carignano, S. Inhomogeneous chiral symmetry breaking in dense neutron-star matter. *Eur. Phys. J. A* **2016**, *52*, 57. [CrossRef]
37. Mueller, E.J. Pseudogaps in strongly interacting Fermi gases. *Rep. Prog. Phys.* **2017**, *80*, 104401. [CrossRef]
38. Harada, M.; Sasaki, C.; Takemoto, S. Enhancement of quark number susceptibility with an alternative pattern of chiral symmetry breaking in dense matter. *Phys. Rev. D* **2010**, *81*, 016009. [CrossRef]
39. Wang, H.; Chudnovskiy, A.L.; Gorsky, A.; Kamenev, A. SYK superconductivity: quantum Kuramoto and generalized Richardson models. *Phys. Rev. Res.* **2020**, *2*, 033025. [CrossRef]
40. Sasaki, C.; Lee, H.K.; Paeng, W.G.; Rho, M. Conformal anomaly and the vector coupling in dense matter. *Phys. Rev. D* **2011**, *84*, 034011. [CrossRef]
41. Ma, Y.L.; Harada, M.; Lee, H.K.; Oh, Y.; Park, B.Y.; Rho, M. Dense baryonic matter in the hidden local symmetry approach: Half-skyrmions and nucleon mass. *Phys. Rev. D* **2013**, *88*, 014016. [CrossRef]
42. Nielsen, H.B.; Rho, M.; Wirzba, A.; Zahed, I. Color anomaly in a hybrid bag model. *Phys. Lett. B* **1991**, *269*, 389. [CrossRef]
43. Lee, H.J.; Min, D.P.; Park, B.Y.; Rho, M.; Vento, V. The proton spin in the chiral bag model: Casimir contribution and Cheshire cat principle. *Nucl. Phys. A* **1999**, *657*, 75. [CrossRef]
44. Berry, D.K.; Caplan, M.E.; Horowitz, C.J.; Huber, G.; Schneider, A.S. 'Parking-garage' structures in nuclear astrophysics and cellular biophysics. *Phys. Rev. C* **2016**, *94*, 055801. [CrossRef]
45. Park, B.Y.; Paeng, W.G.; Vento, V. The Inhomogeneous phase of dense skyrmion matter. *Nucl. Phys. A* **2019**, *989*, 231. [CrossRef]
46. Canfora, F. Ordered arrays of baryonic tubes in the Skyrme model in $(3+1)$ dimensions at finite density. *Eur. Phys. C* **2018**, *78*, 929. [CrossRef]
47. Canfora, F.; Lagos, M.; Vera, A. Crystals of superconducting baryonic tubes in the low energy limit of QCD at finite density. *Eur. Phys. J. C* **2020**, *80*, 697. [CrossRef]
48. Hu, Y.; Jain, J.K. Kohn-Sham theory of the fractional quantum Hall effect. *Phys. Rev. Lett.* **2019**, *123*, 176802. [CrossRef]
49. Zhang, P.; Kimm, K.; Zou, L.; Cho, Y.M. Re-interpretation of Skyrme theory: New topological structures. *arXiv* **2017**, arXiv:1704.05975.
50. Gudnason, S.B.; Nitta, M. Fractional skyrmions and their molecules. *Phys. Rev. D* **2015**, *91*, 085040. [CrossRef]
51. Sulejmanpasic, T.; Shao, H.; Sandvik, A.; Unsal, M. Confinement in the bulk, deconfinement on the wall: infrared equivalence between compactified QCD and quantum magnets. *Phys. Rev. Lett.* **2017**, *119*, 091601. [CrossRef]
52. Annala, E.; Gorda, T.; Kurkela, A.; Nättilä, J.; Vuorinen, A. Evidence for quark-matter cores in massive neutron stars. *Nat. Phys.* **2020**, *16*, 907–910. [CrossRef]
53. Ma, Y.L.; Rho, M. The sound speed and core of massive compact stars: A manifestation of hadron-quark duality. *arXiv* **2021**, arXiv:2104.13822.
54. Schäfer, T.; Wilczek, F. Continuity of quark and hadron matter. *Phys. Rev. Lett.* **1999**, *82*, 3956. [CrossRef]
55. Cherman, A.; Jacobson, T.; Sen, S.; Yaffe, L.G. Higgs-confinement phase transitions with fundamental representation matter. *arXiv* **2020**, arXiv:2007.08539.
56. Rho, M.; Ma, Y.L. Manifestation of hidden symmetries in baryonic matter: From finite nuclei to neutron stars. *Mod. Phys. Lett. A* **2021**, *36*, 2130012. [CrossRef]
57. Hong, D.K.; Rho, M.; Yee, H.U.; Yi, P. Dynamics of baryons from string theory and vector dominance. *J. High Energy Phys.* **2007**, *2007*, 063. [CrossRef]
58. Kitano, R.; Matsudo, R. Vector mesons on the wall. *J. High Energy Phys.* **2021**, *2021*, 1–27. [CrossRef]
59. Brown, G.E.; Harada, M.; Holt, J.W.; Rho, M.; Sasaki, C. Hidden local field theory and dileptons in relativistic heavy ion collisions. *Prog. Theor. Phys.* **2009**, *121*, 1209–1236. [CrossRef]
60. Braun-Munzinger, P.; Koch, V.; Schäfer, T.; Stachel, J. Properties of hot and dense matter from relativistic heavy ion collisions. *Phys. Rep.* **2016**, *621*, 76–126. [CrossRef]
61. Schäfer, T.; Baym, G. From nuclear to unnuclear physics. *arXiv* **2021**, arXiv:2109.06924.
62. Georgi, H. Unparticle physics. *Phys. Rev. Lett.* **2007**, *98*, 221601. [CrossRef]

symmetry MDPI

Review

Introduction to Renormalization Theory and Chiral Gauge Theories in Dimensional Regularization with Non-Anticommuting γ_5

Her mès Bélusca-Maïto [1], Amon Ilakovac [1], Paul Kühler [2], Marija Mađor-Božinović [1], Dominik Stöckinger [2,*] and Matthias Weißwange [2]

[1] Department of Physics, University of Zagreb, Bijenička cesta 32, HR-10000 Zagreb, Croatia
[2] Institut für Kern- und Teilchenphysik, TU Dresden, Zellescher Weg 19, DE-01069 Dresden, Germany
* Correspondence: dominik.stoeckinger@tu-dresden.de

Abstract: This review provides a detailed introduction to chiral gauge theories, renormalization theory, and the application of dimensional regularization with the non-anticommuting BMHV scheme for γ_5. One goal was to show how chiral gauge theories can be renormalized despite the spurious breaking of gauge invariance and how to obtain the required symmetry-restoring counterterms. A second goal was to familiarize the reader with the theoretical basis of the renormalization of chiral gauge theories, the theorems that guarantee the existence of renormalized chiral gauge theories at all orders as consistent quantum theories. Relevant topics include BPHZ renormalization, Slavnov–Taylor identities, the BRST formalism, and algebraic renormalization, as well as the theorems guaranteeing that dimensional regularization is a consistent regularization/renormalization scheme. All of these, including their proofs and interconnections, are explained and discussed in detail. Further, these theoretical concepts are illustrated in practical applications with the example of an Abelian and a non-Abelian chiral gauge theory. Not only the renormalization procedure for such chiral gauge theories is explained step by step, but also the results of all counterterms, including the symmetry-restoring ones, necessary for the consistent renormalization, are explicitly provided.

Keywords: renormalization; chiral gauge theories; dimensional regularization

PACS: 11.10.Gh; 11.15.-q; 12.15.Lk; 12.38.Bx

1. Introduction

Except for gravity, all known fundamental particles and interactions in nature are described by quantum gauge theories. The Standard Model (SM) of particle physics combines the theories for electromagnetic, weak, and strong interactions. It is based on the gauge group SU(3) × SU(2) × U(1) and includes fermionic fields describing spin 1/2 quarks and leptons and bosonic fields describing the Higgs boson and electroweak symmetry breaking.

Exact solutions for quantum gauge theories rarely exist. Often, SM predictions can be successfully evaluated in a perturbative treatment. Based on known exact solutions of the free non-interacting quantum field theory, higher-order corrections can be evaluated step by step. The higher-order corrections lead to Feynman diagrams with closed loops and momentum integrations, which lead to ultraviolet divergences. Therefore, the higher-order amplitudes have to be regularized and renormalized. Equivalently, a mathematically rigorous treatment has to inductively construct higher orders from lower orders, where the construction has to respect fundamental requirements such as causality, Lorentz invariance, and the unitarity of the time evolution. Practical regularization/renormalization prescriptions that agree with such a rigorous approach are called consistent schemes.

For so-called vector gauge theories, in which left-handed and right-handed fermions have the same gauge interactions, an essentially perfect regularization/renormalization framework is provided by dimensional regularization [1–3]. It is not only consistent in the sense above, but it also manifestly preserves the fundamental gauge invariance at all steps of the calculations. Further, a useful practical tool is provided by the validity of the quantum action principle [4], which enables the straightforward study of symmetries and equations of motion on the level of Green functions. Alternative consistent schemes such as analytic renormalization or Pauli–Villars regularization break gauge invariance. For the status of further modern developments of alternative schemes, we refer to Reference [5].

However, a fundamental discovery of elementary particle physics is that electroweak interactions act on chiral fermions, i.e., they treat left-handed and right-handed fermions differently. Accordingly, the SM and all its extensions for potential new physics are chiral gauge theories, in which left-handed and right-handed fermions interact differently with gauge bosons. The presence of such chiral fermions and chiral interactions is manifested through phenomena such as non-conservation of parity and charge conjugation invariance of the weak interactions. Connected with chiral fermions is the possibility of chiral anomalies [6–8], i.e., the possibility that classically conserved currents are not conserved in the full quantum theory. Chiral anomalies lead to observed phenomena such as neutral pion decay into two photons. Chiral gauge theories, however, can only be consistently renormalized if chiral anomalies in currents coupling to gauge fields cancel. Although the cancellation is valid in the SM [9–11], the potential presence of chiral anomalies makes it impossible to define a consistent regularization/renormalization procedure that manifestly preserves all symmetries involving chiral fermions. A particularly transparent analysis can be given in terms of the non-invariance of the fermion path integral measure [12,13].

Within chiral models, dimensional regularization schemes meet the so-called "γ_5-problem", which is a consequence of the fact that γ_5 (similarly, the Levi-Civita tensor $\epsilon_{\mu\nu\rho\sigma}$) is an intrinsically four-dimensional quantity. The three basic properties, anticommutativity of γ_5 with other γ^μ matrices, cyclicity of traces, and the nonzero trace of products of γ_5 with four different γ^μ-matrices, cannot be simultaneously retained without spoiling the consistency of the scheme. The usage of the naive scheme [14], including the γ_5 anticommutativity, is the most common in practical calculations, but it is restricted to subclasses of diagrams [14,15], and within it, the γ_5-matrix is ambiguously defined. Giving up the cyclicity of the trace, one has to introduce a consistent reading prescription defining combinations of reading points for evaluations of noncyclic traces [16–18], which makes the mathematical consistency of higher orders less transparent and questionable. Abandoning the anticommutativity of the γ_5-matrix [1,2,19–21] leads to the mathematically most-rigorously established dimensional regularization scheme, the so-called Breitenlohner–Maison/'t Hooft–Veltman (BMHV) scheme, for which all basic quantum field theory properties were proven to be valid [4,22–24].

Unfortunately, in the BMHV scheme with non-anticommuting γ_5, some of the advantages of dimensional regularization are lost. In particular, gauge invariance is not manifestly valid in chiral gauge theories, reflecting the possibility of anomalies. Even if the actual anomalies cancel, as in the SM, gauge invariance is broken in intermediate steps, and the breaking has to be compensated by a more complicated renormalization procedure. Instead of the typical textbook approach of generating a bare Lagrangian and counterterms by a renormalization transformation of fields and parameters, specific symmetry-restoring counterterms of a more general structure need to be found and included. Several recent works have begun to systematically investigate the practical application of the BMHV scheme to chiral gauge theories and determine such counterterms [25–28]; see also Reference [29] for a compact summary.

The present review provides a detailed introduction into chiral gauge theories, dimensional regularization, renormalization theory, and the application of the BMHV scheme to chiral gauge theories. Its intentions and motivations can be summarized as follows:

- We aimed for a pedagogical review, starting at the level of typical quantum field theory textbooks and containing detailed step-by-step explanations and illustrative examples.
- On a practical level, we show how chiral gauge theories can be renormalized employing the BMHV scheme for γ_5 and how the required symmetry-restoring counterterms can be obtained and used. Thus, we also provide an introduction to the recent literature mentioned above. The general motivation is an increasing need for high-precision (multi-)loop calculations in the SM and beyond and an increasing interest in mathematically rigorous treatments that avoid pitfalls such as inconsistencies, ambiguities, or incorrect results.
- On a conceptual level, we discuss the theoretical basis of the renormalization of chiral gauge theories. The existence of renormalized quantum gauge theories at all orders, together with their physics interpretation, is a major result in theoretical physics. It is based on a large set of complicated theorems and formalisms, ranging from BPHZ theorems on causal and unitary renormalization to Slavnov–Taylor identities and the BRST formalism, the theorems of algebraic renormalization, and to the theorems guaranteeing that dimensional regularization is a consistent regularization/renormalization scheme. All these relevant theorems, their role, and their interconnections are discussed and explained in detail. The proofs are either given or illustrated and explained.
- In line with the pedagogical goals, we used extensive cross-referencing between sections. Wherever possible, introductory sections develop intuition and expectations of later steps, and later sections refer back to simpler, more qualitative explanations and illustrations. In our citations, we cite not only original works, but wherever possible, we also cite textbooks or other reviews, where further details can be found. References to the remarks made at the beginning of this Introduction can be found in the appropriate sections.

In the following, we present an extensive outline of the individual sections.

In Section 2, the basic knowledge necessary for a discussion of chiral gauge theories in dimensional regularization is presented:

- Beginning with key ingredients, first, non-Abelian Yang–Mills gauge theories and spinors, chirality, and chiral fermions are introduced, including required notions from Lie group theory and Poincaré group representations. BRST invariance and a corresponding Slavnov–Taylor identity are discussed in detail already at the classical level. Turning to the quantum level, the notions needed for discussions of Green functions and their generating functionals are introduced. Then, Slavnov–Taylor identities for Green functions and generating functionals are introduced, derived from the path integral and interpreted in detail. The concluding subsection considers the case of an Abelian gauge theory, and simplifications and additionally valid equations compared to the non-Abelian case are shown.

Section 3 gives a detailed introduction to dimensional regularization as a mathematically well-defined regularization procedure, which allows efficient computations and preserves basic properties of quantum field theory:

- As a preview and to set the stage, the general structure of dimensional regularization, renormalization, and the counterterms, as well as corresponding notations are presented. Then, D-dimensional extensions of four-dimensional quantities are discussed, starting with the notion of the quasi D-dimensional space. The core of the method is D-dimensional integrals. After listing their properties relevant for practical calculations, they are mathematically constructed in two ways, using parallel and orthogonal spaces, as well as via Schwinger parametrization. The definition and properties of the metric tensor and its inverse are given. Of particular importance for chiral gauge theories are the definitions and properties of D-dimensional γ matrices. Here, an explicit construction of quasi-D-dimensional γ matrices is provided, which is optimized for the study of chiral gauge theories. The extension to D dimensions leads to the well-known γ_5 problem; this problem is explained, and the BMHV scheme is presented together with its definitions and properties of the γ_5 matrix and the $\epsilon_{\mu\nu\rho\sigma}$ symbol.

- In addition to defining the regularization and constructing its basic elements, the relationship of regularized Feynman rules to Lagrangians in D dimensions via a D-dimensional Gell–Mann–Low formula is discussed. Special emphasis is put on the relation between kinetic terms and corresponding propagators and chiral fermion–gauge boson interactions. As an outlook and somewhat orthogonal topic, the variants HV, CDR, DRED, and FDH of dimensional regularization schemes are briefly discussed. Their distinctions are of particular importance in the context of infrared divergences and in the context of supersymmetric gauge theories.

In Section 4, the quantum action principle and regularized quantum action principle in dimensional regularization are introduced. This is a set of relations between variations of the classical action and variations of the Green functions of the resulting quantum theory, which allow expressing symmetries and symmetry violations of the regularized or renormalized theory:

- First, an instructive, but formal derivation from the path integral is given, sidestepping the need for regularization and renormalization. Then, an exact proof of the regularized quantum action principle within dimensional regularization is presented. This validity constitutes an important advantage of dimensional regularization. Its role is illustrated by proving rigorously the all-order validity of the Slavnov–Taylor identity for QCD and explaining the extent of the validity of supersymmetry in the DRED scheme.

Section 5 is devoted to general renormalization theory, focusing on aspects not yet specific to gauge theories. One goal is to explain the rigorous theorems guaranteeing that the regularization, renormalization, and cancellation of divergences are possible, and physically sensible quantum field theories can be constructed at all orders. A second goal is to analyze conditions for consistent regularization/renormalization procedures and to show how we know that dimensional regularization is one such consistent procedure:

- Renormalization is introduced as a mathematical construction of time-ordered products of free field operators in agreement with the unitarity and causality of the perturbative S-matrix. The "main theorem" of renormalization relates the construction and its ambiguities to reparametrizations. Importantly, the ambiguities and the reparametrizations are local in a well-defined sense. The relationships between the BPH approach and the R-operation, the BPHZ approach and the forest formula, and the usual counterterm approach are explained. Further, analytic regularization is discussed as a conceptually interesting non-dimensional regularization scheme that can facilitate all-order proofs.
- In the second subsection, the main theorem on dimensional regularization is reviewed. First, an extensive discussion of the main statements is given; the most important is the applicability of dimensional regularization as a consistent regularization/renormalization framework. Then, the proof is sketched in detail. The first steps set up Feynman graph theoretical notions, an organization of the loop integrations, and an optimized forest formula. Then, the resulting integrals are investigated in detail, and an inductive proof can be given. All steps are explained and illustrated with examples.

With the fundamentals of regularization and renormalization thus established, Section 6 goes on to consider the case of quantized gauge theories and their renormalization. It focuses on the compatibility of BRST invariance and Slavnov–Taylor identities, which are vital for the correct physical interpretation of gauge theories, and the regularization/renormalization procedure, which may in general spoil symmetries:

- Revisiting first the familiar textbook case of a symmetry-preserving regularization such as in QED or QCD reminds the reader of practically important concepts such as

renormalization transformations and puts into context the symmetry-breaking case, which is the central topic of this review.
- Focusing on this case of interest, the theory of algebraic renormalization is reviewed as the framework in which rigorous and elegant proofs of the renormalizability of gauge theories can be carried out, even if regularization procedures break symmetries. The quantum action principle of BPHZ renormalization emerges as the main theoretical tool of this framework; hence, a brief exposition of this tool is given, and its connection to the quantum action principle in dimensional regularization is explained. The section then illustrates the inductive all-order proof of the restoration of the spuriously broken symmetry by symmetry-preserving finite counterterms. It also includes a brief discussion of anomalies, their cancellation conditions, and an outlook on further applications of algebraic renormalization.
- Finally, coming to the practical goal of this review, the formalism is specialized to dimensional regularization. Here, explicit equations for the computation of symmetry-preserving counterterms are derived and the resulting structure of the counterterm Lagrangian is discussed.

Section 7 gives a detailed illustration of the treatment of chiral gauge theories in the BMHV scheme, using concrete examples:
- It focuses mainly on an Abelian example, a chiral QED model, discusses its structure, symmetry breaking as the result of the scheme, and the required counterterm structure. It explains and compares several ways to determine the required symmetry-restoring counterterms in practical calculations.
- The symmetry restoration is illustrated in detail for the photon self-energy case, where it becomes apparent how the quantum action principle and Ward identities have a crucial practical role in the calculations.
- For the chiral QED model, the calculations are generalized to the full one-loop and the full two-loop level, and the new features arising at the two-loop level are discussed.
- Finally, a detailed comparison of the Abelian chiral QED and a chiral non-Abelian Yang–Mills theory is given at the one-loop level.

2. Setup

In this section, we collect background information on the main theoretical concepts needed to discuss the renormalization of chiral gauge theories in dimensional regularization. We begin with the general notions of Yang–Mills gauge theories and of spinors, γ^μ-matrices, and chirality. On the level of classical field theory, gauge invariance is then extended to BRST invariance, including gauge fixing and Faddeev–Popov ghosts, and BRST invariance is formulated as a Slavnov–Taylor identity (Sections 2.1–2.3). In Section 2.4, the basic objects of quantized field theories, Green functions, and generating functionals are defined. Section 2.5 discusses the role and interpretation of the Slavnov–Taylor identity on the level of the quantum field theory. Finally, Section 2.6 discusses the case of Abelian gauge theories, which involves additional identities. Much of the material of this section can also be found in standard textbooks such as References [30–36].

2.1. Yang–Mills Gauge Theories

We begin by summarizing the construction of general Yang–Mills gauge theories with simple gauge groups such as SU(N) or SO(N) and with generic matter fields.

The first ingredient is the gauge group. It is a Lie group in which all group elements can be written as continuous functions of a certain number N_{gen} of parameters. The Lie group can be associated with a Lie algebra with N_{gen} generators, called t^a, $a = 1 \ldots N_{\text{gen}}$. The generators satisfy the commutation relations:

$$[t^a, t^b] = i f^{ab}_c t^c \tag{1}$$

with antisymmetric structure constants f_c^{ab}. There exists a set of generators for which the structure constants are totally antisymmetric, such that we write $f_c^{ab} \equiv f_{abc} \equiv f^{abc}$. This is the case for sums of simple compact and $U(1)$ subalgebras; see, e.g., [32]. Any set of matrices T^a that satisfy the relation (1) is called a representation of the Lie algebra.

One special representation, the so-called adjoint representation, always exists. It is defined by

$$(T^a_{\text{adj}})_{ij} = -if^{aij} \tag{2}$$

and, thus, a representation in terms of $N_{\text{gen}} \times N_{\text{gen}}$ matrices. The commutation relation (1) is fulfilled because of the Jacobi identity of commutators.

For any representation of the Lie algebra, we can form a representation of the Lie group (at least locally in a region around the identity) by exponentiation:

$$U(\theta^a) = e^{-ig\theta^a T^a} \tag{3}$$

where θ^a are real parameters and where g is the gauge coupling.

Once the Lie group and Lie algebra are defined, we assumed the existence of N_F so-called matter fields $\varphi_i(x), i = 1 \ldots N_F$. We collectively denote them as a tuple $\varphi = (\varphi_i)$. We further assumed that there exists a representation of the Lie algebra in terms of $N_F \times N_F$ matrices T^a, and we define *(local) gauge transformations* of the matter fields as

$$\varphi_i(x) \to U(\theta^a(x))_{ij}\varphi_j(x)\,. \tag{4}$$

The representation may be reducible or irreducible. To simplify the notation, we will often suppress the indices and arguments and write the previous equation as

$$\varphi \to U\varphi\,. \tag{5}$$

Next, we introduce the central elements of Yang–Mills gauge theories: the covariant derivative D_μ and the gauge fields A^a_μ. They are related as

$$D_\mu = \partial_\mu + igT^a A^a_\mu \tag{6}$$

where g is the gauge coupling. As the notation indicates, there is one vector field A^a_μ for each generator $a = 1 \ldots N_{\text{gen}}$. The relation (6) is valid for any representation, and the vector fields A^a_μ are independent of the chosen representation. It is often useful to define the matrix-valued and representation-dependent gauge field $A_\mu \equiv T^a A^a_\mu$.

The fundamental requirement is that, under a gauge transformation, the covariant derivative behaves as

$$D_\mu\varphi \to UD_\mu\varphi\,. \tag{7}$$

This is valid if and only if the matrix-valued gauge field transforms as

$$A_\mu \to UA_\mu U^{-1} - \frac{1}{ig}[\partial_\mu U]U^{-1}\,. \tag{8}$$

Finally, the field strength tensor can be defined as

$$F_{\mu\nu} = \frac{1}{ig}[D_\mu, D_\nu]\,. \tag{9}$$

With this definition, the field strength tensor is matrix-valued and dependent on the chosen representation. We can decompose it as $F_{\mu\nu} = T^a F^a_{\mu\nu}$ and evaluate the previous definition with the result:

$$F^a_{\mu\nu} = \partial_\mu A^a_\nu - \partial_\nu A^a_\mu - g f^{abc} A^b_\mu A^c_\nu \,. \tag{10}$$

Here, we see that the field strength tensors $F^a_{\mu\nu}$ are independent of the chosen representation and are generalizations of the field strength tensor of electrodynamics.

At this point, we collect all gauge transformations in compact and matrix-valued form as

$$\varphi \to U\varphi \,, \tag{11a}$$
$$D_\mu \varphi \to U D_\mu \varphi \,, \tag{11b}$$
$$A_\mu \to U A_\mu U^{-1} - \frac{1}{ig}[\partial_\mu U] U^{-1} \,, \tag{11c}$$
$$F_{\mu\nu} \to U F_{\mu\nu} U^{-1} \,, \tag{11d}$$

where the last equation directly follows from the definition (9). We also record the gauge transformations for the fundamental fields in more explicit form, by taking the parameters θ^a to be infinitesimal, as

$$\varphi \to \varphi - ig\theta^a T^a \varphi \,, \tag{12a}$$
$$A_\mu \to A_\mu + \partial_\mu \theta - ig[\theta, A_\mu] \,, \tag{12b}$$
$$A^a_\mu \to A^a_\mu + \partial_\mu \theta^a + g f^{abc} \theta^b A^c_\mu \,, \tag{12c}$$

where we also set $\theta = T^a \theta^a$. The last of the previous equations is particularly important. It holds universally for any representation. It also contains the gauge coupling g. This is at the heart of the universality of the gauge coupling, i.e., the physical statement that one single gauge coupling governs all interactions of the gauge bosons with other gauge bosons and with any matter fields. Note that this statement relies on the assumption of a simple non-Abelian gauge group.

The renormalizable gauge-invariant Lagrangian for this Yang–Mills theory can be written as

$$\mathcal{L}_{\text{inv}} = \mathcal{L}_{\text{YM}} + \mathcal{L}_{\text{mat}} \,, \tag{13a}$$
$$\mathcal{L}_{\text{YM}} = -\frac{1}{4} F^{a\mu\nu} F^a_{\mu\nu} \,, \tag{13b}$$
$$\mathcal{L}_{\text{mat}} = \mathcal{L}_{\text{mat}}(\varphi, D_\mu \varphi) \,. \tag{13c}$$

The concrete form of the matter field Lagrangian depends on details such as the spin of the matter field and interactions between different matter fields.

2.2. Chiral Fermions

In this subsection, we introduce the next ingredient: chiral fermions. A fundamental discovery of elementary particle physics is that electroweak interactions fundamentally act on chiral fermions, i.e., they treat left-handed and right-handed fermions differently. Chiral fermions are also fundamental building blocks in many extensions of the Standard Model, such as grand unified theories or supersymmetry.

Here, we will first summarize general properties of four-component, or Dirac or Majorana spinors in four dimensions, and then, define the notion of chirality in this context. Thereafter, we also introduce the two-component Weyl/van der Waerden spinor notation, which allows an efficient understanding of many important relationships. We will then collect such relationships.

2.2.1. General Representation-Independent Relations for γ-Matrices and Four-Spinors

Spinors are defined via their properties under Lorentz transformations. Therefore, we begin with the reminder that a Lorentz transformation of ordinary 4-vectors is defined by a matrix $\Lambda^{\mu}{}_{\nu}$, which leaves scalar products of 4-vectors invariant. Infinitesimal Lorentz transformations are given by matrices of the form $\Lambda^{\mu}{}_{\nu} = \delta^{\mu}{}_{\nu} + \omega^{\mu}{}_{\nu}$ with an infinitesimal, antisymmetric matrix $\omega_{\mu\nu}$. A representation of the Lorentz group $U(\Lambda)$ is (at least locally) defined by specifying

$$U(\delta + \omega) = 1 - \frac{i}{2}\omega_{\mu\nu}J^{\mu\nu} \tag{14}$$

with generators $J^{\mu\nu}$ that must satisfy the commutation relations of the corresponding Lie algebra:

$$[J^{\mu\nu}, J^{\rho\sigma}] = i(g^{\nu\rho}J^{\mu\sigma} - g^{\mu\rho}J^{\nu\sigma} + g^{\mu\sigma}J^{\nu\rho} - g^{\nu\sigma}J^{\mu\rho})\,. \tag{15}$$

Now, we can turn to spinors. The basic building blocks of four-component spinor theory are the γ^{μ}-matrices. They are 4×4 matrices satisfying the defining Clifford algebra relation:

$$\{\gamma^{\mu}, \gamma^{\nu}\} = 2g^{\mu\nu}\mathbb{1}\,. \tag{16}$$

Here and everywhere else, we use the mostly minus metric. The fundamental importance of these matrices is that they generate a representation of the Lorentz group. Indeed, setting

$$S^{\mu\nu} = \frac{i}{4}[\gamma^{\mu}, \gamma^{\nu}]\,, \tag{17}$$

one can show that these $S^{\mu\nu}$ satisfy the required commutation relations (15). Hence, we can now define the notion of a four-component (Dirac or Majorana) spinor: a four-component spinor ψ is an object whose Lorentz transformation properties are given by

$$\psi \overset{\Lambda=\delta+\omega}{\longrightarrow} \left(1 - \frac{i}{2}\omega_{\mu\nu}S^{\mu\nu}\right)\psi\,. \tag{18}$$

In addition to the γ^{μ}-matrices, the γ_5 matrix and projection operators $P_{L,R}$ are defined as

$$\gamma_5 = i\gamma^0\gamma^1\gamma^2\gamma^3 = -\frac{i}{4!}\epsilon_{\mu\nu\rho\sigma}\gamma^{\mu}\gamma^{\nu}\gamma^{\rho}\gamma^{\sigma}\,, \qquad P_{L,R} = \frac{1}{2}(\mathbb{1} \mp \gamma_5)\,, \tag{19}$$

with the totally antisymmetric Levi-Civita (pseudo-)tensor $\epsilon_{\mu\nu\rho\sigma}$ with $\epsilon_{0123} = -1$. These matrices satisfy the additional equations:

$$\{\gamma^{\mu}, \gamma_5\} = 0\,, \qquad (\gamma_5)^2 = \mathbb{1}\,, \qquad (P_{L,R})^2 = P_{L,R}\,, \qquad P_L P_R = 0\,. \tag{20}$$

Though not required in general, in many representations (including the chiral representation introduced below), the relations

$$(\gamma^{\mu})^{\dagger} = \gamma^0\gamma^{\mu}\gamma^0\,, \qquad (\gamma^{\mu})^* = \gamma^2\gamma^{\mu}\gamma^2\,, \qquad (\gamma^{\mu})^T = -C^{-1}\gamma^{\mu}C\,, \qquad C = i\gamma^0\gamma^2 \tag{21}$$

hold. In particular, γ^2 is the only imaginary matrix. We will assume these relations in the following.

For any four-spinor ψ, we can define an adjoint spinor $\bar{\psi}$ and a charge-conjugated spinor ψ^C by

$$\bar{\psi} = \psi^{\dagger}\gamma^0\,, \qquad \psi^C = C\bar{\psi}^T\,. \tag{22}$$

In this way, ψ^C is also a four-spinor satisfying the transformation rule (18), and $\bar{\psi}$ transforms with the inverse matrix. One can show that bilinear expressions such as $\bar{\psi}_1\psi_2$, $\bar{\psi}_1\gamma^{\mu}\psi_2$ transform as Lorentz scalars and four-vectors, respectively.

2.2.2. Chirality and Chiral Fermions

At the level of four-component spinors, the concept of chirality is related to the γ_5 matrix and the projectors $P_{L,R}$. Let us define for any four-spinor ψ so-called left-handed and right-handed spinors by

$$\psi_L = P_L\psi, \qquad \psi_R = P_R\psi. \tag{23}$$

Then, we can make three observations:

- The matrix $\gamma_5^2 = \mathbb{1}$. Hence, the eigenvalues of γ_5 are ± 1.
- The spinors ψ_L and ψ_R are eigenspinors of γ_5 with eigenvalues $-1, +1$, respectively.
- The matrix γ_5 and the projectors $P_{L,R}$ commute with the Lorentz generators $S^{\mu\nu}$.

Hence, the left-handed and right-handed spinors are proper spinors in the sense of Equation (18), and they form two distinct invariant subspaces of the Lorentz representation: the representation defined by Equations (17) and (18) is reducible.

We refer to the eigenvalue of γ_5 as chirality; the left-handed and right-handed spinors are chiral, or chirality eigenstates. In view of the above, chirality is a Lorentz-invariant property, and its existence is linked to the structure of the Lorentz group representation theory. For the general analysis, we refer to Reference [37] and, in particular, the textbooks by Weinberg, Srednicki, and Ryder [31,35,38]. The spaces of the left-handed and right-handed spinors each define an irreducible representation of the Lorentz group; these are the simplest nontrivial representations, which are commonly known as the $(\frac{1}{2},0)$ and $(0,\frac{1}{2})$ representations.

Slightly reformulating the previous statements, we may say that the left-handed and right-handed spinors have different, independent Lorentz transformation properties. Hence, in a Lorentz-invariant field theory, left-handed or right-handed spinor fields may appear independently. Specifically, gauge theories may be constructed in which left-handed or right-handed spinor fields appear with different gauge group representations. This is precisely what happens in the case of the electroweak interactions. Chiral fermions are the fermions described by such field theories based on chiral spinor fields.

2.2.3. Chiral Representation and Two-Component Spinor Formalism

Although many important relationships hold independently of any specific representation of the γ^μ-matrices, it is useful to introduce here the so-called chiral representation, which is given as follows by 2×2 block matrices:

$$\gamma^\mu = \begin{pmatrix} 0 & \sigma^\mu \\ \overline{\sigma}^\mu & 0 \end{pmatrix}, \quad \gamma^5 = \begin{pmatrix} -1 & 0 \\ 0 & 1 \end{pmatrix}, \quad P_L = \begin{pmatrix} 1 & 0 \\ 0 & 0 \end{pmatrix}, \quad P_R = \begin{pmatrix} 0 & 0 \\ 0 & 1 \end{pmatrix}. \tag{24}$$

This representation uses the Pauli matrices:

$$\sigma^1 = \begin{pmatrix} 0 & 1 \\ 1 & 0 \end{pmatrix}, \quad \sigma^2 = \begin{pmatrix} 0 & -i \\ i & 0 \end{pmatrix}, \quad \sigma^3 = \begin{pmatrix} 1 & 0 \\ 0 & -1 \end{pmatrix} \tag{25}$$

and the following four-vectors of 2×2 matrices:

$$\sigma^\mu = (1, \sigma^k), \qquad \overline{\sigma}^\mu = (1, -\sigma^k). \tag{26}$$

In this representation of γ-matrices, the Lorentz generators (17) take the form:

$$S^{\mu\nu} = \begin{pmatrix} \frac{i}{4}(\sigma^\mu\overline{\sigma}^\nu - \sigma^\nu\overline{\sigma}^\mu) & 0 \\ 0 & \frac{i}{4}(\overline{\sigma}^\mu\sigma^\nu - \overline{\sigma}^\nu\sigma^\mu) \end{pmatrix}. \tag{27}$$

The block structure of all these matrices makes manifest that the Lorentz representation is reducible and that the left-handed and right-handed spinor spaces are invariant under Lorentz transformations.

This block structure of the chiral representation suggests introducing individual two-component spinors for the left-handed and right-handed parts. In the following, we will briefly introduce the corresponding two-component spinor formalism, which allows a very transparent formulation for many important and useful equations.

We mention that a systematic theory of the Lorentz group representations automatically leads first to such 2-component spinors as the natural spinors for the $(\frac{1}{2},0)$ and $(0,\frac{1}{2})$ representations and that, in such a context, the 4-component spinors appear as secondary objects. We also refer to the review [39] for an excellent account of 2-spinors and relationships between formalisms and relationships between different conventions.[1]

To avoid confusion, in the remainder of the present subsection, we will always denote 4-component spinors with capital Greek letters such as Ψ and 2-component spinors with lower-case Greek letters such as χ, η. The relationship between a 4-component spinor Ψ and 2-component spinors is given by the decomposition:

$$\Psi = \begin{pmatrix} \chi_\alpha \\ \bar{\eta}^{\dot{\alpha}} \end{pmatrix}. \tag{28}$$

Here, the indices $\alpha = 1,2$ and $\dot{\alpha} = 1,2$ and χ_α and $\bar{\eta}^{\dot{\alpha}}$ are two distinct two-component spinors. For the two-component spinors, we define Hermitian conjugation as

$$\bar{\chi}^{\dot{\alpha}} = (\chi^\alpha)^\dagger, \qquad \overline{\chi}_{\dot{\alpha}} = (\chi_\alpha)^\dagger, \tag{29}$$

and the raising and lowering of indices as

$$\chi_\alpha = \epsilon_{\alpha\beta}\chi^\beta, \qquad \chi^\alpha = \epsilon^{\alpha\beta}\chi_\beta, \qquad \overline{\chi}_{\dot{\alpha}} = \epsilon_{\dot{\alpha}\dot{\beta}}\overline{\chi}^{\dot{\beta}}, \qquad \overline{\chi}^{\dot{\alpha}} = \epsilon^{\dot{\alpha}\dot{\beta}}\overline{\chi}_{\dot{\beta}}, \tag{30}$$

with the antisymmetric symbol:

$$\epsilon_{\alpha\beta} = -\epsilon_{\beta\alpha}, \quad \epsilon_{\dot{\alpha}\dot{\beta}} = -\epsilon_{\dot{\beta}\dot{\alpha}}, \quad \epsilon^{\alpha\beta} = \epsilon_{\beta\alpha}, \quad \epsilon^{\dot{\alpha}\dot{\beta}} = \epsilon_{\dot{\beta}\dot{\alpha}}, \quad \epsilon^{12} = 1, \quad \epsilon^{\dot{1}\dot{2}} = 1. \tag{31}$$

The Lorentz transformations of the original 4-spinors induce how the 2-spinors transform. For an infinitesimal Lorentz transformation matrix $\Lambda = \delta + \omega$, we can define the 2×2 matrix:

$$M(\delta+\omega)_\alpha{}^\beta \equiv 1 - \frac{i}{2}\omega_{\mu\nu}\left(\frac{i}{4}(\sigma^\mu\bar{\sigma}^\nu - \sigma^\nu\bar{\sigma}^\mu)\right)_\alpha{}^\beta \tag{32}$$

in accordance with the general Equation (14). The explicit form of $S^{\mu\nu}$ in Equation (27) shows that this is the transformation matrix for two-spinors χ_α. The matrix M is a general complex invertible matrix with $\det(M) = 1$, i.e., M is an element of the group SL(2,$\mathbb{C}$). Elementary computations involving raising and lowering of indices and inspection of $S^{\mu\nu}$ show that, in total, the four kinds of two-spinors transform as follows:

$$\chi_\alpha \to (M)_\alpha{}^\beta\, \chi_\beta, \tag{33a}$$
$$\bar{\eta}^{\dot{\alpha}} \to (M^{-1\dagger})^{\dot{\alpha}}{}_{\dot{\beta}}\, \bar{\eta}^{\dot{\beta}}, \tag{33b}$$
$$\chi^\alpha \to \chi^\beta\, (M^{-1})_\beta{}^\alpha, \tag{33c}$$
$$\bar{\eta}_{\dot{\alpha}} \to \bar{\eta}_{\dot{\beta}}\, (M^\dagger)^{\dot{\beta}}{}_{\dot{\alpha}}. \tag{33d}$$

These relations highlight explicitly that the four types of spinors have four different Lorentz transformation rules. The efficiency of the two-component spinor formalism is

[1] In our presentation we have chosen to start from the 4-component spinors despite the fundamental nature of 2-component spinors. Our most important reason is that we aim to consider DReg, where there is the γ_5-problem which precisely means that the treatment of chirality and specifically 2-component spinors is problematic, while the treatment of ordinary γ^μ-matrices remains possible.

strongly related to this use of the index notation to denote the different Lorentz representations. The four representations are different, but not all inequivalent: The fact that $\epsilon^{\alpha\beta}M_\beta{}^\gamma\,\epsilon_{\gamma\delta}=(M^{-1})_\delta{}^\alpha$ shows that the spinors χ_α and χ^α transform in equivalent (i.e., unitarily related) transformations—the $(\frac{1}{2},0)$ representation. Analogously, the representations for $\bar\eta^{\dot\alpha}$ and $\bar\eta_{\dot\alpha}$ are both equivalent to the general $(0,\frac{1}{2})$ representation.

The Lorentz transformation properties also suggest the following definitions for an index-free notation for spinor products:

$$\chi\eta=\chi^\alpha\eta_\alpha\,,\qquad\qquad \bar\chi\bar\eta=\bar\chi_{\dot\alpha}\bar\eta^{\dot\alpha}\,,\tag{34a}$$

$$\chi\sigma^\mu\bar\eta=\chi^\alpha\sigma^\mu_{\alpha\dot\alpha}\bar\eta^{\dot\alpha}\,,\qquad\qquad \bar\chi\bar\sigma^\mu\eta=\bar\chi_{\dot\alpha}\bar\sigma^{\mu\dot\alpha\alpha}\eta_\alpha\,.\tag{34b}$$

The expressions in the first line are clearly Lorentz-invariant scalar quantities, and a calculation shows that the expressions in the second line transform as Lorentz four-vectors. The index-free notation and the conventions to denote the matrix indices of the σ^μ and $\bar\sigma^\mu$-matrices in this way reflect the Lorentz transformation properties of all these objects.

As announced, we will now use the two-component formalism to write useful spinor relations in a transparent way. We begin with the spinors and their conjugates:

$$\Psi=\begin{pmatrix}\chi_\alpha\\ \bar\eta^{\dot\alpha}\end{pmatrix},\qquad \overline{\Psi}=(\eta^\alpha\;\bar\chi_{\dot\alpha})\,,\qquad \Psi^C=\begin{pmatrix}\eta_\alpha\\ \bar\chi^{\dot\alpha}\end{pmatrix},\qquad \overline{\Psi^C}=(\chi^\alpha\;\bar\eta_{\dot\alpha})\,,\tag{35}$$

i.e., these conjugations simply exchange the two-component spinors and the index positions. Chiral spinors take the forms:

$$\Psi_L=\begin{pmatrix}\chi_\alpha\\ 0\end{pmatrix},\qquad \overline{\Psi_L}=(0\;\bar\chi_{\dot\alpha})\,,\qquad \Psi_R=\begin{pmatrix}0\\ \bar\eta^{\dot\alpha}\end{pmatrix},\qquad \overline{\Psi_R}=(\eta^\alpha\;0)\,.\tag{36}$$

Examples of useful bilinear expressions for anticommuting spinors (which allow rearrangements such as $\chi\eta=\eta\chi$ in view of $\eta^\alpha\chi_\alpha=-\chi_\alpha\eta^\alpha$) are

$$\overline{\Psi}_1P_L\Psi_2 = \overline{\Psi^C_2}P_L\Psi^C_1=\eta_1\chi_2,\tag{37}$$

$$\overline{\Psi}_1P_R\Psi_2 = \overline{\Psi^C_2}P_R\Psi^C_1=\bar\chi_1\bar\eta_2,\tag{38}$$

$$\overline{\Psi}_1\gamma^\mu P_L\Psi_2 = \overline{\Psi^C_2}(-P_L\gamma^\mu)\Psi^C_1=\bar\chi_1\bar\sigma^\mu\chi_2=-\chi_2\sigma^\mu\bar\chi_1,\tag{39}$$

$$\overline{\Psi}_1\{1,\gamma_5,\gamma^\mu,\gamma^\mu\gamma_5\}\Psi_2 = \overline{\Psi^C_2}\{1,\gamma_5,-\gamma^\mu,-\gamma_5\gamma^\mu\}\Psi^C_1.\tag{40}$$

Using the Hermiticity relations for two-spinors:

$$\overline{\psi}^{\dot\alpha}=(\psi^\alpha)^\dagger,\quad \overline{\psi}_{\dot\alpha}=(\psi_\alpha)^\dagger,\quad (\psi_1\psi_2)^\dagger=\overline{\psi}_2\overline{\psi}_1,\quad (\psi_1\sigma^\mu\overline{\psi}_2)^\dagger=\psi_2\sigma^\mu\overline{\psi}_1,\tag{41}$$

directly leads to the following equations for Hermitian conjugation of four-component bilinears:

$$(\overline{\Psi}_1P_L\Psi_2)^\dagger = \overline{\Psi}_2P_R\Psi_1,\tag{42}$$

$$(\overline{\Psi}_1\gamma^\mu P_L\Psi_2)^\dagger = \overline{\Psi}_2P_R\gamma^\mu\Psi_1,\tag{43}$$

$$(\overline{\Psi}_1\gamma^\mu P_R\Psi_2)^\dagger = \overline{\Psi}_2P_L\gamma^\mu\Psi_1,\tag{44}$$

$$(\overline{\Psi}_1\{1,\gamma_5,\gamma^\mu,\gamma^\mu\gamma_5\}\Psi_2)^\dagger = \overline{\Psi}_2\{1,-\gamma_5,\gamma^\mu,-\gamma_5\gamma^\mu\}\Psi_1.\tag{45}$$

At this point, we stress again that all equations of this section are valid in strictly four-dimensional Minkowski spacetime. Later, we will use dimensional regularization in which two-component spinors are not directly defined. However, all equations for four-component spinors written in this section have been written in such a way that they remain valid on the D-dimensional regularized level.

2.3. *BRST Invariance and Slavnov–Taylor Identity*

Though the construction of the Yang–Mills Lagrangian (13) is elegant and predictive, the Lagrangian cannot directly be quantized. On the level of canonical quantization, the canonical conjugate momentum field corresponding to A_0^a identically vanishes; on the level of path integral quantization, the naively defined path integral is ill-defined due to the integration over infinitely many gauge equivalent field configurations.

The well-known proposal by Faddeev and Popov modifies the path integral definition of the quantum theory by separating off this divergent factor [40]. Via a clever manipulation, the path integral can then be written in terms of a modified Lagrangian, which contains a gauge-fixing term, as well as terms with Faddeev–Popov ghost fields. The interactions of the Faddeev–Popov ghosts are determined by the choice of the gauge fixing. This path integral formulation also allows deriving Slavnov–Taylor identities, which could then be used in the first proofs of the renormalizability of Yang–Mills theories, as discussed later in Section 6.

Historically, it was observed afterwards that the resulting Faddeev–Popov Lagrangian is invariant under a new symmetry, the so-called BRST invariance [41–44]. Here, we will directly start with this BRST invariance, which can be intrinsically motivated and which provides an efficient formalism for setting up the quantization of Yang–Mills theories. Our presentation has similarities to the presentation of the Kugo/Ojima formalism in Reference [45] and the presentations of the BRST and Batalin/Vilkovisky formalisms in References [32,46].

The main idea is that the concept of local gauge invariance means that physics is described by equivalence classes. Precisely speaking on the classical level, field configurations that are related by local gauge transformations by definition describe the same physical state. The BRST formalism implements this idea in an elegant way. It first introduces the notion of ghost number N_{gh}. All fields introduced so far have a vanishing ghost number, but we shall introduce objects with positive or negative ghost numbers later. The BRST formalism further postulates the existence of an operator s, the BRST operator, which acts on classical fields and has the following properties and interpretations:

- It generalizes gauge invariance in the sense that: a field configuration X with ghost number zero is "physical" if

$$sX = 0\,. \tag{46}$$

- It generalizes gauge transformations and gauge equivalence in the sense that: two "physical" field configurations X_1, X_2 with ghost number zero are physically equivalent if some Y exists with

$$X_1 = X_2 + sY\,. \tag{47}$$

  As a side note, objects X, which are total BRST transformations,

$$X = sY\,, \tag{48}$$

  are, therefore, "unphysical" in the sense that they are equivalent to the trivial field configuration where all fields vanish (even if they also satisfy $sX = 0$).
- It is nilpotent:

$$s^2 = 0\,, \tag{49}$$

  and this nilpotency is important for the consistency of the previous two relations.

- In general, s acts as a fermionic differential operator, which increases the ghost number by one. Specifically, on products of fermionic and bosonic expressions F_i, B_i, it satisfies the product rules corresponding to a so-called graded algebra:

$$s(B_1B_2) = (sB_1)B_2 + B_1(sB_2), \tag{50a}$$
$$s(F_1B_2) = (sF_1)B_2 - F_1(sB_2), \tag{50b}$$
$$s(F_1F_2) = (sF_1)F_2 - F_1(sF_2). \tag{50c}$$

In order to define an operator with these properties, one first introduces ghost fields $c^a(x)$, which are scalar fields with fermionic statistics and ghost number $+1$. As for the gauge fields, there is one such ghost field for each gauge group generator $a = 1 \ldots N_{\text{gen}}$, and we can also write $c = T^a c^a$ with representation matrices T^a. On the ordinary fields, the BRST operator is then defined as an infinitesimal gauge transformation (see Equation (12)), but with the replacement $\theta^a \to c^a$:

$$sA_\mu(x) = \partial_\mu c(x) - ig[c(x), A_\mu(x)], \tag{51a}$$
$$sA^a_\mu(x) = \partial_\mu c^a(x) + g f^{abc} c^b(x) A^c_\mu(x) = (D_\mu c(x))^a, \tag{51b}$$
$$s\varphi(x) = -igc(x)\varphi(x). \tag{51c}$$

Here, we also used the covariant derivative acting on ghost fields, which is defined by using the adjoint representation for the generators. The BRST transformation of the ghost fields themselves is defined via the structure constants of the Lie algebra:

$$sc^a(x) = \frac{1}{2} g f^{abc} c^b(x) c^c(x), \tag{52a}$$
$$sc(x) = -igc(x)^2. \tag{52b}$$

In this way, the BRST operator is indeed nilpotent if it acts on any combination of these fields, and it clearly generalizes the original gauge transformations.

In this formalism, introducing gauge fixing and associated ghost interaction terms becomes very natural and transparent. The existence of two further kinds of fields is postulated, the antighosts $\bar{c}^a$ and the Nakanishi–Lautrup auxiliary fields B^a (with ghost numbers -1 and 0, respectively). From the present point of view, these fields essentially have the sole purpose of allowing the formulation of a gauge fixing. They form a so-called BRST doublet, which means the following very simple BRST transformations:

$$s\bar{c}^a(x) = B^a(x), \tag{53a}$$
$$sB^a(x) = 0, \tag{53b}$$

which are again consistent with nilpotency. It is known that introducing such a BRST doublet does not change the cohomology classes of the BRST operator [47]. In terms of the interpretation specified above, this means that introducing the BRST doublet does not not change the physical content of the theory.

With these ingredients, we can discuss Lagrangians of the type:

$$\mathcal{L}_{\text{fix,gh}} = s[\bar{c}^a X^a] \tag{54}$$

with some ghost number zero object X^a. Evaluating the BRST transformation on the right-hand side produces terms of ghost number zero, which are allowed terms in a Lagrangian. Given the interpretations listed above, such Lagrangians are "unphysical" since they are total BRST transformations. Similarly, adding such a Lagrangian to the original gauge-invariant Yang–Mills Lagrangian $\mathcal{L}_{\text{inv}} + \mathcal{L}_{\text{fix,gh}}$ does not change the physical content.

Hence, we may use this possibility to design a Lagrangian of this type that can be used for gauge fixing, allowing straightforward quantization of the theory. The common choice is

$$\begin{aligned}\mathcal{L}_{\text{fix,gh}} &= s\left[\bar{c}^a\left((\partial^\mu A^a_\mu) + \frac{\xi}{2}B^a\right)\right] \\ &= B^a(\partial^\mu A^a_\mu) + \frac{\xi}{2}(B^a)^2 - \bar{c}^a\partial^\mu(D_\mu c)^a\,. \end{aligned} \tag{55}$$

The B-fields are auxiliary fields in the sense that they have no kinetic term and have purely algebraic equations of motion. They can, hence, be eliminated by their equations of motion:

$$B^a = -\frac{1}{\xi}\partial^\mu A^a_\mu, \tag{56}$$

$$\mathcal{L}_{\text{fix,gh}} = -\frac{1}{2\xi}(\partial^\mu A^a_\mu)^2 - \bar{c}^a\partial^\mu(D_\mu c)^a\,. \tag{57}$$

In this way, the Lagrangian contains the usual ξ-dependent gauge fixing term, and the way it was constructed led to corresponding ghost kinetic terms and ghost–antighost–gauge boson interactions. The result of this construction is the same as the result of the Faddeev–Popov approach.

Before turning to quantization, there is one final useful extension of the classical Lagrangian. We note that most of the BRST transformations are local products of fields, i.e., constitute nonlinear field transformations. In a non-Abelian gauge theory, the only exceptions are the BRST transformations $s\bar{c}^a$ and sB^a, which are linear or zero. In an Abelian theory (where f^{abc} would vanish), also the BRST transformations of c^a and A^a_μ would be linear. In the quantized theory, such field products will define composite operators that require dedicated renormalization. It is useful to introduce "sources" for these composite operators, i.e. classical fields $\rho^{a\mu}(x)$, $\zeta^a(x)$, $Y_i(x)$,[2] which couple to the composite operators in the Lagrangian. We, therefore, define

$$\mathcal{L}_{\text{ext}} = \rho^{a\mu}sA^a_\mu + \zeta^a sc^a + Y_i s\varphi_i\,. \tag{58}$$

Each source has a negative ghost number such that the Lagrangian has in total a zero ghost number, and each source has the opposite statistics of the original field, such that the Lagrangian is bosonic. The dimensions of the sources are such that the Lagrangian has dimension four. Specifically, the sources $\rho^{a\mu}$ are fermionic with ghost number -1 and dimension 3, and the sources ζ^a are bosonic with ghost number -2 and dimension 4. By convention, the BRST transformation of all sources vanishes.

In total, we can then define the full classical Lagrangian, which will be the basis of quantization, as follows:

$$\mathcal{L}_{\text{cl}} = \mathcal{L}_{\text{inv}} + \mathcal{L}_{\text{fix,gh}} + \mathcal{L}_{\text{ext}}\,. \tag{59}$$

Each of the three parts is individually BRST-invariant. The first part is the gauge-invariant physical Lagrangian. It depends only on ordinary fields, on which BRST transformations act as gauge transformations. The second part contains the gauge fixing and ghost terms, which allow quantization of the theory. Together, they are a total BRST transformation and, hence, BRST-invariant and unphysical. The third part is BRST-invariant in view of the nilpotency $s^2 = 0$. In total,

$$s\mathcal{L}_{\text{cl}} = 0\,. \tag{60}$$

[2] These sources are not quantized and not integrated over in the path integral. These sources are also called "external sources" or "external fields" or "antifields". One may also regard them as local, x-dependent parameters of the Lagrangian.

The same statement can be rewritten in functional form. Defining the classical action:

$$\Gamma_{\text{cl}} = \int d^4x \mathcal{L}_{\text{cl}} \tag{61}$$

allows rewriting Equation (60) as the Slavnov–Taylor identity:

$$\mathcal{S}(\Gamma_{\text{cl}}) = 0 \tag{62}$$

with the Slavnov–Taylor operator:

$$\mathcal{S}(\mathcal{F}) = \int d^4x \left(\frac{\delta \mathcal{F}}{\delta \rho^{a\mu}(x)} \frac{\delta \mathcal{F}}{\delta A^a_\mu(x)} + \frac{\delta \mathcal{F}}{\delta \zeta^a(x)} \frac{\delta \mathcal{F}}{\delta c^a(x)} + \frac{\delta \mathcal{F}}{\delta Y_i(x)} \frac{\delta \mathcal{F}}{\delta \varphi_i(x)} + B^a(x) \frac{\delta \mathcal{F}}{\delta \bar{c}^a(x)} \right). \tag{63}$$

The Slavnov–Taylor identity (62) is the ultimate reformulation of the gauge invariance of the classical action after introducing gauge fixing, ghost terms, and external sources for composite operators. This identity will be a crucial ingredient in the renormalization procedure.[3]

2.4. Green Functions in Quantum Field Theory

In this subsection, we introduce the basic notation for quantum field theory required for our discussion of higher orders and regularization and renormalization. We considered a generic quantum field theory with dynamical fields $\phi_i(x)$ (these may be the gauge fields, matter fields, or ghost or antighost fields introduced in earlier subsections) and a Lagrangian $\mathcal{L}$.

Fundamental objects of the full, interacting quantized theory are Green functions, i.e., time-ordered expectation values of Heisenberg picture field operators ϕ_i^H in the full vacuum $|\Omega\rangle$ of the interacting theory:

$$G_{i_1 \ldots i_n}(x_1, \ldots, x_n) = \langle \Omega | T \phi^H_{i_1}(x_1) \ldots \phi^H_{i_n}(x_n) | \Omega \rangle \,. \tag{64}$$

We also considered Green functions involving composite local operators $\mathcal{O}$:

$$\begin{aligned} G^{k_1 \ldots k_m}_{i_1 \ldots i_n}(y_1, \ldots, y_m, x_1, \ldots, x_n) &= \langle \Omega | T \mathcal{O}^H_{k_1}(y_1) \ldots \mathcal{O}^H_{k_m}(y_m) \phi^H_{i_1}(x_1) \ldots \phi^H_{i_n}(x_n) | \Omega \rangle \\ &\equiv \langle T \mathcal{O}_{k_1}(y_1) \ldots \mathcal{O}_{k_m}(y_m) \phi_{i_1}(x_1) \ldots \phi_{i_n}(x_n) \rangle \,. \end{aligned} \tag{65}$$

Here, ϕ denotes a generic quantum field, and the above expressions may contain different kinds of such fields. Where unambiguous, we shall write $\phi_{i_1}(x_1) \equiv \phi_{i_1}$. The second line here introduces an alternative short-hand notation for such Green functions, where the explicit symbols for the vacuum state and for the Heisenberg picture are suppressed. We will often use this short-hand notation in the following.

Generally, Green functions are important since they encapsulate the essential information of a given quantum field theory. We briefly remark how they particularly allow constructing important observable quantities. The physical rest masses of one-particle states are reflected in the poles of momentum-space two-point functions, as a result of the Källen–Lehmann representation. S-matrix elements for scattering processes between

[3] We remark that the choice of gauge fixing used in the present review is not the only option. Other options include physical gauges such as the axial gauge, where no ghosts are required, or the background field gauge; see, e.g., References [32,34] for textbook discussions. Of particular interest for the present discussion is the application of the background field gauge to the electroweak SM, which includes chiral fermions (and electroweak symmetry breaking) [48]. Later, in Section 6.2.4, we will further comment on the proofs of renormalizability and physical properties such as charge universality in these different gauges. The central point of the present review is the application of the BMHV scheme for non-anticommuting γ_5 to chiral gauge theories. Here, it is noteworthy that this application is essentially unchanged regardless of whether the gauge fixing of the main text or the background field gauge is used. The corresponding discussion and the required computation of symmetry-restoring counterterms were carried out in Reference [27]. The main technical difference to the formalism presented here is that the dominant role of the Slavnov–Taylor identity is replaced by a Ward identity reflecting gauge invariance with respect to background fields; the overall logic and detailed calculational steps are essentially the same.

asymptotically free states are obtained via the Lehmann–Symanzik–Zimmermann reduction formalism, which can be derived from Haag–Ruelle scattering theory (see, e.g., the textbooks by Srednicki and Peskin/Schroeder [33,35] and the monograph by Duncan [49] for a particularly detailed account). We note here an important subtlety. Green functions are particularly defined in momentum-space for off-shell momenta, while physical observables are related to the on-shell limits, where Green functions may develop infrared divergences. In the present review, we will not discuss the specifics of the on-shell limits of Green functions.

A very useful tool for general discussions is the generating functional $Z(J,K)$ for the most-general Green functions with elementary fields and composite operators. It can be written by introducing sources (or "external fields", i.e., fields that always remain classical and never are quantized) $J_i(x)$ for the elementary fields and $K_i(x)$ for the composite operators such that

$$\begin{aligned} &G^{k_1\ldots k_m}_{i_1\ldots i_n}(y_1,\ldots,y_m,x_1,\ldots,x_n) \\ &\quad = \frac{1}{Z(0,0)} \left. \frac{\delta^{m+n} Z(J,K)}{\delta i K_{k_1}(y_1)\ldots\delta i K_{k_m}(y_m)\ldots\delta i J_{i_1}(x_1)\ldots\delta i J_{i_n}(x_n)\ldots} \right|_{J=K=0}. \end{aligned} \tag{66}$$

In perturbation theory, the Green functions are given by Feynman diagrams obtained from the well-known Gell–Mann–Low formula. Specifically, in perturbation theory, the Lagrangian is split as $\mathcal{L} = \mathcal{L}_{\text{free}} + \mathcal{L}_{\text{int}}$, where the free part $\mathcal{L}_{\text{free}}$ is bilinear in the quantum fields, allowing quantization as a free field theory. This quantization then leads to free field operators, which we denote as ϕ_i without the superscript, and to a free vacuum $|0\rangle$. The Gell–Mann–Low formula for the perturbative evaluation of Green functions then yields an explicit construction of the generating functional:

$$Z(J,K) = \frac{\langle 0|T\exp(i\int d^4x(\mathcal{L}_{\text{int}} + J_i\phi_i + K_i\mathcal{O}_i))|0\rangle}{\langle 0|T\exp(i\int d^4x\mathcal{L}_{\text{int}})|0\rangle}. \tag{67}$$

The evaluation of this formula via Wick contractions leads to Feynman rules and Feynman diagrams. In Equation (67), we also introduce a short-hand notation, which we will often use: all appearing fields and sources $J_i, \phi_i, K_i, \mathcal{O}_i$ and the Lagrangian $\mathcal{L}_{\text{int}}$ have the spacetime argument x, which is suppressed. Further, there is a summation over the index i, and the summation range extends over all quantum fields in the term $J_i\phi_i$ and over all composite operators with sources in the term $K_i\mathcal{O}_i$.

Another representation of the generating functional is given by the path integral:

$$Z(J,K) = \int \mathcal{D}\phi\, e^{i\int d^4x(\mathcal{L}+J_i\phi_i+K_i\mathcal{O}_i)}, \tag{68}$$

where $\mathcal{D}\phi$ is the measure of the integration over all field configurations and the quantities in the exponent are number-valued fields (either sources or path integral integration variables). The same short-hand notation suppressing the arguments is used. We stress that both Equations (67) and (68) are formal and not yet fully defined: the literal application of the Gell–Mann–Low formula leads to divergences unless the theory is regularized, and the path integral formula requires a precise definition of the path integral measure. Both formulas will become well-defined via the process of regularization and renormalization (this process can also be regarded as a constructive definition of the path integral measure).

The full Green functions discussed so far are described by the most-general Feynman diagrams, which are allowed to contain several disconnected components. It is possible to define a second generating functional Z_c that directly generates only connected Green functions, i.e., the sums of connected Feynman diagrams. The relation is given by

$$Z(J,K) = e^{iZ_c(J,K)}. \tag{69}$$

For a proof that this generates precisely the connected Green functions, see, e.g., References [50,51].[4]

For renormalization, one-particle irreducible (1PI) Feynman diagrams are most useful since they are the smallest building blocks that suffice to discuss ultraviolet divergences and counterterms. The corresponding 1PI Green functions can also be generated by a generating functional. This 1PI generating functional is called Γ, or effective action. It is defined by a Legendre transform of Z_c, which replaces the sources by classical fields.

In order to prepare for the introduction of this 1PI generating functional Γ, we make two remarks: First, we note that there is a mapping between the sources J_i and expectation values of field operators ϕ_i. Specifically, the first derivatives of the generating functional Z_c have the special interpretation as the expectation values of the field operators:

$$\phi_i^{\text{class}}(x) \equiv \frac{\delta Z_c}{\delta J_i(x)} = \langle \phi_i(x) \rangle^{J,K} . \tag{70}$$

In contrast to Equation (66), we have not set the sources to zero. Each choice of the sources $J_i(x)$ (for fixed $K_i(x)$), thus, defines expectation values of the quantum field operators. These expectation values are number-valued, "classical" fields $\phi_i^{\text{class}}(x)$. We may regard these classical fields as functionals of the sources $J_i(x)$ (for fixed $K_i(x)$), or we may invert the relationship and regard the sources as functionals of the classical fields. In the following, we will always assume that the vacuum expectation values of the operators ϕ_i vanish. Here, this means that $J = 0$ is mapped to $\phi^{\text{class}} = 0$ and vice versa (for $K = 0$):

$$\left. \frac{\delta Z_c}{\delta J_i(x)} \right|_{J=K=0} = 0 . \tag{71}$$

The second remark is the following: In the classical limit, the path integral is dominated by the classical field configuration minimizing the classical action. Hence, in the classical limit ("cl.lim.") and up to an irrelevant constant, we have

$$Z(J,K) = e^{iZ_c(J,K)} \overset{\text{cl.lim.}}{\longrightarrow} \left. e^{i(\Gamma_{\text{cl}}(\phi^{\text{class}},K) + \int d^4x J_i \phi_i^{\text{class}})} \right|_{0 = \frac{\delta \Gamma_{\text{cl}}}{\delta \phi^{\text{class}}} \pm J} , \tag{72}$$

where $\Gamma_{\text{cl}} = \int d^4x (\mathcal{L} + K_i \mathcal{O}_i)$ is the classical action (including source terms for composite operators) and where the $\pm$ signs apply for bosonic/fermionic fields ϕ, respectively.

This motivates the definition of a new functional Γ via the analogous, exact relation:

$$Z_c(J,K) = \Gamma(\phi^{\text{class}}, K) + \left. \int d^4x J_i \phi_i^{\text{class}} \right|_{J = \mp \frac{\delta \Gamma}{\delta \phi^{\text{class}}}} . \tag{73}$$

This relation is a Legendre transformation, which can be inverted to

$$\Gamma(\phi^{\text{class}}, K) = Z_c(J,K) - \left. \int d^4x J_i \phi_i^{\text{class}} \right|_{\phi^{\text{class}} = \frac{\delta Z_c}{\delta J}} . \tag{74}$$

In the Legendre transformation, the sources K_i for composite operators act as spectators, such that the relation:

$$\frac{\delta \Gamma(\phi^{\text{class}}, K)}{\delta K_i(x)} = \frac{\delta Z_c(J,K)}{\delta K_i(x)} \tag{75}$$

holds.

[4] The conventions for the generating functionals differ slightly between most references. Our conventions are essentially the same as in Reference [33], except that our connected functional $Z_c = -E$ there.

The functional Γ defined in this way has two very important properties. First, it is equal to the classical action plus quantum corrections, i.e.,

$$\Gamma(\phi^{\text{class}}, K) = \Gamma_{\text{cl}}(\phi^{\text{class}}, K) + \mathcal{O}(\hbar)\,, \tag{76}$$

where we reinstate explicit powers of $\hbar$ to count the number of loops. This justifies the name "effective action". Second, Γ generates one-particle irreducible (1PI) Green functions. For the full proofs of these statements, see, e.g., the textbooks by Zinn-Justin or Itzykson/Zuber [50,51], and for detailed discussions including subtleties in cases with spontaneous symmetry breaking, see, e.g., the textbooks by Weinberg or Brown [32,52].

Let us introduce further useful notation related to Green functions and Γ. First, in the following and in general, we simplify the notation for Γ and write only ϕ_i instead of ϕ_i^{class} for its arguments if no misunderstanding is possible.

Next, we introduce the notation for specific 1PI Green functions. Such concrete 1PI Green functions in position-space are obtained from derivatives of Γ with respect to the classical fields as

$$\Gamma_{\phi_i\phi_j\ldots}(x_1, x_2, \ldots) = \frac{\delta\Gamma}{\delta\phi_i(x_1)\delta\phi_j(x_2)\ldots}\bigg|_{\phi=0} = -i\langle\phi_i(x_1)\phi_j(x_2)\ldots\rangle^{\text{1PI}}\,. \tag{77}$$

In terms of Feynman diagrams, $i\Gamma_{\phi_i\phi_j\ldots}$ corresponds to the set of 1PI diagrams with the indicated external fields. When passing to momentum-space via the Fourier transform, we split off a δ-function corresponding to momentum conservation; symbolically:

$$\Gamma_{\phi_i\phi_j\ldots}\bigg|^{\text{F.T.}}(p_1, p_2, \ldots) = \Gamma_{\phi_i\phi_j\ldots}(p_1, p_2, \ldots)(2\pi)^4\delta^{(4)}(\textstyle\sum_{j=1}^n p_j)\,. \tag{78}$$

Equations (72) and (76) show that, naturally, the source terms for composite operators combine with the Lagrangian; hence, it is motivated to absorb these source terms into the Lagrangian. This is precisely what was performed in Section 2.3 for certain important operators corresponding to nonlinear BRST transformations; see Equation (58). In this way, the renormalization of such composite operators is fully integrated into the standard renormalization procedure.

Sometimes, special operators need to be considered only in the simpler context of single operator insertions. Let $\mathcal{O}$ be such an operator and $K_{\mathcal{O}}$ the corresponding source, treated as in Equations (67) or (68) or absorbed into the Lagrangian. The sources for all remaining operators are collectively called K. Then, for single insertions of $\mathcal{O}$, a special notation is defined:

$$\mathcal{O}(x)\cdot Z(J,K) = \frac{\delta Z(J,K,K_{\mathcal{O}})}{\delta(iK_{\mathcal{O}}(x))}\bigg|_{K_{\mathcal{O}}=0}, \tag{79a}$$

$$\mathcal{O}(x)\cdot \Gamma(\phi,K) = \frac{\delta \Gamma(\phi,K,K_{\mathcal{O}})}{\delta K_{\mathcal{O}}(x)}\bigg|_{K_{\mathcal{O}}=0}. \tag{79b}$$

For particular 1PI Green functions with a single operator insertion, we can write

$$(\mathcal{O}(x)\cdot\Gamma)_{\phi_i\phi_j\ldots}(x_1, x_2, \ldots) = -i\langle\mathcal{O}(x)\phi_i(x_1)\phi_j(x_2)\ldots\rangle^{\text{1PI}}\,. \tag{80}$$

In terms of Feynman diagrams, $i(\mathcal{O}(x)\cdot\Gamma)_{\phi_i\phi_j\ldots}$ corresponds to 1PI diagrams with the indicated external fields and one insertion of a vertex corresponding to $i\mathcal{O}(x)$, where the factor i results as usual from the exponential function in the Gell–Mann–Low formula (67).

An important consequence is the lowest-order behavior of the operator insertion into Γ:

$$\mathcal{O}\cdot\Gamma(\phi) = \mathcal{O}^{\text{class}} + \mathcal{O}(\hbar)\,, \tag{81}$$

where $\mathcal{O}^{\text{class}}$ is the classical field product corresponding to the operator $\mathcal{O}$. This is in line with the interpretation of Γ as the effective action.

2.5. Slavnov–Taylor Identities for Green Functions and Their Interpretation

In Section 2.3, we introduced BRST invariance as a substitute for gauge invariance in the presence of a gauge fixing, and we found the BRST-invariant classical action. The question is now: How is this BRST invariance reflected in the full quantum theory? The most-general answer is that the off-shell Green functions introduced in Section 2.4 satisfy so-called Slavnov–Taylor identities. Here, we provide a formal derivation of these Slavnov–Taylor identities. This derivation is simple and elegant and allows an efficient understanding and interpretation of the structure of the Slavnov–Taylor identities. It is however formal in the sense that it ignores the procedure of regularization and renormalization; hence, we will later, in Section 6, need to discuss how this procedure might change the identities. There, we will also discuss the important role of the Slavnov–Taylor identities in establishing the renormalizability of Yang–Mills theories, including the decoupling of unphysical degrees of freedom and the unitarity of the physical S-matrix.

We start from the BRST invariance of the classical action, which was already expressed by Equation (60) and rewritten as the Slavnov–Taylor identity (62). Here, we rewrite it as an invariance relation:

$$\Gamma_{\text{cl}}(\phi, K) = \Gamma_{\text{cl}}(\phi + \delta\phi, K) \tag{82}$$

where ϕ denote all dynamical fields (A_μ, φ_i, c, $\bar{c}$, B) and K denote all sources (ρ^μ, Y_i, ζ) and where the field transformations are given as

$$\delta\phi = \theta s\phi \tag{83}$$

with an infinitesimal fermionic parameter θ such that $\delta\phi$ always has the same bosonic/fermionic statistics as ϕ itself. Equation (82) is meant at first order in θ, and at this order, it is clearly equivalent to both Equations (60) and (62).

Now, we use this invariance as a starting point and derive the Slavnov–Taylor identities for the generating functional (68) in the path integral formulation. We assumed that the path integral measure is invariant under the same symmetry transformation $\phi \to \phi + \delta\phi \equiv \phi'$ and, therefore, write

$$\begin{aligned} Z(J, K) &= \int \mathcal{D}\phi' \, e^{i(\Gamma_{\text{cl}}(\phi', K) + \int d^4x J_i \phi'_i)} \\ &= \int \mathcal{D}\phi \, e^{i(\Gamma_{\text{cl}}(\phi, K) + \int d^4x J_i \phi_i + J_i \delta\phi_i)} . \end{aligned} \tag{84}$$

The variation $\delta\phi$ only appears in the exponent. We can expand the right-hand side at first order in $\delta\phi$ and subtract it from the left-hand side to obtain

$$0 = \int \mathcal{D}\phi \left(\int d^4x J_i \delta\phi_i \right) e^{i(\Gamma_{\text{cl}}(\phi, K) + \int d^4x J_i \phi_i)} . \tag{85}$$

This is already one basic version of the Slavnov–Taylor identity. We can rewrite it in several ways to familiarize us with its interpretation:

- A first way is to replace the path integral with its interpretation as an operator expectation value, in line with Equation (70). Then, we obtain

$$0 = \int d^4x J_i \langle \delta\phi_i \rangle^{J,K} . \tag{86}$$

This can be further rewritten by replacing the sources J_i in terms of derivatives of Γ, the effective action or generating functional of 1PI Green functions, via the Legendre transform (73) such that

$$0 = \int d^4x \langle \delta\phi_i \rangle^{J,K} \frac{\delta\Gamma}{\delta\phi_i}, \tag{87}$$

where, again, the sum over all fields i is implied and where the order of the factors was exchanged to compensate the $\pm$ signs in the relation for J_i in Equation (73). Both of these equations have the forms of typical infinitesimal invariance relations. We may also rewrite the previous equation as

$$\Gamma(\phi, K) = \Gamma\Big(\phi + \langle\delta\phi\rangle^{J,K}, K\Big), \tag{88}$$

valid to first order in the variation. This equation is directly analogous to the starting point (82). It clarifies the interpretation of the Slavnov–Taylor identity as an invariance relation for the full effective action Γ under symmetry transformations given by $\langle\delta\phi_i\rangle^{J,K}$. An important distinction can now be made about these symmetry transformations. In general, the $\delta\phi_i$ are nonlinear products of fields (i.e., composite operators), and generally, the expectation value of a product is different from the product of expectation values. In other words, the symmetry transformations may receive nontrivial quantum corrections. Hence, the symmetry transformation in Equation (88) is in general different from the classical expression $\delta\phi_i$ (which becomes $\delta\phi_i^{\text{class}}$ using the more explicit notation of the previous section), which one might have expected to appear. Only in the case where all $\delta\phi_i$ are linear in the dynamical fields, the symmetry relation (88) corresponds to the same invariance as Equation (82).

- A second way to rewrite the Slavnov–Taylor identity (85) is by taking derivatives with respect to the sources as in Equation (66) to obtain identities for specific Green functions. In this way, Equation (85) leads to infinitely many identities of the kind:

$$\begin{aligned} 0 &= \delta\langle T\phi_{i_1}(x_1)\ldots\phi_{i_n}(x_n)\rangle^{J,K} \\ &\equiv \langle T(\delta\phi_{i_1}(x_1))\ldots\phi_{i_n}(x_n)\rangle^{J,K} + \ldots + \langle T(\phi_{i_1}(x_1))\ldots\delta\phi_{i_n}(x_n)\rangle^{J,K}, \end{aligned} \tag{89}$$

where the first line is defined as an abbreviation for the second line and the uniform + signs of all terms are correct because the transformation δ as defined by Equation (83) is of a bosonic nature. In these identities, Green functions involving ordinary fields ϕ_i and the symmetry transformation composite operators $\delta\phi_i$ appear. In this form, Slavnov–Taylor identities may be checked explicitly by computing Feynman diagrams for such Green functions. We can illustrate this with a simple, but important example. Taking the Yang–Mills theory of the previous subsections with fermionic matter fields ψ, we can consider $\delta\langle\bar{c}\psi_i\bar{\psi}_k\rangle$ and use the BRST transformations in Equations (51c) and (53) to obtain

$$0 = \langle TB\psi_i\bar{\psi}_k\rangle + ig\langle T\bar{c}(c\psi)_i\bar{\psi}_k\rangle - ig\langle T\bar{c}\psi_i(\bar{\psi}c)_k\rangle \tag{90}$$

where the brackets indicate local composite operators. The auxiliary field B will effectively be replaced by $\partial^\mu A_\mu$ via Equation (56). In Abelian QED, the ghosts are free and can be factored out of the matrix elements. Hence, in QED, this identity simply leads to the familiar Ward identity between the electron self-energy and the electron–electron–photon vertex function. In non-Abelian Yang–Mills theories, the identity also relates the fermion self-energy and the fermion–fermion–gauge boson three-point function, but the relationship is more complicated and involves nontrivial composite operators, which need to be renormalized.

- A final way to rewrite the Slavnov–Taylor identity is to write it as functional equations for the generating functionals Z, Z_c, or Γ. Since we coupled the nonlinear classical symmetry transformation (83) to the sources K in the classical action (58), the expectation values of nonlinear composite operators appearing in the previous equations may be rewritten in terms of functional derivatives with respect to K. A slight technical complication is that there are also linear symmetry transformations that we have not coupled to sources, such as the BRST transformations of the $\bar{c}$ and B fields. Precisely, we can, therefore, replace the nonlinear $\delta\phi_i$ by $\delta/\delta(iK_i)$ in the Slavnov–Taylor identity (85), but the linear $\delta\phi_i$ remain. If we express the path integral in terms of the connected functional, Equation (85) takes the schematic form:

$$0 = \int d^4x \sum_{\delta\phi_i=\text{nonlinear}} J_i \frac{\delta Z_c(J,K)}{\delta K_i} + \int d^4x \sum_{\delta\phi_i=\text{linear}} J_i \langle \delta\phi_i \rangle^{J,K}, \tag{91}$$

where the expectation value in the last term really is a linear combination of the expectation values of fundamental fields, i.e., a linear combination of ϕ_j^{class} as used in Equation (74) and, thus, equal to what we mean by $\delta\phi_i^{\text{class}}$, where the index ^{class} will be dropped again. The previous equation can be rewritten as an equation for the 1PI functional Γ by replacing the sources J_i via the Legendre transformation to Γ and by using that the sources K are unaffected by the Legendre transformation, as expressed by Equation (75). In this way, we obtain

$$0 = \int d^4x \sum_{\delta\phi_i=\text{nonlinear}} \frac{\delta\Gamma(\phi,K)}{\delta K_i} \frac{\delta\Gamma(\phi,K)}{\delta\phi_i} + \int d^4x \sum_{\delta\phi_i=\text{linear}} \delta\phi_i \frac{\delta\Gamma(\phi,K)}{\delta\phi_i}. \tag{92}$$

This is literally the same equation as the Slavnov–Taylor identity for the classical action with the Slavnov–Taylor operator (63), but rewritten for the full effective action:

$$\mathcal{S}(\Gamma) = 0. \tag{93}$$

This explains the reason why we rewrote the BRST invariance of the classical action in Section 2.3 as the Slavnov–Taylor identity using Equation (63): This equation has the potential of remaining valid without modification in the full quantum theory, provided the above formal manipulations survive the regularization and renormalization procedure.

Finally, we comment on the validity of our derivation. The derivation assumed the classical action to be symmetric, the path integral to be well-defined, and the path integral measure to be invariant under the symmetry. A full treatment must define the quantum theory via the procedure of regularization and renormalization, which may be viewed as a constructive definition of the path integral and its measure and which might change the action, e.g., by counterterms. An essential result of algebraic renormalization theory (see below in Section 6) is that the above derivations are essentially correct up to local terms in the following sense: If the above Slavnov–Taylor identity (93) is valid at some given loop order, then at the next loop order, it can at most be violated by a local functional of the fields. Hence, there is a chance that any such local violation can be canceled by adding local, symmetry-restoring counterterms. If this is possible, the Slavnov–Taylor identity indeed can be established at all orders in the renormalized theory.

In the present review, we mainly work in dimensional regularization. In this context, the above derivation acquires a more literal meaning. In Section 4, we will discuss the so-called regularized quantum action principle, which essentially states that all derivations remain literally valid in dimensional regularization if all quantities are defined via regularized Feynman diagrams in $D \neq 4$ dimensions. In that case, however, it becomes questionable whether the D-dimensional version of the classical action satisfies the same symmetry (82) as the original four-dimensional version. If this is not the case, there is again a violation of the Slavnov–Taylor identity at the regularized level, which needs to be studied and which may be canceled by introducing symmetry-restoring counterterms.

2.6. Peculiarities of Abelian Gauge Theories

So far, the discussions above focused on the non-Abelian case. However, there are some peculiarities in the Abelian case that will be highlighted in this subsection (we assume the absence of spontaneous symmetry breaking). Obviously, in an Abelian gauge theory, there are less interactions than in the non-Abelian case, with corresponding implications for higher-order corrections. However, there are also less restrictions by the gauge group, which leads to the need for an additional symmetry condition to ensure a consistent renormalization of the Abelian coupling constant, as discussed below. For further information of Abelian theories in this context, we refer the reader to [42,53–56], where they focused, in contrast to the present section, on the Abelian case with spontaneous symmetry breaking, whereas the more general case of the Standard Model and extensions was discussed in [57–59]. For a general overview, we refer to the textbook by Piguet/Sorella [47].

Starting with the classical Lagrangian of the Abelian gauge theory of quantum electrodynamics, using the notation of Section 2.3, we may write it in the same form as in Equation (59), this time, however, with

$$\mathcal{L}_{\text{inv}} = i\overline{\psi}_i \not{D}_{ij}\psi_j - \frac{1}{4}F^{\mu\nu}F_{\mu\nu}\,, \tag{94}$$

with the covariant derivative $D^{\mu}_{ij} = \partial^{\mu}\delta_{ij} + ieQ_i\delta_{ij}A^{\mu}$ and the field strength tensor $F_{\mu\nu} = \partial_{\mu}A_{\nu} - \partial_{\nu}A_{\mu}$, with the gauge-fixing and ghost Lagrangian:

$$\mathcal{L}_{\text{fix,gh}} = s\left[\bar{c}\left((\partial^{\mu}A_{\mu}) + \frac{\xi}{2}B\right)\right] = B(\partial^{\mu}A_{\mu}) + \frac{\xi}{2}B^2 - \bar{c}\partial^{\mu}\partial_{\mu}c\,, \tag{95}$$

with $B = -(\partial_{\mu}A^{\mu})/\xi$ and with the Lagrangian of the external sources:

$$\mathcal{L}_{\text{ext}} = \rho^{\mu}sA_{\mu} + \bar{R}^{i}s\psi_i + R^{i}s\overline{\psi}_i\,, \tag{96}$$

where we used the concrete name R^i for the matter field sources instead of the generic name Y_i of Section 2.3. The classical action is then, again, given by (61).

The BRST transformations in the Abelian case, already used in (95), are provided by

$$sA_{\mu}(x) = \partial_{\mu}c(x)\,, \tag{97a}$$
$$s\psi_i(x) = -ieQ_ic(x)\psi_i(x)\,, \tag{97b}$$
$$s\overline{\psi}_i(x) = ieQ_ic(x)\overline{\psi}_i(x)\,, \tag{97c}$$
$$sc(x) = 0\,, \tag{97d}$$
$$s\bar{c}(x) = B(x)\,, \tag{97e}$$
$$sB(x) = 0\,. \tag{97f}$$

It can be seen that, in the Abelian case, except from the BRST transformations for the fermions ψ_i and $\overline{\psi}_i$, all other BRST transformations are linear in dynamical fields. Recall that, for a linear classical symmetry of the form:

$$\delta\phi_i(x) = v_i(x) + \int d^4y\; t_{ij}(x,y)\phi_j(y) \tag{98}$$

with number-valued kernel t_{ij}, its expectation value is identical to the classical symmetry transformation (see also the discussions around Equations (88), (91) and (92)), i.e.,

$$\langle\delta\phi_i(x)\rangle^{J,K} = v_i(x) + \int d^4y\; t_{ij}(x,y)\langle\phi_j(y)\rangle^{J,K} = \delta\phi_i^{\text{class}}. \tag{99}$$

Hence, on the basis of Equations (87) and (88) from Section 2.5, the full effective quantum action Γ is invariant under such linear classical symmetries as they do not receive nontrivial quantum corrections. In other words, linear symmetry transformations of the classical action Γ_{cl} are automatically symmetry transformations of the full effective quantum action Γ.

In particular, the BRST transformation of the photon A_μ is linear, and hence, sA_μ does not receive quantum corrections, and the expectation value $\langle sA_\mu \rangle^{J,K}$ is identical to the classical expression $(sA_\mu)^{\text{class}}$.

Further, R^i and $\bar{R}^i$ are external sources, and the Abelian Faddeev–Popov ghost and antighost completely decouple from the rest of the theory (cf. (95)). Hence, neither R^i and $\bar{R}^i$ nor the ghost c and antighost $\bar{c}$ can occur in loops; they can only appear as external legs, as there are no corresponding interactions and the external sources are not dynamical fields, and thus cannot propagate. Consequently, none of the Abelian BRST transformations obtain quantum corrections, or in other words, in the Abelian case, the BRST transformations do not renormalize.

In a theory with a non-Abelian simple gauge group G with gauge coupling g, the generators T^a are uniquely determined by choosing a representation. For this reason, the couplings of all matter fields to the gauge fields $\propto gT^a$ and of all gauge boson self-interactions $\propto gf^{abc}$ are uniquely determined up to one common, universal gauge coupling g.

In contrast to this, in an Abelian gauge theory, every diagonal matrix would be a representation of the corresponding Lie algebra. Thus, the corresponding charges Q_i of the respective fermions could in principle be arbitrary real numbers. Group theory alone would allow these charges to obtain quantum corrections, i.e., they could renormalize, and could thus even take different values at every order in the perturbation theory. Hence, due to the fact that the group structure of an Abelian gauge group is not as powerful as the one of a non-Abelian gauge group, the Abelian couplings need to be determined, in all orders, by an additional symmetry condition to the full effective quantum action, either by the local Ward identity or by the so-called antighost equation.

The special simplicity of Abelian gauge theories and the existence of additional all-order identities is technically reflected in several field derivatives of the classical action. We begin with the antighost equation:

$$\frac{\delta \Gamma_{\text{cl}}}{\delta c(x)} = \Box \bar{c}(x) + \partial_\mu \rho^\mu(x) - ieQ_i \bar{R}^i(x)\psi_i(x) + ieQ_i \overline{\psi}_i(x) R^i(x) \,. \tag{100}$$

Additionally, varying Γ_{cl} with respect to the antighost and the external source of the photon yields

$$\frac{\delta \Gamma_{\text{cl}}}{\delta \bar{c}(x)} = -\Box c(x) \,, \qquad \frac{\delta \Gamma_{\text{cl}}}{\delta \rho_\mu(x)} = sA^\mu(x) = \partial^\mu c(x) \,, \tag{101}$$

which can be combined to obtain the so-called ghost equation:

$$\left(\frac{\delta}{\delta \bar{c}} + \partial_\mu \frac{\delta}{\delta \rho_\mu} \right) \Gamma_{\text{cl}} = 0 \,. \tag{102}$$

The gauge fixing condition is obtained by varying Γ_{cl} with respect to the Nakanishi–Lautrup field B:

$$\frac{\delta \Gamma_{\text{cl}}}{\delta B(x)} = \xi B(x) + \partial_\mu A^\mu(x) \,. \tag{103}$$

Importantly, it can be seen that all of the above Equations (100)–(103) are linear in dynamical fields, e.g., $\delta\Gamma_{\text{cl}}/\delta c(x) = $ (linear expression). In contrast, all other functional derivatives of the classical action:

$$\frac{\delta\Gamma_{\text{cl}}}{\delta R^i(x)} = s\overline{\psi}_i(x) = ieQ_i c(x)\overline{\psi}_i(x)\,, \tag{104a}$$

$$\frac{\delta\Gamma_{\text{cl}}}{\delta\psi_i(x)} = i\partial_\mu\overline{\psi}_i(x)\gamma^\mu + eQ_i\overline{\psi}_i(x)\not{A}(x) + ieQ_i\bar{R}^i(x)c(x)\,, \tag{104b}$$

$$\frac{\delta\Gamma_{\text{cl}}}{\delta\bar{R}^i(x)} = s\psi_i(x) = -ieQ_i c(x)\psi_i(x)\,, \tag{104c}$$

$$\frac{\delta\Gamma_{\text{cl}}}{\delta\overline{\psi}_i(x)} = i\not{\partial}\psi_i(x) - eQ_i\not{A}(x)\psi_i(x) - ieQ_iR^i(x)c(x)\,, \tag{104d}$$

are nonlinear in dynamical fields. The special feature of linear Equations (100)–(103) is that there are no quantum corrections expected, which could spoil these linear relations. Indeed, for loop corrections, we need interactions and, thus, at least three dynamical fields, which is not the case here.

Hence, we may require that these identities hold at all orders as part of the definition of the theory, meaning that they also hold for the full effective quantum action Γ, i.e.,[5]

$$\frac{\delta\Gamma}{\delta c(x)} \overset{!}{=} \frac{\delta\Gamma_{\text{cl}}}{\delta c(x)}, \qquad \frac{\delta\Gamma}{\delta\bar{c}(x)} \overset{!}{=} \frac{\delta\Gamma_{\text{cl}}}{\delta\bar{c}(x)}, \qquad \frac{\delta\Gamma}{\delta\rho_\mu(x)} \overset{!}{=} \frac{\delta\Gamma_{\text{cl}}}{\delta\rho_\mu(x)}, \qquad \frac{\delta\Gamma}{\delta B(x)} \overset{!}{=} \frac{\delta\Gamma_{\text{cl}}}{\delta B(x)}. \tag{105}$$

The charges Q_i of all fields explicitly occur in the antighost Equation (100), and (105), thus, fixes the charges of the fields to all orders.

Additionally, we can derive the aforementioned Ward identity (which, in the present case without spontaneous symmetry breaking and in the presence of the identities (105), is equivalent to the Slavnov–Taylor identity). Starting with the Slavnov–Taylor identity for the Abelian case:

$$0 = \mathcal{S}(\Gamma) = \int d^4x \left(\frac{\delta\Gamma}{\delta\bar{R}^i}\frac{\delta\Gamma}{\delta\psi_i} + \frac{\delta\Gamma}{\delta R^i}\frac{\delta\Gamma}{\delta\overline{\psi}_i} + \frac{\delta\Gamma}{\delta\rho^\mu}\frac{\delta\Gamma}{\delta A_\mu} + B\frac{\delta\Gamma}{\delta\bar{c}} \right), \tag{106}$$

cf. (63) for the non-Abelian case, varying it with respect to the Faddeev–Popov ghost $c(x)$, i.e.,

$$\begin{aligned}
0 &= \frac{\delta\mathcal{S}(\Gamma)}{\delta c(x)} \\
&= \int d^4y \Bigg[\left(\frac{\delta}{\delta c(x)}\frac{\delta\Gamma}{\delta\bar{R}^i(y)} \right)\frac{\delta\Gamma}{\delta\psi_i(y)} + \frac{\delta\Gamma}{\delta\bar{R}^i(y)}\left(\frac{\delta}{\delta c(x)}\frac{\delta\Gamma}{\delta\psi_i(y)} \right) \\
&\qquad + \left(\frac{\delta}{\delta c(x)}\frac{\delta\Gamma}{\delta R^i(y)} \right)\frac{\delta\Gamma}{\delta\overline{\psi}_i(y)} + \frac{\delta\Gamma}{\delta R^i(y)}\left(\frac{\delta}{\delta c(x)}\frac{\delta\Gamma}{\delta\overline{\psi}_i(y)} \right) \\
&\qquad + \left(\frac{\delta}{\delta c(x)}\frac{\delta\Gamma}{\delta\rho^\mu(y)} \right)\frac{\delta\Gamma}{\delta A_\mu(y)} - \frac{\delta\Gamma}{\delta\rho^\mu(y)}\left(\frac{\delta}{\delta c(x)}\frac{\delta\Gamma}{\delta A_\mu(y)} \right) \\
&\qquad + B\left(\frac{\delta}{\delta c(x)}\frac{\delta\Gamma}{\delta\bar{c}(y)} \right) \Bigg] \\
&= \int d^4y \Bigg[\left(\frac{\delta}{\delta\bar{R}^i(y)}\frac{\delta\Gamma}{\delta c(x)} \right)\frac{\delta\Gamma}{\delta\psi_i(y)} - \frac{\delta\Gamma}{\delta\bar{R}^i(y)}\left(\frac{\delta}{\delta\psi_i(y)}\frac{\delta\Gamma}{\delta c(x)} \right) \\
&\qquad + \left(\frac{\delta}{\delta R^i(y)}\frac{\delta\Gamma}{\delta c(x)} \right)\frac{\delta\Gamma}{\delta\overline{\psi}_i(y)} - \frac{\delta\Gamma}{\delta R^i(y)}\left(\frac{\delta}{\delta\overline{\psi}_i(y)}\frac{\delta\Gamma}{\delta c(x)} \right) \\
&\qquad - \left(\frac{\delta}{\delta\rho^\mu(y)}\frac{\delta\Gamma}{\delta c(x)} \right)\frac{\delta\Gamma}{\delta A_\mu(y)} - B\left(\frac{\delta}{\delta\bar{c}(y)}\frac{\delta\Gamma}{\delta c(x)} \right) \Bigg] \\
&= -ieQ_i\psi_i(x)\frac{\delta\Gamma}{\delta\psi_i(x)} + ieQ_i\bar{R}^i(x)\frac{\delta\Gamma}{\delta\bar{R}^i(x)} + ieQ_i\overline{\psi}_i(x)\frac{\delta\Gamma}{\delta\overline{\psi}_i(x)} \\
&\quad - ieQ_iR^i(x)\frac{\delta\Gamma}{\delta R^i(x)} - \partial_\mu\frac{\delta\Gamma}{\delta A_\mu(x)} - \Box B(x)\,,
\end{aligned} \tag{107}$$

---

5 In case of an Abelian gauge theory with spontaneous symmetry breaking, not all of these identities are valid, but one may introduce background fields which allow obtaining a valid local Ward identity and/or an Abelian antighost equation, see Refs. [54,56].

where we used the fact that fermionic objects anticommute and that $\delta/\delta c(x)$ is a fermionic functional derivative. After the third equality, we moved $\delta/\delta c(x)$ past the other respective functional derivative and utilized the antighost Equation (100) (which is valid to all orders, see Equation (105)). We dropped the penultimate term of the second equality, as the RHS of the antighost Equation (100) does not contain a term depending on A_μ. Rearranging the last line, we obtain the functional form of the local Abelian Ward identity:

$$\left(\partial_\mu \frac{\delta}{\delta A_\mu(x)} + ieQ_i \sum_\Psi (-1)^{n_\Psi} \Psi(x) \frac{\delta}{\delta \Psi(x)}\right)\Gamma = -\Box B(x), \tag{108}$$

with $\Psi \in \{\psi_i, \overline{\psi}_i, R^i, \bar{R}^i\}$ and $n_\Psi \in \{0,1,0,1\}$. The well-known Ward identity for the relation of the electron self-energy and the electron–electron–photon interaction vertex may then be deduced from this equation. Further discussions will be made later in Section 7 for the example of chiral QED. Again, the charges Q_i of all fields are fixed as (108) is established to all orders. Consequently, the above statements imply a nonrenormalization of the field charges Q_i, which means that a single counterterm is sufficient to renormalize the Abelian coupling to all orders of the perturbation theory, thus guaranteeing a consistent renormalization of the coupling constant.

The above identities, viewing them as part of the definition of the theory, constrain the regularization and renormalization procedure. On the one hand, symmetry-preserving (field and parameter) renormalization constants are constrained by the equations (meaning, in particular, that certain combinations such as the gauge fixing term or terms such as $\bar{R}^i s\psi_i$ do not renormalize). On the other hand, the local Ward identity (108) particularly will be of interest in determining symmetry-restoring counterterms. It can be used to interpret the breaking and restoration of the Slavnov–Taylor identity.

3. Dimensional Regularization

In a perturbative quantum field theory, Feynman diagrams with closed loops correspond to higher orders in $\hbar$. They, hence, represent genuine quantum corrections and are of fundamental interest. Such loop diagrams, however, are known to give rise to ultraviolet (UV) divergences, which need to be handled. The reason for this can easily be understood by imagining a loop made of a propagator with coinciding end points. Since the propagator is a distribution, one may expect this object to be ill-defined, as is the product of distributions at the same spacetime point in general. In fact, such loops correspond to the exchange of virtual particles, whose momenta are integrated over and which may run up to infinity, hence the possibility of divergent integrals in momentum-space. In essence, the purpose of renormalization is to remove all divergences and assign a meaning to such ill-defined expressions and ultimately to define physically meaningful results.

In practice, this means that we first need to isolate the aforementioned divergences before they can be subtracted. In the typical setting, isolating divergences is achieved via regularization, while their subtraction is performed via counterterms, which are added to the Lagrangian. The entire procedure constitutes the renormalization. Hence, in order to obtain meaningful results at the quantum level, i.e., including higher-order corrections, one needs regularization and renormalization, as already mentioned at the end of Section 2.5.

There are several regularization schemes; here, we focus on dimensional regularization (DReg). In this present Section 3 and the subsequent Section 4, we provide an overview of the main properties of DReg and of how to perform calculations using this regularization procedure.

Dimensional regularization and its variants are the most-common regularization schemes in relativistic quantum field theories. These schemes have several key advantages that make them particularly useful in practical, concrete computations. The structure of integrals in formally D dimensions is essentially unchanged, allowing efficient integration techniques. The divergent terms appear as $1/(D-4)$ poles and can be isolated in a transparent way. Lorentz invariance and gauge invariance of non-chiral gauge theories is essentially kept manifest. Furthermore, fundamental properties such as equivalence to

BPHZ renormalization, consistency with the unitarity and causality of quantum field theory, and consistent applicability at all orders are rigorously established. The key disadvantage is the problematic treatment of the γ_5 matrix and the $\epsilon_{\mu\nu\rho\sigma}$ symbol. As a result, gauge invariance is manifestly broken in chiral gauge theories. The treatment of such theories is the main topic of the present review.

The previous statements are discussed in detail later in Section 5. That section will explain that, based on DReg, local counterterms exist that can subtract the UV divergences. It will also explain how the regularization/counterterm/renormalization procedure in DReg amounts to a rigorous and physically sensible construction of higher orders. Then, in Section 6, we will consider DReg applied to gauge theories and see that (under certain conditions where chiral gauge anomalies are absent) the Slavnov–Taylor identity can be established at all orders in the renormalized, finite theory. In case DReg breaks the symmetry in intermediate steps, the existence of symmetry-restoring counterterms is then guaranteed.

The basic idea of DReg is to replace the 4-dimensional spacetime and the 4-dimensional momentum-space by formally D-dimensional ones, with parametrization $D = 4 - 2\epsilon$. In this way, all integrals become formally D-dimensional. DReg was put forward in several works by 't Hooft and Veltman [1], by Bollini and Giambiagi [2], and by Ashmore [3]. Specifically, Reference [1] already highlighted all key advantages and disadvantages mentioned above and showed how to compute 1-loop and 2-loop Feynman diagrams using DReg.

In the following, we begin the section by introducing our notation for the dimensionally regularized and renormalized effective quantum action and schematically sketch its construction. This provides a short overview of the general structure of dimensional regularization and renormalization (Section 3.1).

Then, we will explain what the properties of D-dimensional integrals are and how these integrals can be consistently defined (Section 3.2). Together with the integrals, many other quantities have to be formally continued to D dimensions, in particular momenta, vector fields, metric tensors, and γ matrices. Section 3.3 will focus on such quantities and delineate to what extent a purely D-dimensional treatment is correct and at which points a distinction of four-dimensional and D-dimensional quantities needs to be made in the calculations. In particular, it introduces the BMHV scheme for non-anticommuting γ_5.

Section 3.4 describes an important feature of DReg, which is not shared by all regularization methods: the precise expressions of regularized Feynman diagrams in D dimensions may be encoded in a formally D-dimensional Lagrangian, from which Feynman rules are obtained in the usual way. This relation is obviously useful in the study of symmetries of regularized Feynman diagrams since the properties of diagrams can be obtained from the properties of the regularized Lagrangian. In Section 3.5, we discuss several variants of DReg such as regularization by dimensional reduction and further subvariants. We will discuss the relationships between the variants on the level of the regularized Lagrangians and on the level of Green functions and S-matrix elements.

3.1. General Structure of Dimensional Regularization and Renormalization

Before we discuss the properties of D-dimensional integrals and how to formally continue certain quantities to D dimensions, and thus perform calculations in DReg, we briefly introduce our notation with respect to the dimensionally regularized and renormalized effective quantum action, the key quantity of the theory, and sketch its construction.

As mentioned above, UV divergences in loop integrals are isolated as $1/(D-4)$ poles in DReg. These divergences must be subtracted using counterterms in order to renormalize the theory. In general, such counterterms may not only contain these UV divergent but also finite contributions.[6] Here we sketch the renormalization procedure and introduce useful notation.

[6] The general counterterm structure of a dimensionally regularized theory using the BMHV scheme makes use of further subdivisions of counterterms. This will be presented in Section 6 and illustrated in a practical example in Section 7.

The perturbative expansion is organized in terms of orders in $\hbar$, equivalent to orders in loops. The classical action of order $\hbar^0$ defining the theory is denoted $S_0 \equiv \Gamma_{\text{cl}}$; the counterterm action is denoted as S_{ct}; the sum of the two is called the bare action S_{bare}. In the following, symbols without an upper index denote all-order quantities, while for perturbative expressions, an upper index i labels quantities of precisely order i, whereas quantities up to and including order i are labeled with an upper index (i). Using this notation, the bare and the counterterm actions may be written as

$$S_{\text{bare}} = S_0 + S_{\text{ct}}, \qquad S_{\text{ct}} = \sum_{i=1}^{\infty} S_{\text{ct}}^i, \qquad S_{\text{ct}}^{(i)} = \sum_{j=1}^{i} S_{\text{ct}}^j. \tag{109}$$

In dimensional regularization and renormalization, the perturbative construction of the effective action is performed iteratively at each order of $\hbar$, i.e., at each loop order, starting from the tree-level action S_0. Then, a counterterm action S_{ct}^i needs to be constructed at each higher order $i \geq 1$, which has to satisfy the two conditions that the renormalized theory is UV-finite and in agreement with all required symmetries.

The subrenormalized quantum action of order i is denoted by

$$\Gamma_{\text{subren}}^i \tag{110}$$

and obtained at order i by using Feynman rules from the tree-level action and counterterms up to order $i-1$. The counterterms S_{ct}^i to be constructed at the order i are subdivided into singular counterterms (which, by definition, contain only pole terms in $(D-4)$ and are denoted by subscript $_{\text{sct}}$) and finite counterterms (finite in the limit $D \to 4$ and denoted by subscript $_{\text{fct}}$). By constructing and including singular counterterms of the order i, we obtain

$$\lim_{D \to 4} \left(\Gamma_{\text{subren}}^i + S_{\text{sct}}^i \right) = \text{finite}, \tag{111}$$

which determines the singular counterterms unambiguously. If necessary, we may then also include additional finite counterterms. Once the finite counterterms are determined, we obtain

$$\Gamma_{\text{DRen}}^i \equiv \Gamma_{\text{subren}}^i + S_{\text{sct}}^i + S_{\text{fct}}^i. \tag{112}$$

This quantity Γ_{DRen}^i is finite and essentially renormalized, but it may still contain the variable $\epsilon = (4-D)/2$ and so-called evanescent quantities, which vanish in strictly $D = 4$ dimensions. Thus, the completely renormalized quantum action is obtained by taking the limit $D \to 4$ and setting all evanescent quantities to zero. This procedure is denoted by[7]

$$\Gamma^i \equiv \underset{D \to 4}{\text{LIM}}\, \Gamma_{\text{DRen}}^i. \tag{113}$$

Some comments on the finite counterterms are in order. They can have two purposes. On the one hand, it may happen that regularized quantum corrections spoil a symmetry of the theory, such that, e.g., the Slavnov–Taylor identity is invalid on the level of Equation (111). If the symmetry is part of the definition of the theory, finite counterterms must be found and added such that the symmetry is valid on the renormalized level (113). The purpose of counterterms is then not solely to remove UV divergences, but also to restore symmetries if necessary (and if possible). If no finite counterterms can be found that restore the symmetry, the symmetry is lost. This situation is called an anomaly, or anomalous symmetry breaking. It signals an irreconcilable clash of the symmetry and the quantum theory. If the symmetry is part of the definition of the theory or required for

[7] We will sometimes synonymously refer to the completely renormalized and 4-dimensional quantum action as Γ_{ren}^i, i.e. $\Gamma^i \equiv \Gamma_{\text{ren}}^i$, in order to emphasize that it is completely renormalized.

the consistency of the theory, the theory must be abandoned. The later Section 6.2.3 will provide a detailed discussion of the symmetry restoration using finite counterterms.

On the other hand, the finite counterterms can also be used in order to fulfil certain renormalization conditions. In general, the choice of the finite counterterms (beyond symmetry restoration) is called a renormalization scheme. Popular examples of renormalization schemes are on-shell or (modified) minimal subtraction schemes. In the present review we will not further discuss renormalization schemes. For textbook-level discussions of this important topic we refer to the books by Böhm/Denner/Joos and Srednicki [34,35].[8]

Finally, we reiterate that we only sketched the general procedure and introduced the notation, but we have not yet proven that this procedure actually works. This will be performed in the later Sections 5 and 6, and it is exemplarily illustrated in Section 7 for the case where finite symmetry-restoring counterterms are required.[9]

3.2. Integrals in D Dimensions

In this subsection, we will discuss momentum integrations in DReg. As explained above, in DReg, we replace four-dimensional spaces by formally D-dimensional ones. In this way, all integrals become formally D-dimensional, and we can schematically write for the loop integration measure:

$$\int \frac{d^4k}{(2\pi)^4} \to \mu^{4-D} \int \frac{d^Dk}{(2\pi)^D}, \tag{114}$$

where μ denotes a new, artificial mass scale, the dimensional regularization scale. Though the basic idea [1–3] is simple, care is needed to avoid incorrect or inconsistent results. After the first detailed discussions in Reference [22,61,62], very systematic definitions and analyses of D-dimensional integrals were given by Breitenlohner and Maison [4] and by Collins [63].

3.2.1. Quasi-D-Dimensional Space

Before discussing integrals, we discuss the simpler concept of a D-dimensional space. Let us denote the original four-dimensional Minkowski space as 4S and the formal, or quasi-D-dimensional space as QDS. The question is which properties QDS can have and what its relationship to the original space 4S can be.

Clearly, even on the regularized level, we need the usual properties of linear combinations. If two momenta p^μ and q^μ are elements of QDS, then also $ap^\mu + bq^\mu$ is an element of QDS for any real or complex a and b, with the usual properties of linear combinations. Hence, QDS must constitute a proper mathematical vector space. However, there do not exist mathematical vector spaces with dimensionality D if D is a non-integer real or complex number.

[8] Although the main focus of the review is on the renormalization of Green functions, we provide here a remark on the extraction of physical S-matrix elements via LSZ reduction as mentioned in Section 2.4. LSZ reduction involves the need for so-called wave function renormalization, which ties in with the discussion of finite counterterms and renormalization schemes. In order to obtain properly normalized S-matrix elements, Green functions need to be divided by $\sqrt{z_i}$ for each external line, where z_i is the residue of the corresponding two-point function at the pole corresponding to the rest mass of the considered external particle i. This may be automatically achieved by choosing an on-shell renormalization scheme for renormalized fields, where all such residues are equal to unity; see, e.g., the discussion in Reference [34]. If a different renormalization scheme is chosen, the wave function factor $\sqrt{z_i}$ may be different from unity and needs to be explicitly taken into account, such as in the scheme proposed in Reference [60] for the electroweak Standard Model. In practical computations in DReg, it is actually often possible to carry out the renormalization program only partially, such that quantum fields remain unrenormalized and the residue factors $\sqrt{z_i}$ remain divergent. After LSZ reduction and proper wave function renormalization, nevertheless finite and correct S-matrix elements can be obtained.

[9] In textbooks and in practical computations, counterterms are often obtained by applying a so-called renormalization transformation onto the tree-level action. Section 4.3 and, in more generality, Section 6.1 will also explain under which conditions this procedure is possible.

The crucial observation [61] is that, on the regularized level, we need to accept that arbitrary sets of momentum vectors may have to be treated as linearly independent. Hence, we need to accept that Q*D*S must actually be an infinite-dimensional vector space. Correspondingly, what we call *D*-dimensional momentum vectors are actually elements of Q*D*S with infinitely many components (of course, in the case of physical momenta, only four of them will be nonzero). It turns out to be possible to define objects and operations on Q*D*S with the desired properties, which resemble *D*-dimensional behavior, justifying the name quasi-*D*-dimensional space.

An important consequence for practical applications is that the original space 4S is always a subspace of Q*D*S:

$$4\text{S} \subset \text{Q}D\text{S}\,, \tag{115}$$

regardless of whether $D > 4$, or $D < 4$, or D is complex, assuming the opposite relation leads to mathematical inconsistencies, which will be discussed in the context of dimensional reduction below in Section 3.5.

3.2.2. Properties of *D*-Dimensional Integrals

Now, we turn to integrals over functions of vectors defined on Q*D*S. Clearly, the plethora of successful calculations and available multi-loop techniques (see, e.g., the book [64]) provides ample evidence of the existence of *D*-dimensional integrals and of the consistency of their evaluations. Still, as stressed in Reference [63], it is important to establish the existence of *D*-dimensional integrals in general and to prove the uniqueness of the results. In the literature, different constructive definitions have been proposed. Here, we will describe the construction by Collins [63], which extends the earlier work by Wilson [61].

We begin by listing important properties of *D*-dimensional integration given in Reference [63]. It is generally sufficient to discuss the case of the Euclidean metric. *D*-dimensional Minkowski spacetime can then be treated as one fixed time dimension combined with $(D-1)$-dimensional Euclidean space, and in quantum field theory applications, Minkowski space integrals can be converted to Euclidean space integrals via Wick rotation. Depending on the context, either Minkowski space or Euclidean space notation can be more convenient. For the following integrals, we assumed Euclidean space, with the Euclidean metric for scalar products of vectors:

Property (a) Linearity: for all functions $f_{1,2}$ and coefficients a, b:

$$\int d^D k \Big(a f_1(\vec{k}) + b f_2(\vec{k}) \Big) = a \int d^D k f_1(\vec{k}) + b \int d^D k f_2(\vec{k})\,. \tag{116}$$

Property (b) Translation invariance: for all vectors $\vec{p}$ ∈Q*D*S:

$$\int d^D k f_1(\vec{k} + \vec{p}) = \int d^D k f_1(\vec{k})\,. \tag{117}$$

Property (c) Scaling: for all numbers s,

$$\int d^D k f_1(s\vec{k}) = s^{-D} \int d^D k f_1(\vec{k})\,. \tag{118}$$

Property (d) The *D*-dimensional Gaussian integral in *D*-dimensional Euclidean metric has the value:

$$\int d^D k e^{-\vec{k}^2} = \pi^{D/2}\,. \tag{119}$$

Using D-dimensional spherical coordinates to evaluate this rotationally symmetric integral, $\int d^Dk \to \int d^{D-1}\Omega \int_0^\infty dk k^{D-1} e^{-k^2}$, implies the result for the surface of a D-dimensional sphere:

$$\Omega_D \equiv \int d^{D-1}\Omega = \frac{2\pi^{D/2}}{\Gamma(D/2)} \tag{120}$$

which depends on the well-known Γ-function defined as $\Gamma(z) = \int_0^\infty t^{z-1}e^{-t}dt$ for $\mathrm{Re}(z) > 0$ and by analytic continuation otherwise.

Remark: Properties a, b, c, and d may also be viewed as axioms on the integration. Taken together, they uniquely fix the integration [61].

Property (e) Commutation with differentiation:

$$\frac{\partial}{\partial \vec{p}} \int d^Dk f_1(\vec{k}, \vec{p}) = \int d^Dk \frac{\partial}{\partial \vec{p}} f_1(\vec{k}, \vec{p}). \tag{121}$$

Property (f) Partial integration: The previous equation, together with translation invariance (117), implies the possibility for partial integration:

$$\int d^Dk \frac{\partial}{\partial \vec{k}} f_1(\vec{k}) = 0. \tag{122}$$

Property (g) Two different integrations can be interchanged:

$$\int d^Dp \int d^Dk f(\vec{p}, \vec{k}) = \int d^Dk \int d^Dp f(\vec{p}, \vec{k}). \tag{123}$$

Property (h) If an integral is finite in four dimensions, the D-dimensional version is analytic in a region for D around $D = 4$ and in the external momenta, and it reproduces the original value for $D = 4$.

Remark: The explicit construction of References [61,63] guarantees the existence of the D-dimensional integration and allows establishing general properties. Uniqueness together with existence implies "consistency" in the sense that one initial expression in DReg will always lead to one unique final expression, no matter how and in which order the calculational steps are organized.

3.2.3. Uniqueness and Construction of D-Dimensional Integrals Using Parallel and Orthogonal Spaces

For the full proofs of the properties listed above and for further properties, we refer to Reference [63]. In the following, we summarize the uniqueness proof and then sketch the integral constructions of References [61,63].

We begin with the uniqueness of the D-dimensional integral. It is sufficient to assume the Euclidean metric, such that scalar products are given by $\vec{p} \cdot \vec{k} = p_1k_1 + p_2k_2 + \ldots$ for D-dimensional vectors $\vec{p}, \vec{k}$. Reference [61] starts from the observation that any function of the form $f(\vec{p}_1 \cdot \vec{k}, \ldots, \vec{p}_n \cdot \vec{k}, \vec{k}^2)$ can be obtained from suitable combinations of the derivatives of the generating function:

$$g(s, \vec{p}, \vec{k}) \equiv e^{-s\vec{k}^2 + \vec{p}\cdot\vec{k}}. \tag{124}$$

Indeed, derivatives with respect to $\vec{p}$ and s generate arbitrary polynomials in all components of $\vec{k}$ and $\vec{k}^2$, multiplied by $g(s, \vec{p}, \vec{k})$. Ignoring convergence questions, any function can be sufficiently approximated in this way.

Using linearity (116), it is sufficient to prove the uniqueness of the integral over the generating function $g(s, \vec{p}, \vec{k})$. Using translation invariance to complete the square, scaling, and the D-dimensional Gaussian integral, we obtain

$$\int d^D k g(s, \vec{p}, \vec{k}) \overset{(117)}{=} \int d^D k e^{-s\vec{k}^2 + \vec{p}^2/4s} \tag{125a}$$

$$\overset{(118)}{=} s^{-D/2} e^{\vec{p}^2/4s} \int d^D k e^{-\vec{k}^2} \tag{125b}$$

$$\overset{(119)}{=} s^{-D/2} e^{\vec{p}^2/4s} \pi^{D/2} \,. \tag{125c}$$

The integral over the generating function is uniquely fixed given the four properties (116)–(119), establishing the general uniqueness of the integral.

Now, we sketch the D-dimensional integral construction proposed by References [61,63]. Suppose the function $f(\vec{p}_1 \cdot \vec{k}, \dots \vec{p}_n \cdot \vec{k}, \vec{k}^2)$ is to be integrated over $\vec{k}$, and we take seriously that all these vectors are elements of QDS, which is actually infinite-dimensional. The result will depend on the n "external momenta" $\vec{p}_1 \dots \vec{p}_n$, and these span a subspace, which is at most n-dimensional. The basic idea is then that the space of all $\vec{k}$ can be split into a "parallel" space and an "orthogonal" space. The parallel space is defined such that it contains all n external vectors $\vec{p}_1 \dots \vec{p}_n$. It has a finite, integer dimensionality n_p. Once the parallel space is fixed, we can uniquely decompose any loop momentum and its scalar products as

$$\vec{k} = \vec{k}_{\|} + \vec{k}_{\perp} \qquad \vec{p}_i \cdot \vec{k} = \vec{p}_i \cdot \vec{k}_{\|} \qquad \vec{k}^2 = \vec{k}_{\|}^2 + \vec{k}_{\perp}^2 \,. \tag{126}$$

For this reason, the $\vec{k}$ dependence of the integrand may be abbreviated as

$$f(\vec{p}_1 \cdot \vec{k}, \dots \vec{p}_n \cdot \vec{k}, \vec{k}^2) \equiv f(\vec{k}_{\|}, \vec{k}_{\perp}^2) \,, \tag{127}$$

i.e., the $\vec{k}$ dependence is separated: the vector $\vec{k}_{\|}$ appears explicitly, but it is an element of a finite-dimensional vector space where ordinary integrals are defined. The orthogonal components appear only as the square $\vec{k}_{\perp}^2$. This is the crucial simplification, which allows the two-step definition, where first, the integral is split as

$$\int d^D k f(\vec{p}_1 \cdot \vec{k}, \dots \vec{p}_n \cdot \vec{k}, \vec{k}^2) \equiv \int d^{n_p} k_{\|} \int d^{D-n_p} k_{\perp} f(\vec{k}_{\|}, \vec{k}_{\perp}^2) \tag{128}$$

and second, the $D - n_p$-dimensional integral on the right-hand side is defined via spherical coordinates, using Equation (120):

$$\int d^{D-n_p} k_{\perp} f(\vec{k}_{\|}, \vec{k}_{\perp}^2) \equiv \Omega_{D-n_p} \int_0^{\infty} dk k^{D-n_p-1} f(\vec{k}_{\|}, k^2) \,. \tag{129}$$

In these two steps, the original D-dimensional integral is defined in terms of a series of ordinary integrals in one dimension and in n_p dimensions. The effect of the regularization becomes manifest with the D-dependence in the exponent, which governs the behavior of the integrand at large k and at small k. If the function f has at most a power-like divergence at large/small k, there is a range of D for which the k-integral is well defined. Its value for arbitrary D is then defined by analytical continuation.

Reference [63] provides detailed discussions of the independence of the choice of the parallel space and its dimensionality n_p, of the analytical continuation in the variable D, and of more general integrals.

We will now discuss the computation of such integrals with two examples, which will illustrate several important general points. The examples are (we, again, work in Euclidean space and use a dimensionless integration variable $\vec{k}$)

$$I_{\rm all}(D) = \int d^D k \vec{k}^2 \delta(\vec{k}^2 - 1)\,, \tag{130a}$$

$$I_1(D) = \int d^D k k_1^2 \delta(\vec{k}^2 - 1)\,. \tag{130b}$$

Both integrals only depend on the dimensionality D. In both cases, we essentially integrate over the surface of the unit sphere, in the first case multiplied by $\vec{k}^2$ and in the second case multiplied by k_1^2. Since no direction is special, the second integral would not change if we replaced k_1^2 by any other k_i^2 with a fixed index i. We will discover a useful relationship between the two integrals.

The first integral may immediately be computed by treating the entire $\vec{k}$ as $\vec{k}_\perp$. We can apply the definition (129) and evaluate the integral as

$$I_{\rm all}(D) = \frac{\Omega_D}{2}\,. \tag{131}$$

For the second integral, we treat the first component as special and align the parallel space along this first component (the explicit component k_1 might also be regarded as the scalar product $\vec{p} \cdot \vec{k}$ with a vector that happens to be $\vec{p} = (1, 0, 0, \ldots)$). Then, the integral becomes by definition

$$I_1(D) = \int_{-\infty}^{\infty} dk_1 k_1^2 \int d^{D-1} k_\perp \delta(\vec{k}_\perp^2 - (1 - k_1^2))\,. \tag{132}$$

The $D-1$-dimensional integral is now of the same type as $I_{\rm all}$ except in reduced dimensionality, and it is only nonzero if $|k_1| \leq 1$. Applying standard substitutions, we obtain

$$I_1(D) = \int_{-1}^{1} dk_1 k_1^2 \frac{\Omega_{D-1}}{2} (1 - k_1^2)^{(D-3)/2}\,. \tag{133}$$

The remaining integral can be related to the definition of the Beta function $B(3/2, (D-1)/2)$ by the substitution $x = k_1^2$, and the result is

$$I_1(D) = \frac{\Omega_{D-1}}{2} \frac{\Gamma(3/2)\Gamma(D/2 - 1/2)}{\Gamma(D/2 + 1)}\,. \tag{134}$$

As announced, these results illustrate important general points:

- The result (134) can be simplified by using the explicit result $\Gamma(3/2) = \sqrt{\pi}/2$, the recursion relation $z\Gamma(z) = \Gamma(z+1)$, and the explicit result for Ω_D in Equation (120). After simplification, we obtain

$$I_1(D) = \frac{\Omega_D}{2D}\,, \tag{135}$$

  where the $(D-1)$-dimensional surface volume is replaced by the D-dimensional one.
- As a result, we simply obtain the relation:

$$I_{\rm all}(D) = D I_1(D)\,, \tag{136}$$

  which agrees with the naive expectation from a D-dimensional space with D vector components despite the construction of QDS as an infinite-dimensional vector space.

- These two integrals I_{all} and I_1 and their relationships will allow defining metric tensors on the quasi-D-dimensional space QDS with appropriate properties resembling D-dimensional behavior.
- Similar relationships are also the essence of the proof of the independence of the choice of the parallel space in defining the integrals [63].

3.2.4. Construction of D-Dimensional Loop Integrals via Schwinger Parametrization

In addition to the integral construction via parallel and orthogonal spaces, we also sketch a second way to construct D-dimensional integrals. This second way was carried out and used, in particular, in References [4,22]. It also realizes the four basic properties of linearity, translation invariance, scaling, and the generalization of the Gaussian integral (116)–(119), but otherwise, it is formulated specifically for loop integrals in Minkowski space quantum field theory. It is based on the well-understood Schwinger parametrization, which has been developed for arbitrary loop integrals and used, e.g., in BPHZ renormalizability proofs in References [65–67] and in the context of analytical regularization [68]. For general accounts, see also the books [64,69]. We present here, first, a simple example and, then, indicate the general case.

The example is a standard one-loop two-point function with loop integrand

$$\frac{i^2 e^{i(u_1^\mu (k+p)_\mu + u_2^\mu k_\mu)}}{[(k+p)^2 - m^2 + i\varepsilon][k^2 - m^2 + i\varepsilon]} \equiv \frac{i^2 e^{i(u_1^\mu (k+p)_\mu + u_2^\mu k_\mu)}}{D_1 D_2} \tag{137}$$

with loop integration momentum k and external momentum p, two equal masses, and the customary $+i\varepsilon$ prescription. We also allowed for a generating function in the numerator similar to Equation (124) with two vector-like parameters u_1^μ, u_2^μ such that the derivatives at $u_{1,2} = 0$ can generate arbitrary polynomials of propagator momenta in the numerator. The Schwinger parametrization, or α-parametrization, uses the following replacement for generic propagators:

$$\frac{1}{[p^2 - m^2 + i\varepsilon]^\nu} = \frac{1}{i^\nu \Gamma(\nu)} \int_0^\infty d\alpha \alpha^{\nu-1} e^{i(p^2 - m^2 + i\varepsilon)\alpha}, \tag{138}$$

which is derived by substitution and by using the definition of the Γ function. In this way, the integrand (137) becomes

$$\int_0^\infty d\alpha_1 d\alpha_2 e^{i(D_1\alpha_1 + D_2\alpha_2)} e^{i(u_1\cdot(k+p) + u_2\cdot k)} \tag{139}$$

and the appearing exponent is a quadratic polynomial in the loop momentum which, up to the factor i, can be written as[10]

$$k^2 M + 2k^\mu J_\mu + K + K', \tag{140}$$

or, by completing the square, as

$$k'^2 M - J^2 M^{-1} + K + K', \tag{141}$$

[10] Note that in this particular case, the quantity M is a number, while in the general case of multiloop integrals M will be a matrix.

with

$$k'_\mu = k_\mu + M^{-1} J_\mu\,, \tag{142a}$$

$$M = \alpha_1 + \alpha_2\,, \tag{142b}$$

$$J_\mu = p_\mu \alpha_1 + \frac{1}{2}(u_1 + u_2)_\mu\,, \tag{142c}$$

$$K = p^2 \alpha_1 + u_1 \cdot p\,, \tag{142d}$$

$$K' = (i\varepsilon - m^2)(\alpha_1 + \alpha_2)\,. \tag{142e}$$

Using this rearrangement in the exponent, the loop integral over k becomes essentially a Gaussian integral over $e^{ik'^2 M}$. Using translation invariance and the scaling property (117) and (118) and employing the Minkowski metric, we obtain

$$\int \frac{d^D k}{(2\pi)^D} e^{i(k'^2 + i\varepsilon)M} = (4\pi)^{-D/2} i^{1-D/2} M^{-D/2}\,. \tag{143}$$

The previous steps transform the integrand (137) into a product of a purely Gaussian integrand and a remainder, which does not depend on the integration momentum. This leads to the following definition:

$$\int \frac{d^D k}{(2\pi)^D} \frac{i^2 e^{i(u_1\cdot(k+p) + u_2 \cdot k)}}{D_1 D_2} = (4\pi)^{-D/2} i^{1-D/2} \times \int_0^\infty d\alpha_1 d\alpha_2 M^{-D/2} e^{i(-J^2 M^{-1} + K + K')}\,. \tag{144}$$

In this way, the D-dimensional integral is defined in terms of two standard integrals over $\alpha_{1,2}$. The integrand depends on $\alpha_{1,2}$ via the exponential function and via the term $M^{-D/2}$, where the D dependence enters.

This example can be generalized to arbitrary loop integrals, and it may be generalized to numerator polynomials in the integration momentum. We provide here the result for the general case of a 1PI graph G with L loops, loop momenta k_i, I internal lines with momenta ℓ_k, a generating function with parameters u_k, and a derivative operator $Z(-i\partial/\partial_u)$ with respect to all the u_k in the numerator (see, e.g., [4,64]):

$$\mathcal{T}_G = \int d^D k_1 \ldots d^D k_L Z(-i\partial/\partial_u) \frac{i^I e^{iu_k \cdot \ell_k}}{D_1 \ldots D_I}\bigg|_{u=0}\,. \tag{145}$$

Selecting specific choices of the operator Z and setting $u = 0$ after taking the derivative produce specific numerators. Going through similar steps as before, the integrand can be rearranged into the form of pure Gaussian integrals, leading to the result and D-dimensional definition:

$$\mathcal{T}_G = c_D^L \int_0^\infty d\alpha_1 \ldots d\alpha_I Z(-i\partial/\partial_u) \mathcal{U}^{-D/2} e^{iW}\bigg|_{u=0}\,, \tag{146a}$$

$$c_D = i^{1-D/2} (4\pi)^{-D/2}\,. \tag{146b}$$

By definition, the variables u have to be set to zero before performing the α integration. The formula clearly corresponds to the one-loop example, where $L = 1$, $I = 2$, and $Z = 1$, and

$$\mathcal{U} = M = \alpha_1 + \alpha_2\,, \tag{147a}$$

$$W = \frac{p^2 \alpha_1 \alpha_2 - \alpha_1 u_2 \cdot p + \alpha_2 u_1 \cdot p - \frac{1}{4}(u_1 + u_2)^2}{\mathcal{U}} + K'\,. \tag{147b}$$

In the general case, the quantities in the result (146a) have the following properties:

- $\mathcal{U}$ is the so-called Symanzik polynomial in the αs of degree L. All its terms have a unity coefficient; hence, inside the α-integration range, $\mathcal{U}$ is positive.
- The ultraviolet divergences (including subdivergences) of the original loop integral are mapped to the singularities of the α integrals at small α. As some of the αs approach zero, $\mathcal{U}$ vanishes with a certain power-like behavior, depending on the original power counting of the Feynman diagram. The D-dependence of $\mathcal{U}^{-D/2}$ then effectively regularizes the divergences.
- The exponent W is a rational function in the αs and depends on the external momenta, the masses, and the u_k variables.

The definition of the general loop integral (145) via Equation (146a) provides not only a second constructive definition of D-dimensional integration (which is of course equivalent to the one in Section 3.2.2 thanks to the uniqueness theorem), but it also provides a starting point for practical computations and allows rigorous proofs of renormalizability and further renormalization properties [4,22].

For completeness, we present here briefly the full computation of the one-loop example (144) for the scalar numerator case where $u_{1,2} = 0$. With the substitutions $\alpha = \alpha_1 + \alpha_2$ and $\beta = \alpha_1/\alpha$, we obtain

$$(144) = (4\pi)^{-D/2} i^{1-D/2} \int_0^\infty d\alpha \int_0^1 d\beta \alpha^{1-D/2} e^{-i\alpha Q(\beta)} \tag{148}$$

with

$$Q(\beta) = -p^2\beta(1-\beta) + m^2 - i\varepsilon\,. \tag{149}$$

The α-integration is given by the Γ function up to a substitution, so we obtain the final expression:

$$(144) = -i(4\pi)^{-D/2}\Gamma(2-D/2) \int_0^1 d\beta Q(\beta)^{D/2-2}\,, \tag{150}$$

which is the well-known one-dimensional integral representation of the result.

3.3. *Metric Tensors, γ Matrices, and Other Covariants in D Dimensions*

In this subsection, we will discuss covariant objects used in DReg calculations, such as momentum vectors k^μ, vector fields $A^\mu(x)$, γ^μ-matrices, and the metric tensor $g^{\mu\nu}$. We first provide a summary of the basic properties, which are often sufficient in practical calculations. Afterwards, we will give details on the explicit construction of the required objects on the quasi-D-dimensional space QDS. As in the case of integrals, the explicit construction is important to guarantee the consistency of the calculational rules.

In the context of Equation (115), we have seen that the original four-dimensional Minkowski space is necessarily a subspace of QDS. Hence, strictly four-dimensional objects always exist in addition to the quasi-D-dimensional ones, and we will discuss the relevant relationships. At the end of the subsection, we will discuss the objects γ_5 and $\epsilon_{\mu\nu\rho\sigma}$, which are tied to strictly four dimensions.

3.3.1. Properties of D-Dimensional Covariants and γ-Matrices

We begin with the main properties that can be used in the calculations:

- Vectors or more general objects X^μ on QDS with upper indices such as k^μ, $A^\mu(x)$, γ^μ, and $g^{\mu\nu}$ can be defined by the explicit values of their components. The index μ takes infinitely many values and runs from $0, 1, 2, \ldots$ to infinity.
- Indices can be lowered and raised with the D-dimensional metric tensor $g_{\mu\nu}$ and $g^{\mu\nu}$ as

$$X_\mu = g_{\mu\nu} X^\nu \qquad\qquad X^\mu = g^{\mu\nu} X_\nu\,. \tag{151}$$

We reiterate that we used a mostly minus metric.

- The D-dimensional metric tensor with a mostly minus signature satisfies the expected relations:

$$g^{\mu\nu} = g_{\mu\nu} = \begin{cases} +1 & \text{for } \mu = \nu = 0 \\ -1 & \text{for } \mu = \nu = 1, 2, \ldots \\ 0 & \text{for} \mu \neq \nu \end{cases} \tag{152a}$$

$$g_{\mu\nu} g^{\mu\nu} = D \,. \tag{152b}$$

These two relations extend the most-important and -obvious properties of the metric tensor to D dimensions. They, however, seem contradictory since the indices take infinitely many values, and naively, one might expect the contraction in the second equation to diverge. The solution is to regard a contraction with the lower index $g_{\mu\nu}$ as a linear mapping, acting on upper index quantities, instead of defining it via summation over explicit index values. Below, we show in detail how this idea reconciles the two equations (152) and gives meaning to general lower index quantities.

- Contraction with $g_{\mu\nu}$ commutes with D-dimensional integration, as, e.g., in

$$g_{\mu\nu} \int d^D k k^\mu k^\nu f(k) = \int d^D k g_{\mu\nu} k^\mu k^\nu f(k) = \int d^D k k^2 f(k) \,, \tag{153}$$

and if a tensor $T^{\mu\nu}$ has only a finite number of nonvanishing entries, the expected result with an explicit summation is obtained:

$$g_{\mu\nu} T^{\mu\nu} = \sum_{\mu,\nu=0}^{\infty} g_{\mu\nu} T^{\mu\nu} = T^{00} - \sum_{i=1}^{\infty} T^{ii} \,. \tag{154}$$

- The γ^μ-matrices may also be defined on QDS, i.e., for $\mu = 0, 1, 2, \ldots$ up to infinity, such that they satisfy the basic relations:

$$\{\gamma^\mu, \gamma^\nu\} = 2 g^{\mu\nu} \mathbb{1} \,, \qquad \gamma_\mu \gamma^\mu = D \mathbb{1} \,. \tag{155}$$

A representation exists that satisfies the same relations for complex conjugation, Hermitian conjugation, and charge conjugation as the ones of Equations (21) also for all μ. Hence, it is also possible to define spinors on QDS and to use the definitions (22) for adjoint and charge-conjugated spinors in D dimensions.

As a result, the following relations hold for bilinear expressions of anticommuting spinors on QDS:

$$\bar{\psi}_1 \Gamma \psi_2 = \overline{\psi_2^C} \Gamma^C \psi_1^C \qquad \text{with} \qquad \Gamma^C = -C \Gamma^T C \tag{156a}$$

$$(\bar{\psi}_1 \Gamma \psi_2)^\dagger = \bar{\psi}_2 \bar{\Gamma} \psi_1 \qquad \text{with} \qquad \bar{\Gamma} = \gamma^0 \Gamma^\dagger \gamma^0 \tag{156b}$$

and

$$\{1, \gamma_5, \gamma^\mu, \gamma^\mu \gamma_5\}^C = \{1, \gamma_5, -\gamma^\mu, -\gamma_5 \gamma^\mu\}, \tag{157a}$$

$$\overline{\{1, \gamma_5, \gamma^\mu, \gamma^\mu \gamma_5\}} = \{1, -\gamma_5, \gamma^\mu, -\gamma_5 \gamma^\mu\}. \tag{157b}$$

For more details on the γ_5 matrix, see Section 3.3.3.

- The quasi-D-dimensional space actually is infinite-dimensional and, hence, contains the original four-dimensional Minkowski space, as expressed in Equation (115). On the level of covariants, we therefore can define the purely 4-dimensional metric tensor $\bar{g}^{\mu\nu}$ by the 4-dimensional entries $\bar{g}^{00} = -\bar{g}^{ii} = +1$ for $i = 1, 2, 3$ and $\bar{g}^{\mu\nu} = 0$ in all other cases. This tensor acts as a projector on the original Minkowski space. It also allows defining a complementary projector, the metric tensor of the $(D-4)$-

dimensional complement as $\hat{g}^{\mu\nu} = g^{\mu\nu} - \bar{g}^{\mu\nu}$. In summary, all these tensors satisfy the following equations:

$$D\text{-dim.}: \; g^{\mu\nu} = \bar{g}^{\mu\nu} + \hat{g}^{\mu\nu} \qquad 4\text{-dim.}: \; \bar{g}^{\mu\nu} \qquad (D-4)\text{-dim.}: \; \hat{g}^{\mu\nu} \tag{158}$$

with the dimensionalities expressed by

$$g_{\mu\nu}g^{\mu\nu} = D\,, \qquad \bar{g}_{\mu\nu}\bar{g}^{\mu\nu} = 4\,, \qquad \hat{g}_{\mu\nu}\hat{g}^{\mu\nu} = D-4 \tag{159}$$

and the following contraction rules, expressing the projection and subspace relationships:

$$\bar{g}_{\mu\nu}\bar{g}^{\nu\rho} = \bar{g}_{\mu\nu}g^{\nu\rho} = g_{\mu\nu}\bar{g}^{\nu\rho} = \bar{g}_{\mu}^{\rho}\,, \tag{160a}$$
$$\hat{g}_{\mu\nu}\hat{g}^{\nu\rho} = \hat{g}_{\mu\nu}g^{\nu\rho} = g_{\mu\nu}\hat{g}^{\nu\rho} = \hat{g}_{\mu}^{\rho}\,, \tag{160b}$$
$$\bar{g}_{\mu\nu}\hat{g}^{\nu\rho} = \hat{g}_{\mu\nu}\bar{g}^{\nu\rho} = 0\,. \tag{160c}$$

- Since the metric tensors $\bar{g}^{\mu\nu}$ and $\hat{g}^{\mu\nu}$ act as projectors on the four-dimensional and $(D-4)$-dimensional subspaces, we can generally decompose any vector X^μ as

$$X^\mu = \bar{X}^\mu + \hat{X}^\mu \qquad \bar{X}^\mu = \bar{g}^\mu_{\;\nu}X^\nu \qquad \hat{X}^\mu = \hat{g}^\mu_{\;\nu}X^\nu\,, \tag{161}$$

such that, e.g., squares and scalar products behave as

$$X^2 = \bar{X}^2 + \hat{X}^2 \qquad X_\mu Y^\mu = \bar{X}_\mu\bar{Y}^\mu + \hat{X}_\mu\hat{Y}^\mu \qquad \bar{X}_\mu\hat{Y}^\mu = 0\,. \tag{162}$$

Similar relationships can be defined for tensors in obvious ways.

- As in Equation (161), we can define four-dimensional and $(D-4)$-dimensional versions $\bar{\gamma}^\mu$ and $\hat{\gamma}^\mu$, respectively, which satisfy

$$\{\gamma^\mu, \bar{\gamma}^\nu\} = \{\bar{\gamma}^\mu, \bar{\gamma}^\nu\} = 2\bar{g}^{\mu\nu}\mathbb{1} \qquad \gamma_\mu\bar{\gamma}^\mu = \bar{\gamma}_\mu\bar{\gamma}^\mu = 4\,\mathbb{1}\,, \tag{163a}$$
$$\{\gamma^\mu, \hat{\gamma}^\nu\} = \{\hat{\gamma}^\mu, \hat{\gamma}^\nu\} = 2\hat{g}^{\mu\nu}\mathbb{1}\,, \qquad \gamma_\mu\hat{\gamma}^\mu = \hat{\gamma}_\mu\hat{\gamma}^\mu = (D-4)\mathbb{1}\,, \tag{163b}$$
$$\{\bar{\gamma}^\mu, \hat{\gamma}^\nu\} = 0\,, \qquad \bar{\gamma}_\mu\hat{\gamma}^\mu = 0\,. \tag{163c}$$

Traces of γ-matrices are defined such that

$$\mathrm{Tr}(\mathbb{1}) = 4 \qquad \mathrm{Tr}(\gamma^\mu) = 0\,. \tag{164}$$

With these relations, all other traces of products of γ-matrices can be calculated.

- The properties of γ_5 and $\epsilon_{\mu\nu\rho\sigma}$ are discussed below in Section 3.3.3.
- Generally, objects (covariants or operators) that vanish in purely four dimensions are called evanescent. Examples of evanescent objects are all contractions with $\hat{g}^{\mu\nu}$ such as $\hat{g}^{\mu\nu}$ itself, $\hat{\gamma}^\mu$, or products such as $\hat{\gamma}^\mu\hat{\gamma}^\nu$, $\hat{\gamma}^\mu\bar{\gamma}^\nu$. Later, we will see that many objects related to γ_5 or related to Fierz identities are also evanescent.

3.3.2. Construction of D-Dimensional Covariants and γ Matrices

Now, we describe how objects may be defined that satisfy these relations. The main difficulties are to define the lower index metric tensor and its contraction rules and the γ^μ-matrices. We essentially follow Collins [63] in the construction of all these quantities.

As mentioned above, at first sight, it appears difficult to reconcile the different properties (152) of the D-dimensional metric tensor $g_{\mu\nu}$. The basic idea is that, fundamentally, tensors with lower indices can be viewed as multilinear forms, i.e., mappings of objects with upper indices to numbers. In the case of infinite-dimensional vector spaces, it is not always sufficient to specify their component values. For the Euclidean metric and for a general tensor T with components T^{ij}, Collins proposed the definition of $\delta_{ij}T^{ij}$ as an abbreviation of a mapping $\delta(T)$. This mapping can be defined via a D-dimensional integral [63]:

$$\delta_{ij}T^{ij} = \delta(T) = A\int d^D k T^{ij} k_i k_j \delta(\vec{k}^2 - 1) \tag{165}$$

with normalization constant $A = D\Gamma(D/2)/\pi^{D/2}$. For the integration momentum, we simply take $k_i = k^i$ such that $\delta^{ij}k_i k_j = \vec{k}^2$. The crucial point is that, by definition, the index contraction is performed before evaluating the integral. As a special case, the definition also contains a definition of the individual components:

$$\delta_{ij} = A\int d^D k k_i k_j \delta(\vec{k}^2 - 1)\,. \tag{166}$$

The calculations of the integrals in Equations (130) leading to Equation (136) then show that

$$\delta^{ij}\delta_{ij} = D\,, \tag{167a}$$

$$\delta_{ij} = \delta^{ij}\,. \tag{167b}$$

The first of these relations demonstrates the effective D-dimensional behavior of the metric tensor, and the second holds componentwise and shows that the individual components have the usual values. However, the equations also show, again, that contraction with δ_{ij} is not defined by summation over explicit component values, but via the integral (165), where contraction and integration cannot be interchanged. Clearly,

$$\sum_{i,j=1}^{\infty} \delta^{ij}\delta_{ij} = \infty \tag{168}$$

in contrast to the correct Equation (167a).

By treating the space-like components of $g_{\mu\nu}$ analogously to the definition of δ_{ij} discussed above, it is clear that we can define a metric tensor that indeed fulfills the announced Equations (152). General tensor contractions of the form $T^{\mu\nu}g_{\mu\nu}$ are defined via integrals such as Equation (165) and not via explicit summation over component values; in general, summation over indices does not commute with integration (which here defines contraction). The exception are cases of tensors with only a finite number of nonvanishing components, in which case, Equations (152a) and (167b) immediately establish the relation (154). In addition, the definition via an integral benefits from the fact that different D-dimensional integrations can be interchanged; see Equation (123). Therefore, $g_{\mu\nu}$ may be pulled inside or outside integrals as exemplified in Equation (153). In this way, we established all desired properties of the D-dimensional metric tensor by explicit construction.

Next, we discuss the construction of γ^μ-matrices that satisfy the formally D-dimensional relations (155). We define them similarly to Reference [63]. We start from any standard representation for the usual four-dimensional γ^μ-matrices such as the representation (24) and denote these 4×4-matrices now as $\gamma^\mu_{[4]}$, $\mu = 0,1,2,3$. The usual four-dimensional γ_5-matrix is now denoted as $\gamma_{[4]5} = i\gamma^0_{[4]}\gamma^1_{[4]}\gamma^2_{[4]}\gamma^3_{[4]}$. We assumed a representation such as (24) in which the properties (21) hold, such that only γ^2 is imaginary and all others are real.

Then, the formally D-dimensional γ^μ-matrices can be defined as infinite-dimensional block matrices. Adapting the construction of Reference [63], we first set for $\mu = 0,1,2,3$

$$\gamma^\mu = \begin{pmatrix} \gamma^\mu_{[4]} & 0 & 0 & \cdots \\ 0 & \gamma^\mu_{[4]} & 0 & \cdots \\ 0 & 0 & \gamma^\mu_{[4]} & \cdots \\ & & \cdots & \end{pmatrix} \qquad (\mu = 0,1,2,3)\,, \tag{169}$$

where each entry corresponds to a 4×4 submatrix. To construct γ^μ with $\mu > 3$, we define the intermediate matrices $\hat{\gamma}_{(4^k)}$ by

$$\hat{\gamma}_{(4)} = \gamma_{[4]5} \qquad \hat{\gamma}_{(4^{k+1})} = \begin{pmatrix} \hat{\gamma}_{(4^k)} & 0 & 0 & 0 \\ 0 & -\hat{\gamma}_{(4^k)} & 0 & 0 \\ 0 & 0 & -\hat{\gamma}_{(4^k)} & 0 \\ 0 & 0 & 0 & \hat{\gamma}_{(4^k)} \end{pmatrix} \qquad (k \geq 1)\,. \tag{170}$$

In this way, $\hat{\gamma}_{(4^k)}$ is a real, Hermitian, 4^k-dimensional matrix, which consists of $\pm\gamma_{[4]5}$-blocks on the diagonal and which satisfies $(\hat{\gamma}_{(4^k)})^2 = 1$. Using these matrices, we define, for any $\mu \geq 4$, the $2^{2\mu+1}$-dimensional real, anti-Hermitian block matrix:

$$\gamma^\mu_{(2(2\mu+1))} = \begin{pmatrix} 0 & \hat{\gamma}_{(4^\mu)} \\ -\hat{\gamma}_{(4^\mu)} & 0 \end{pmatrix} \qquad (\mu \geq 4) \tag{171}$$

and finally, the infinite-dimensional block matrix:

$$\gamma^\mu = \begin{pmatrix} \gamma^\mu_{(2^{2\mu+1})} & 0 & \cdots \\ 0 & \gamma^\mu_{(2^{2\mu+1})} & \\ \vdots & & \ddots \end{pmatrix} \qquad (\mu \geq 4)\,. \tag{172}$$

The γ^μ-matrices defined in Equations (169) and (172) satisfy all properties announced in Section 3.3.1; with the exception of the commutation relations of γ_5 (see below), these are identical to the purely four-dimensional properties listed in Equations (16), (19), (21) and (22).

We note that the construction of Reference [63] is different in that the Hermiticity/reality/charge conjugation properties of the γ^μ-matrices are different from Equation (21). Our construction corresponds essentially to a subset of the γ^μ-matrices of Reference [63].

3.3.3. Definition of γ_5 and $\epsilon_{\mu\nu\rho\sigma}$ in DReg

A particularly problematic issue is the definition γ_5 and the $\epsilon_{\mu\nu\rho\sigma}$ symbol in DReg;the issue is often referred to as the "γ_5-problem of DReg". In four dimensions, three properties hold for the γ_5-matrix and traces:

$$\{\gamma_5, \gamma^\mu\} = 0, \tag{173a}$$

$$\mathrm{Tr}(\gamma_5\gamma^\mu\gamma^\nu\gamma^\rho\gamma^\sigma) = -4i\epsilon^{\mu\nu\rho\sigma}, \tag{173b}$$

$$\mathrm{Tr}(\Gamma_1\Gamma_2) = \mathrm{Tr}(\Gamma_2\Gamma_1)\,. \tag{173c}$$

The last equality means that the traces are cyclic. In $D \neq 4$ dimensions, it is inconsistent to require these properties simultaneously, and one has to give up one of them. To exhibit the problem, we consider the trace $t_{\mu_1\ldots\mu_4} = \mathrm{Tr}(\gamma_{\mu_1}\cdots\gamma_{\mu_4}\gamma_5)$ and employ the following series of steps, making use of Equation (173).

$$\begin{aligned} Dt_{\mu_1\ldots\mu_4} &= \mathrm{Tr}\big(\gamma^\alpha\gamma_\alpha\gamma_{\mu_1}\cdots\gamma_{\mu_4}\gamma_5\big) \\ &= \mathrm{Tr}\big((2\gamma^\alpha g_{\alpha\mu_1} - \gamma^\alpha\gamma_{\mu_1}\gamma_\alpha)\cdots\gamma_{\mu_4}\gamma_5\big) \\ &= \ldots \\ &= 8t_{\mu_1\ldots\mu_4} + \mathrm{Tr}\big(\gamma^\alpha\gamma_{\mu_1}\cdots\gamma_{\mu_4}\gamma_\alpha\gamma_5\big) \\ &= (8-D)t_{\mu_1\ldots\mu_4}\,. \end{aligned} \tag{174}$$

In the first step, the D-dimensional contraction rule is used, leading to the factor D; in the intermediate steps, the γ^μ anticommutation rule is used four times, leading to the factor of eight. In the last step, cyclicity and the anticommutation relation (173a) are used to relate all terms to the initial trace. The outcome is that

$$(4-D)t_{\mu_1\ldots\mu_4} = 0\,; \tag{175}$$

hence, either $D = 4$ or the trace must vanish. In other words, for $D \neq 4$, two of the Equations (173) imply that the third equation is wrong. In order to set up a consistent regularization that allows a continuous limit to four dimensions, we need both $D \neq 4$ and a nonvanishing trace at the same time, and therefore, we need to give up the validity of some of the Equations (173).

As a result, there is a plethora of proposals for how to treat γ_5. The standard one, which is known to be mathematically well-defined and consistent, is the so-called BMHV scheme [1,4]. This scheme gives up the anticommutation property of γ_5; it is consistent in the sense that it is compatible with the unitarity and causality of quantum field theory, but it does not manifestly lead to the correct conservation/non-conservation properties of currents and does not manifestly preserve the gauge invariance of chiral gauge theories.

In the BMHV scheme, γ_5 is defined in the identical way as in four dimensions:

$$\gamma_5 = i\gamma^0\gamma^1\gamma^2\gamma^3\,. \tag{176}$$

This clearly treats the first, original four dimensions differently from the remaining $(D-4)$ dimensions. Accordingly, we obtain the modified anticommutation relations:

$$\{\gamma^\mu, \gamma_5\} = \{\hat{\gamma}_\mu, \gamma_5\} = 2\hat{\gamma}_\mu\gamma_5\,, \tag{177a}$$

$$\{\bar{\gamma}_\mu, \gamma_5\} = 0\,, \tag{177b}$$

$$[\hat{\gamma}_\mu, \gamma_5] = 0\,, \tag{177c}$$

where, as in Equation (163), the split $\gamma^\mu = \bar{\gamma}^\mu + \hat{\gamma}^\mu$ into the four-dimensional and $(D-4)$-dimensional parts was used. Only the original matrices $\bar{\gamma}^\mu$ fully anticommute with γ_5. In this way, D-dimensional Lorentz invariance is effectively broken by the regularization. Similarly, this modification leads to a breaking of gauge invariance in chiral gauge theories on the regularized level in DReg. This is clearly a drawback and a central topic of the present review.

Similarly, the Levi-Civita $\epsilon_{\mu\nu\rho\sigma}$ symbol, defined as a fully antisymmetric object with four indices, is only well defined in purely four dimensions. Hence, using the split notation, we may write, as stressed in Reference [4],

$$\epsilon_{\mu\nu\rho\sigma} = \bar{\epsilon}_{\mu\nu\rho\sigma}\,, \qquad \hat{\epsilon}_{\mu\nu\rho\sigma} = 0\,, \tag{178}$$

and rewrite the definition of γ_5 as

$$\gamma_5 = -\frac{i}{4!}\bar{\epsilon}_{\mu\nu\rho\sigma}\bar{\gamma}^\mu\bar{\gamma}^\nu\bar{\gamma}^\rho\bar{\gamma}^\sigma\,, \tag{179}$$

with the sign convention:

$$\epsilon^{0123} = -\epsilon_{0123} = +1\,, \tag{180}$$

which was already used in Equation (19). In practical computations, often, combinations of two ϵ-symbols appear. The following four-dimensional identity remains valid:

$$\bar{\epsilon}^{\mu\nu\rho\sigma}\bar{\epsilon}_{\alpha\beta\gamma\delta} = -\bar{g}^{\mu}{}_{\alpha}\bar{g}^{\nu}{}_{\beta}\bar{g}^{\rho}{}_{\gamma}\bar{g}^{\sigma}{}_{\delta} \pm \ldots \tag{181}$$

where the dots denote 23 further similar terms leading to total antisymmetrization in the indices. Some calculations, e.g., the prescription by Larin [70] propose to elevate this identity to the level of D dimensions, i.e., to assume the validity of the corresponding identity with formally D-dimensional metric tensors, i.e., effectively without the bars. Let us remark that such a D-dimensional identity can ultimately lead to inconsistencies in the sense that one initial expression could lead to different answers. To make this inconsistency explicit, we denote the right-hand side of Equation (181) in D dimensions as $p^{\mu\nu\rho\sigma}_{\alpha\beta\gamma\delta}$. Then, consider the product of four ϵ-symbols:

$$\epsilon^{\mu\nu\rho\sigma}\epsilon_{\alpha\beta\gamma\delta}\epsilon_{\mu\nu\rho\sigma}\epsilon^{\alpha\beta\gamma\delta}\,. \tag{182}$$

This can be evaluated in two ways with the two results:

$$\text{either } p^{\mu\nu\rho\sigma}_{\alpha\beta\gamma\delta}p^{\alpha\beta\gamma\delta}_{\mu\nu\rho\sigma} \qquad \text{or } p^{\mu\nu\rho\sigma}_{\mu\nu\rho\sigma}p^{\alpha\beta\gamma\delta}_{\alpha\beta\gamma\delta}\,. \tag{183}$$

In strictly four dimensions, both expressions give $24^2 = 576$, so there is no inconsistency. However, assuming the validity of these equations in D dimensions and using D-dimensional metric tensors in the contractions, the two results are different:

$$\text{either } 24D(D-1)(D-2)(D-3) \qquad \text{or } [D(D-1)(D-2)(D-3)]^2\,. \tag{184}$$

Hence, in an amplitude involving such contractions of ϵ-symbols, the result is ambiguous, except for the leading poles in $1/(D-4)$. For this reason, in a fully consistent treatment, only the four-dimensional version of the identity (181) is valid [4].

In view of the drawbacks of the BMHV scheme, many alternative versions of DReg have been proposed in the literature. For instance, Reference [14] proposed that a fully anticommuting γ_5 may be used in certain Feynman graphs, in spite of the inconsistency between the Equation (173) mentioned above. Similarly, References [71,72] derived that, in specific applications, the correct results can be also be obtained using simpler schemes with anticommuting γ_5. A well-known review of the situation was given by Jegerlehner [15], where further arguments were presented that the "naive" anticommuting γ_5 may be used in many cases. Kreimer et al. [17] proposed a different kind of alternative to BMHV: out of the three Equations (173), the cyclicity of the trace is given up, but the anticommutativity is kept. In this case, special attention must be paid to "subdiagram consistency", as described in Reference [72]: "It should give unique results independently of whether some diagram is considered as a subdiagram, and independently of the order in which subdiagrams are calculated. Otherwise subdivergences could not be properly subtracted in multiloop diagrams." Reference [17] introduced so-called "reading-point" prescriptions to deal with this difficulty.

All these alternative proposals have in common that their general applicability to all cases has not been established; hence, the all-order proofs of the renormalizability properties of, e.g., References [4,22,73,74] do not apply to them.

We also briefly comment on two recent investigations of the γ_5-problem in alternatives to DReg. Reference [75] considered dimensional schemes in various slightly different implementations (e.g., the so-called four-dimensional helicity (FDH) scheme discussed in more detail below in Section 3.5) from the point of view of practical one- and two-loop calculations. At the two-loop level, there is no single scheme that stands out as computationally most efficient. Reference [76] considered strictly four-dimensional schemes as alternatives to dimensional regularization, in the hope that these schemes might offer practical advantages with respect to the treatment of γ_5. The considered class of schemes is wide and general, but contains only schemes that do not break gauge invariance as immediately as, e.g., the Pauli–Villars scheme. This reference showed clearly that all these schemes have very similar problems for γ_5 as dimensional schemes. The reason is that, in those schemes, the regularization is essentially performed by replacement rules, and those replacement rules do not necessarily commute with applying, e.g., the cyclicity of traces.

3.4. Relation to the Lagrangian in D Dimensions

This subsection is devoted to a seemingly simple statement, which, however, constitutes another important advantage of DReg. DReg can already be formulated at the level of the Lagrangian, and regularized Feynman diagrams can literally be obtained from a D-dimensional version of the Gell–Mann–Low formula with a D-dimensional Lagrangian. This fact allows a very efficient investigation of the properties of regularized Green functions. Examples are the all-order proof of the regularized quantum action principle (see Section 4.2) and the textbook derivation of renormalization group β functions and anomalous dimensions from divergences in the counterterm Lagrangian (see, e.g., the textbook by Srednicki [35]).

The explicit construction of formally D-dimensional objects in DReg provides all objects needed to formulate a D-dimensional Lagrangian. Fields $\phi(x)$ are defined as functions of D-dimensional vectors x^μ, i.e., of elements of the quasi-D-dimensional space QDS. Metric tensors, derivatives, vector fields, and γ-matrices have all been extended to D dimensions as well. The construction of γ-matrices implies also a definition of D-dimensional extensions of four-spinor fields (which have infinitely many components in view of Equation (169)). For this reason, any Lagrangian of a four-dimensional quantum field theory involving such fields can be naturally extended to D dimensions.[11]

If a Lagrangian involves the γ_5-matrix or the $\epsilon_{\mu\nu\rho\sigma}$ symbol, e.g., in the case of chiral fermion interactions, an extension to D dimensions remains possible, but the D-dimensional version involves, e.g., γ_5 with its modified anticommutation relations (177). Hence, in such cases, the resulting D-dimensional Lagrangian will not be invariant under formally D-dimensional Lorentz transformations. This, however, does not preclude the application of DReg. In particular, even in such cases, it remains true that four-dimensional Lorentz invariance is manifestly preserved.

This issue illustrates a more general point. Though there is often a preferred choice, the extension of any Lagrangian to D dimensions is, in principle, never unique. It is always possible to change so-called evanescent terms in the Lagrangian, i.e., terms that vanish in four dimensions. If γ_5 is present, this possibility is obvious, e.g., a four-dimensional expression $\overline{\psi}\gamma^\mu P_L\psi$ may be extended to the following three inequivalent D-dimensional choices:

$$\overline{\psi}\gamma^\mu P_L\psi\,, \qquad \text{or } \overline{\psi}P_R\gamma^\mu\psi\,, \qquad \text{or } \overline{\psi}P_R\gamma^\mu P_L\psi\,. \tag{185}$$

In four dimensions, these terms are all equal, but in D dimensions, they are different due to the modified anticommutation relations. However, even independently of γ_5, one may extend, e.g., an interaction term between a vector and a scalar field as

$$\phi^\dagger A^\mu \partial_\mu \phi\,, \qquad \text{or } \phi^\dagger \bar{A}^\mu \bar{\partial}_\mu \phi\,, \tag{186}$$

where the second possibility involves only the purely four-dimensional part of the derivative.

[11] Unfortunately, the 2-component spinor notation described in Section 2.2.3 is not known to be extendable to D dimensions since it is explicitely tied to the representation theory of the 4-dimensional Lorentz group. 2-component spinor Lagrangians need to be rewritten in terms of 4-component spinors before an extension to D dimensions and an application of DReg becomes possible.

Despite the non-uniqueness, clearly, any field theory Lagrangian can be extended to a D-dimensional version. This Lagrangian $\mathcal{L}^{(D)}$ can then be split into a free part and a remainder (the "interaction" part):

$$\mathcal{L}^{(D)} = \mathcal{L}^{(D)}_{\text{free}} + \mathcal{L}^{(D)}_{\text{int}}\,, \tag{187}$$

where the free part must be bilinear in the fields and contain the appropriate kinetic terms. The non-uniqueness affects mainly the "interaction" part; a constraint we will always impose is that the kinetic terms involve strictly D-dimensional derivatives. A reason for this constraint will be illustrated below. It essentially fixes the "free" part of the Lagrangian, such that we may schematically write the free Lagrangian as

$$\mathcal{L}^{(D)}_{\text{free}} = \frac{1}{2}\phi_i \mathcal{D}^{(D)}_{ij}\phi_j \tag{188}$$

with some differential operator $\mathcal{D}^{(D)}_{ij}$ involving D-dimensional derivatives. The notation is meant in a general sense, including the familiar expressions for complex scalar fields, spinor fields, or vector fields. Standard free field theory quantization then leads to the D-dimensional propagators:

$$\mathcal{P}^{(D)}_{jk} = \langle 0|T\phi_j\phi_k|0\rangle \tag{189}$$

which are the Green functions of the differential operators, i.e., which satisfy the inverse relation:

$$\tilde{\mathcal{D}}^{(D)}_{ij}\tilde{\mathcal{P}}^{(D)}_{jk} = i\delta_{ik} \tag{190}$$

in momentum-space in D dimensions.

Let us exemplify these relations and highlight the related subtleties, e.g., for spinor fields, we take the straightforward D-dimensional free Lagrangian $\bar{\psi}(i\gamma^\mu\partial_\mu - m)\psi \equiv \bar{\psi}\mathcal{D}^{(D)}\psi$, leading to the momentum-space propagator:

$$\tilde{\mathcal{P}}^{(D)} = \langle 0|T\psi\bar{\psi}|0\rangle^{\text{F.T.}} = \frac{i}{\not{p} - m} = \frac{i(\not{p} + m)}{p^2 - m^2} \tag{191}$$

where F.T. denotes Fourier transformation of the respective expression (x-arguments are suppressed); the argument of the Fourier transformation is the momentum p; all appearing momenta are D-dimensional, and the $+i\varepsilon$ prescription in the propagator denominator is suppressed. Such propagator Feynman rules lead to loop integrals such as the ones of Section 3.2.4 and denominator structures as in the example (137). The propagator (191) is indeed the inverse of the momentum-space differential operator of the Lagrangian:

$$\tilde{\mathcal{D}}^{(D)} = (\not{p} - m). \tag{192}$$

Taking instead the purely four-dimensional derivative $\bar{\partial}_\mu$ in the free Lagrangian would lead to

$$\langle 0|T\psi\bar{\psi}|0\rangle^{\text{F.T}} = \frac{i}{\bar{\not{p}} - m} = \frac{i(\bar{\not{p}} + m)}{\bar{p}^2 - m^2}\,, \tag{193}$$

which involves only the purely four-dimensional momentum in the denominator. The problem of this choice is that loop integrals would not be regularized; hence, such a choice is not permitted. Similarly, one may propose a recipe where Dirac propagators are regularized as

$$\langle 0|T\psi\bar{\psi}|0\rangle^{\text{F.T}} \to \frac{i(\bar{\not{p}} + m)}{p^2 - m^2}\,, \tag{194}$$

which involves the purely four-dimensional momentum in the numerator and the D-dimensional momentum in the denominator. Such a recipe cannot arise from a D-dimensional Lagrangian; it will not be used, and statements such as the regularized quantum action principle would not necessarily be valid.

As illustrated by this example, the general D-dimensional relationships for the free Lagrangian and the propagators (188)–(190) can always be realized; they will always be assumed, and they are nontrivial.

Once the free Lagrangian is chosen in agreement with the mentioned constraint and the interaction Lagrangian is fixed, D-dimensional regularized Feynman diagrams can be defined via the standard Gell–Mann–Low formula, suitably written in D dimensions. One way to write it is to take the original formula (67) and replace the integrations by D-dimensional ones. In this case, the parameters and fields must have appropriately modified dimensionalities; see, e.g., [77] for a presentation that makes extensive use of this possibility. A second way is to write

$$Z(J,K) = \frac{\langle 0|T\exp\left(i\mu^{D-4}\int d^D x(\mathcal{L}^{(D)}_{\text{int}} + J_i\phi_i + K_i\mathcal{O}_i)\right)|0\rangle}{\langle 0|T\exp\left(i\mu^{D-4}\int d^D x\mathcal{L}^{(D)}_{\text{int}}\right)|0\rangle}, \tag{195}$$

where the regularization scale μ is introduced such that the regularized Lagrangian has mass dimension four. Either way, if the Gell–Mann–Low formula is evaluated via Wick contractions and Fourier transformed, the correct DReg expressions for regularized Feynman diagram amplitudes are obtained. The variant (195) also generates a factor μ^{4-D} accompanying each loop integration, as indicated by Equation (114).

As mentioned in the beginning, this relation between the Lagrangian and regularized Feynman diagrams has important consequences, some of which we will discuss in subsequent sections. Here, we remark that the present discussion allows the possibility that the Lagrangian contains $1/(D-4)$ poles in coefficients; in particular, the discussion is unaffected if the interaction Lagrangian $\mathcal{L}^{(D)}_{\text{int}}$ is defined to include counterterms that are defined order by order to cancel divergences or to restore symmetries.

3.5. *Variants: Dimensional Reduction and* CDR, HV, *and* FDH *Schemes*

DReg as defined so far still leaves room for different options, and there are other variants of dimensional schemes that share the idea of D-dimensional integrals. Here, we give a brief overview of several schemes used in the literature. The overview essentially follows the review [5], and we refer to this review for more details and the original references.

We remark that the following distinction between the schemes does not have much influence on the discussion of chiral fermions and the treatment of γ_5 in DReg. The remarks of Section 3.3.3 apply to all the following schemes, and different alternative treatments of γ_5 have been employed in the literature. In the following discussion, we focus on aspects independent of γ_5.

All the following schemes treat integrals always in D dimensions. They differ in their treatment of vector fields. In order to consistently define the different schemes, it has turned out to be useful [78,79] to introduce the following spaces extending the original four-dimensional space 4S. In Section 3.2.1, we already introduced the quasi-D-dimensional space QDS, on which objects such as formally D-dimensional momenta p^μ and momentum integrations are defined. The explicit construction showed that this space necessarily is infinite-dimensional and contains the original space 4S. Now, we introduce an even bigger space QD_sS (later, $D_s = 4$ will be taken, so this is often called a "quasi-4-dimensional" space). It contains QDS and is formally D_s-dimensional. The relationships are thus

$$4S \subset QDS \subset QD_sS \tag{196}$$

regardless of the values of D and D_s.

Before describing the scheme definitions, we note that vector fields can appear in different roles in Feynman diagrams:

- There are vector fields appearing in propagators in loop diagrams or as propagators or external fields in phase space regions, which lead to infrared, soft, or collinear singularities. We call such vector fields *singular* vector fields. They may be treated in either 4S, Q*D*S, or QD_sS.
- All other vector fields appear outside of 1PI diagrams and outside singular phase space regions. We call them *regular* , and they may be treated differently from singular vector fields.

To motivate the concrete scheme choices, we further list two simple observations:

- Gauge invariance relies on the gauge covariant derivative D_μ, which combines the ordinary derivative (which is always D-dimensional) and vector fields. In order not to directly break gauge invariance on the regularized level, there should be at least a fully D-dimensional covariant derivative. Hence, the singular vector fields should be treated at least as D-dimensional.
- Supersymmetry relies on an equal number of fermionic and bosonic degrees of freedom. The number of spinor degrees of freedom is essentially fixed via $\mathrm{Tr}\mathbb{1} = 4$. Hence, in order not to directly break supersymmetry, singular vector fields should be treated as four-dimensional.

It appears difficult to reconcile the requirements of gauge invariance and supersymmetry, and the different schemes are motivated by focusing on different aspects.

Now, we list the four schemes and refer to Table 1 for a summary:

- Dimensional regularization has two subvariants, called HV and CDR ('t Hooft/Veltman and Conventional Dimensional Regularization). Both variants treat singular vector fields as D-dimensional, i.e. in Q*D*S. This is in line with D-dimensional gauge invariance[12] but leads to a direct breaking of supersymmetry. The HV scheme treats regular vector fields without regularization, i.e. in 4S, and the CDR scheme treats all vector fields in Q*D*S. The space QD_sS is not used.
- The other class of choices is dimensional reduction, originally introduced in the context of supersymmetry [80]. It also has two subvariants, called FDH and DRED (four-dimensional helicity scheme and dimensional reduction). Singular vector fields are treated as D_s-dimensional, and in practical calculations, D_s is eventually set to $D_s = 4$. Hence, singular vector fields are essentially treated as quasi-4-dimensional, but the quasi-4-dimensional space contains the D-dimensional subspace, such that both gauge invariance and supersymmetry are not immediately broken. FDH is analogous to HV and treats regular vector fields as strictly four-dimensional, and DRED treats all vector fields in QD_sS.

Table 1. Treatment of singular and regular vector fields in the four different schemes. The table indicates which metric tensor is to be used in propagator numerators and polarization sums. This table is adapted from References [5,79].

	CDR	HV	FDH	DRED
singular vector field	${g_{[D]}}^{\mu\nu}$	${g_{[D]}}^{\mu\nu}$	${g_{[D_s]}}^{\mu\nu}$	${g_{[D_s]}}^{\mu\nu}$
regular vector field	${g_{[D]}}^{\mu\nu}$	${g_{[4]}}^{\mu\nu}$	${g_{[4]}}^{\mu\nu}$	${g_{[D_s]}}^{\mu\nu}$

12 We stress again that here our definitions of the four schemes only refer to the treatment of vector fields. In principle, in either scheme one would also have different options of treating γ_5, of which the non-anticommuting one is the most rigorous. The agreement with gauge invariance is meant on a superficial level. The existence of a D-dimensional covariant derivative by itself does not prove the all-order preservation of gauge invariance, and clearly gauge invariance of chiral gauge theories can be broken in dimensional schemes. For an example rigorous statement on the preservation of gauge invariance see later Section 4.3.

Technically, the schemes are expressed and summarized by Table 1 by specifying which metric tensor is to be used in propagator numerators or in polarization sums for squared matrix elements. In the table and in the remainder of this subsection, we use a more explicit notation for metric tensors on the different spaces and use the symbols ${g_{[dim]}}^{\mu\nu}$, where *dim* denotes the respective space, i.e., $dim = 4, D, D_s$ or $dim = D-4, D_s - D$. Our previous notation is rewritten as

$$\bar{g}^{\mu\nu} \equiv g^{\mu\nu}_{[4]}, \qquad \hat{g}^{\mu\nu} \equiv g^{\mu\nu}_{[D-4]}, \qquad g^{\mu\nu} \equiv g^{\mu\nu}_{[D]}. \tag{197}$$

The scheme differences for singular vector fields (which are sufficient for 1PI Green functions) can be well explained by comparing the gauge covariant derivatives. In the CDR and HV schemes, a generic covariant derivative is purely D-dimensional:

$${D_{[D]}}^{\mu} = {\partial_{[D]}}^{\mu} + ig{A_{[D]}}^{\mu}, \tag{198}$$

and the regularized vector field ${A_{[D]}}^{\mu}$ plays the role of a D-dimensional gauge field. In contrast, a covariant derivative in the DRED and FDH schemes can be split as

$${D_{[D_s]}}^{\mu} = {\partial_{[D]}}^{\mu} + ig{A_{[D]}}^{\mu} + ig_e{A_{[D_s-D]}}^{\mu}. \tag{199}$$

From a D-dimensional spacetime point of view, only the part ${A_{[D]}}^{\mu}$ acts as a D-dimensional gauge and vector field. In contrast, the field components ${A_{[D_s-D]}}^{\mu}$ are extra fields that behave like scalar fields in D dimensions; they are often referred to as "ϵ-scalars". The behavior under renormalization reflects this difference, and in general, the two coupling constants g_e, g renormalize differently.

In practical calculations, it is often not required to write the covariant derivative as explicitly as in Equation (199). Often, it is sufficient to set $D_s = 4$ and $g_e = g$ such that the vector field in the covariant derivative in DRED and FDH behaves essentially four dimensionally. If this is possible, it constitutes an advantage of these schemes. Specifically in supersymmetric theories, the symmetry leads to $g = g_e$. In general, however, the split (199) is, in principle, always possible and, sometimes, required. In the literature, the split was often useful to understand scheme behaviors, to resolve inconsistencies, and to derive scheme translation rules (for some examples, see Reference [5]).

We now give a brief overview of the theoretical status of the DRED and FDH schemes. For a more practical description with example calculations in all schemes, we refer to Reference [5]. DRED was introduced with the goal to preserve supersymmetry on the regularized level [80,81]. Over time, however, several inconsistencies were reported in the literature. Reference [82] found a mathematical inconsistency in the simultaneous application of four-dimensional and D-dimensional algebra. The inconsistency is very similar to Equations (182) and (184). It turned out that the inconsistency is due to the assumption that the D-dimensional space is a proper subspace of the original four-dimensional space. If one distinguishes between the original 4-dimensional space and the quasi-4-dimensional space QD_sS and uses the relationships (196), the inconsistency is resolved [78].

An important result is the all-order equivalence between all the schemes [83,84] (the proof was given for Green functions without infrared divergences and, hence, does not distinguish CDR/HV or DRED/FDH). For this proof, the split (199) and the independent renormalization of couplings such as g_e and g are essential. In this way, another inconsistency reported in Reference [85] was resolved. In that reference, couplings such as g_e and g were always assumed to be identical, and it was shown that the unitarity of the S-matrix can be violated at higher orders. This necessity of the split (199) and its role for renormalization, finiteness, and unitarity has also been stressed and exemplified by explicit calculations in References [86,87]. In summary, DRED is established as a fully consistent and applicable regularization for UV divergences.

The scheme properties for infrared divergences have also been investigated, in particular focusing on the computation of real and virtual higher-order corrections to physical processes. In the context of such calculations, the different treatments of regular vector fields become important. The schemes HV and, in particular, FDH are motivated by the potential to carry out much of the algebra in strictly four dimensions, allowing, e.g., powerful spinor and helicity methods. It was shown that the CDR, HV, and FDH schemes are equivalent at the next-to-leading (NLO) level, and elegant scheme transition rules have been derived [88–90]. In a parallel development, several references observed an apparent inconsistency in DRED with infrared factorization [91–93]. The resolution of this inconsistency [79,94] is, again, based on the observation that the split (199) and a separate treatment of D-dimensional gauge fields and ϵ-scalars is, in general, necessary. Further higher-order extensions of these analyses were presented in References [95–97].

As discussed in Reference [79], some of the described results were somewhat obscured by the fact that different authors used different names for equivalent schemes and, sometimes, the same names for different schemes: The schemes called DR (dimensional reduction) in References [88–90] are actually equivalent to the FDH scheme [98], but References [91–93] used the term dimensional reduction in the same sense as we define DRED here.

Finally, we comment on the question of supersymmetry preservation. In dimensional regularization (regardless of whether CDR or HV), the number of bosonic and fermionic degrees of freedom on the regularized level is different. This immediately leads to a violation of supersymmetry relations already at the one-loop level. Dimensional regularization may still be used, but specific finite supersymmetry-restoring counterterms have to be added to the Lagrangian. Such counterterms were evaluated and documented in [99–101].

For dimensional reduction (DRED or FDH), many studies have confirmed the compatibility with SUSY and the absence of non-SUSY counterterms. Overviews of the results can be found, e.g., in References [78,83,84,102]. References [78,103] made clear that, in the consistent versions of DRED/FDH, supersymmetry will eventually be broken. The reason is that the regularized Lagrangian is formulated not in the actual four-dimensional space, but in QD_SS, where the Fierz identities do not hold. The quantum action principle in DRED [78] then implies a supersymmetry breaking on the level of Green functions; the reasoning applied in Reference [78] is essentially the same as the strategy described in the present review for restoring gauge invariance in chiral gauge theories. Because of this general statement, the supersymmetry of DRED must be investigated on a case-by-case basis, and it has turned out that, for a large set of relevant multi-loop calculations, supersymmetry is preserved [86,102,104,105].

4. Quantum Action Principle in DReg

If Green functions of a quantum field theory are defined via the path integral (68) or the Gell–Mann–Low formula (67), the properties of Green functions clearly reflect the properties of the underlying Lagrangian. Example properties are the Ward or Slavnov–Taylor identities already discussed in Sections 2.5 and 2.6, which reflect symmetry properties of the Lagrangian.

This section is devoted to a related, but more general relationship—the so-called quantum action principle, specifically the regularized quantum action principle in DReg. This is a very useful relationship, allowing, e.g., rigorous derivations of Slavnov–Taylor identities or their breakings. The quantum action principle might appear obvious or straightforward, and sometimes, its validity is taken for granted. However, actually, its validity and also its precise meaning depend on the chosen regularization and renormalization procedure. For DReg, it was proven in [4] both on the regularized and on the renormalized level; the proof was extended to the consistent version of dimensional reduction in Reference [78]. We remark that there is also a regularization-independent quantum action principle, established in the context of BPHZ-renormalization in References [106–111]. We will discuss it and its relation to the regularized quantum action principle of DReg later in Section 6.2.

Here, in this section, we will begin with a formal derivation to motivate the statement, to highlight its simplicity, and to fix its interpretation (Section 4.1). Then, we will present a full proof of the regularized quantum action principle in DReg (Section 4.2). Finally, Section 4.3 will illustrate how to use this regularized quantum action principle to establish symmetry properties.

4.1. Formal Derivation of the Quantum Action Principle

The quantum action principle is a simple relation between the properties of the Lagrangian and the full Green functions. Here, we will present a formal derivation using the path integral (allowing general dimension D):

$$Z(J,K) = \int \mathcal{D}\phi \, e^{i\int d^Dx(\mathcal{L}+J_i\phi_i)}\,, \tag{200}$$

where possible composite operator terms coupled to sources K have been absorbed into the Lagrangian $\mathcal{L}$. Similar to Section 2.5, we consider a variable transformation:

$$\phi \to \phi + \delta\phi; \tag{201}$$

however, here, we do not assume that the action is invariant, but instead, we allow a change of the Lagrangian:

$$\mathcal{L} \to \mathcal{L} + \delta\mathcal{L}\,. \tag{202}$$

By assuming the path integral measure to be invariant under the transformation, steps analogous to the ones of Section 2.5 lead to

$$0 = \int \mathcal{D}\phi\Big(\int d^Dx\, i(\delta\mathcal{L} + J_i\delta\phi_i)\Big)e^{i\int d^Dx(\mathcal{L}+J_i\phi_i)}. \tag{203}$$

This is the most-important basic version of the quantum action principle.

In an even simpler way, one may derive the following relations for derivatives with respect to an external field $K(x)$ or to a parameter λ appearing in the Lagrangian:

$$\frac{\delta Z(J,K)}{\delta K(x)} = \int \mathcal{D}\phi\Big(\frac{\delta}{\delta K(x)}\int d^Dx\, i\mathcal{L}\Big)e^{i\int d^Dx(\mathcal{L}+J_i\phi_i)}, \tag{204a}$$

$$\frac{\partial Z(J,K)}{\partial \lambda} = \int \mathcal{D}\phi\Big(\frac{\partial}{\partial\lambda}\int d^Dx\, i\mathcal{L}\Big)e^{i\int d^Dx(\mathcal{L}+J_i\phi_i)}. \tag{204b}$$

These are further variants of the quantum action principle.

Similar to Section 2.5, it is instructive to rewrite the quantum action principle in various ways. First, identities for explicit Green functions can be obtained by taking suitable derivatives of the above identities with respect to sources J. In summary, the three variants of the quantum action principle then read as follows:

- Variation of quantum fields: $\delta = \int d^Dx \delta\phi_i(x)\frac{\delta}{\delta\phi_i(x)}$.

  $$i\,\delta\langle T\phi_1\dots\phi_n\rangle = \langle T\phi_1\dots\phi_n\Delta\rangle\,, \tag{205}$$

  where $\Delta = \int d^Dx\,\delta\mathcal{L}$ and the left-hand side is an abbreviation of Green functions involving $\delta\phi_i$ as in Equation (89). Here, and generally in the present section, we use a compact notation and suppress field arguments in a self-explanatory way such that, e.g., $\phi_k \equiv \phi_k(x_k)$, $\int d^Dx J_i\phi_i \equiv \int d^Dx J_i(x)\phi_i(x)$, etc.
- Variation of an external (non-propagating) field $K(x)$:

  $$-i\frac{\delta}{\delta K(x)}\langle T\phi_1\dots\phi_n\rangle = \langle T\phi_1\dots\phi_n\Delta\rangle\,, \tag{206}$$

  with $\Delta = \frac{\delta}{\delta K(x)}\int d^Dx\mathcal{L}$.

- Variation of a parameter λ:

$$-i\frac{\partial}{\partial\lambda}\langle T\phi_1\dots\phi_n\rangle = \langle T\phi_1\dots\phi_n\Delta\rangle\,, \tag{207}$$

with $\Delta = \frac{\partial}{\partial\lambda}\int d^Dx\mathcal{L}$.

An important further way to rewrite the quantum action principle is in terms of the generating functional Γ. By suitable Legendre transformation and expressing $\delta\phi$ in Equation (203) by derivatives with respect to sources K, we obtain, in particular, the form:

$$\mathcal{S}(\Gamma) = \Delta\cdot\Gamma\,, \tag{208}$$

where $\mathcal{S}(\Gamma)$ is a Slavnov–Taylor operator as in Equation (93) or (62) and where $\Delta = \mathcal{S}\left(\int d^Dx\mathcal{L}\right)$. Interestingly, this identity relates the Slavnov–Taylor identity for full Green functions on the LHS with the Slavnov–Taylor identity for the action appearing in the path integral on the RHS.

4.2. Proof of the Quantum Action Principle in DReg

The derivation presented above is only heuristic because the path integral measure $\mathcal{D}\phi$ was assumed to be invariant under the variable transformation. This is precisely the point where the regularization and renormalization enter. Hence, the quantum action principle has to be established separately for each regularization. Here, we consider what is called the regularized quantum action principle in DReg and present its proof. The proof was first given in Reference [4]; here, we follow the presentation of Reference [78], where the proof was extended to dimensional reduction.

Put simply, on the regularized level in DReg, all identities presented above are literally valid, provided all equations are interpreted as identities between Feynman diagrams regularized in DReg in D dimensions. A possible interpretation of this validity is that DReg provides a concrete perturbative definition of the path integral in which the measure is invariant under all field transformations of the form (201).

For the proof, we focus on the most-basic and most-complicated case, Equation (203) equivalently rewritten for explicit Green functions in Equation (205). All other identities can be treated similarly. To precisely formulate the statement, we rewrite Equation (205) as an identity of Feynman diagrams regularized in DReg. As stressed in Section 3.4, the Feynman diagrams regularized in DReg can be obtained from the Gell–Mann–Low formula in D dimensions. We call the regularized Lagrangian simply $\mathcal{L}$, omitting the superscript $^{(D)}$, and split it again as

$$\mathcal{L} = \mathcal{L}_{\text{free}} + \mathcal{L}_{\text{int}}\,, \tag{209}$$

where $\mathcal{L}_{\text{free}}$ determines the propagators in Feynman diagrams and $\mathcal{L}_{\text{int}}$ may contain terms coupling composite operators to sources K; it may contain counterterms involving coefficients with $1/(D-4)$ poles. Then, Equation (205) is rewritten as

$$\sum_{k=1}^{n} i\,\langle T\phi_1\dots(\delta\phi_k)\dots\phi_n\exp(i\textstyle\int d^Dx\mathcal{L}_{\text{int}})\rangle = \langle T\phi_1\dots\phi_n\Delta\exp(i\textstyle\int d^Dx\mathcal{L}_{\text{int}})\rangle \tag{210}$$

with

$$\Delta = \int d^Dx(\delta\mathcal{L}_{\text{free}} + \delta\mathcal{L}_{\text{int}})\,, \tag{211}$$

where both sides of Equation (210) are to be evaluated via Wick contractions in dimensional regularization. This is the statement that needs to be proven.

Let us write down the three parts of Equation (210) at some specific order with N powers of $\mathcal{L}_{\text{int}}$. Each term on the left-hand side becomes

$$\frac{i}{N!}\left\langle T\phi_1\dots(\delta\phi_k)\dots\phi_n\underbrace{(i\int d^Dx_1\mathcal{L}_{\text{int}})\dots(i\int d^Dx_N\mathcal{L}_{\text{int}})}_{N\text{ factors}}\right\rangle \tag{212}$$

and the term involving $\delta\mathcal{L}_{\text{int}}$ on the right-hand side becomes

$$\frac{1}{(N-1)!}\left\langle T\phi_1\dots\phi_n(\int d^Dx\delta\mathcal{L}_{\text{int}})\underbrace{(i\int d^Dx_1\mathcal{L}_{\text{int}})\dots(i\int d^Dx_{N-1}\mathcal{L}_{\text{int}})}_{N-1\text{ factors}}\right\rangle \tag{213}$$

For the term involving $\delta\mathcal{L}_{\text{free}}$, the discussion of Section 3.4 is crucial. The free Lagrangian in DReg contains fully D-dimensional derivative operators and can be schematically written as $\mathcal{L}_{\text{free}}=\frac{1}{2}\phi_i\mathcal{D}^{(D)}_{ij}\phi_j$, such that $\delta\mathcal{L}_{\text{free}}=\delta\phi_i\mathcal{D}^{(D)}_{ij}\phi_j$. Hence, the corresponding term in Equation (210) becomes

$$\frac{1}{N!}\left\langle T\phi_1\dots\overbrace{\phi_k\dots\phi_n(\int d^Dx\overbrace{\delta\phi_i\mathcal{D}^{(D)}_{ij}\phi_j}\overbrace{)\underbrace{(i\int d^Dx_1\mathcal{L}_{\text{int}})\dots(i\int d^Dx_N\mathcal{L}_{\text{int}})}_{N\text{ factors}}}}\right\rangle. \tag{214}$$

Each term must be evaluated using Wick contractions. It will be sufficient to consider all possible kinds of Wick contractions for the special field operator ϕ_j in $\delta\mathcal{L}_{\text{free}}$ as indicated in Equation (214). This field operator can be Wick contracted either with $\delta\phi_i$ at the same spacetime point (Contraction (a)), or with an external field operator ϕ_k (Contraction (b)), or with a field operator inside one of the $\mathcal{L}_{\text{int}}$ factors (Contraction (c)).

For each contraction, we can use the crucial property (190), which means that the Feynman diagram propagators are the inverse of the kinetic operators appearing in the regularized Lagrangian:

$$\mathcal{D}^{(D)}_{ij}\mathcal{P}^{(D)}_{jk}=i\delta_{ik}\,. \tag{215}$$

This relation establishes the relationship between the Lagrangian and Feynman rules and is the core reason why the quantum action principle holds. Using this relation, Contraction (a) produces a single-loop integral over $\mathcal{D}^{(D)}_{ij}$ times the propagator $\mathcal{P}^{(D)}_{jl}$ from ϕ_j to some field ϕ_l within the composite operator $\delta\phi_i$. The loop integrand is, therefore, simply a constant δ_{il}, hence scaleless and, therefore, zero.

Contraction (b) with the external field ϕ_k produces the combination $\mathcal{D}^{(D)}_{ij}\mathcal{P}^{(D)}_{jk}=i\delta_{ik}$. In this way, the $\int d^Dx$ integral is effectively canceled and the external field operator ϕ_k is replaced by $i\delta\phi_k$. Hence, the contractions of Type (b) in Equation (214) yield exactly the same as Equation (212), and we have proven the first required cancellation.

Finally, a contraction of Type (c) between ϕ_j and some field ϕ_l within one of the $\mathcal{L}_{\text{int}}$ factors results in the product $\mathcal{D}^{(D)}_{ij}\mathcal{P}^{(D)}_{jl}\frac{\delta\mathcal{L}_{\text{int}}}{\delta\phi_l}$. Using the inverse relation for the propagators again, we found that all contractions of Type (c) in Equation (214) lead to

$$\frac{i^2N}{N!}\left\langle T\phi_1\dots\phi_n(\int d^Dx\delta\phi_l\tfrac{\delta\mathcal{L}_{\text{int}}}{\delta\phi_l})\underbrace{(i\int d^Dx\mathcal{L}_{\text{int}})\dots(i\int d^Dx\mathcal{L}_{\text{int}})}_{N-1\text{ factors}}\right\rangle. \tag{216}$$

This is precisely the negative of Equation (213). In total, we have, therefore, shown the equality (212) $=$ (213) $+$ (214), and we have established the quantum action principle (210).

In the same way, it is possible to prove all other identities presented in Section 4.1. The essential point in the proof is the possibility to express Feynman diagrams in DReg via the Gell–Mann–Low formula together with the relationship between regularized propagators and the regularized free Lagrangian. Reference [4] gave the proof using the α-representation of all diagrams explained in Section 3.2.4, where the relationship for the propagators is less obvious. Reference [78] extended the proof to the consistent version of dimensional reduction.

4.3. Examples of Applications of the Quantum Action Principle

The quantum action principle is a very powerful tool to study symmetry properties of Green functions. Here, we provide two example applications that illustrate this. The examples are very important in their own right, but they also provide a blueprint for the analysis of chiral gauge theories discussed later.

The first example is gauge invariance in non-chiral gauge theories such as QED or QCD. The gauge-invariant Lagrangian of QCD with one quark flavor is given by

$$\mathcal{L}_{\text{inv}} = -\frac{1}{4}F^{a\mu\nu}F^{a}_{\mu\nu} + \bar{\psi}i\not{D}\psi\,, \tag{217a}$$

$$D_{\mu} = \partial^{\mu} + igT^{a}A^{a}_{\mu}\,, \tag{217b}$$

where the generators T^a correspond to the triplet representation of SU(3) and the field strength tensor is defined as in Equation (9). The full Lagrangian including gauge fixing and ghost terms and source terms for BRST transformations is given by

$$\begin{aligned}\mathcal{L}_{\text{cl}} =& \mathcal{L}_{\text{inv}} + B^{a}(\partial^{\mu}A^{a}_{\mu}) + \frac{\xi}{2}(B^{a})^{2} - \bar{c}^{a}\partial^{\mu}(D_{\mu}c)^{a} \\ &+ \rho^{a\mu}sA^{a}_{\mu} + \zeta^{a}sc^{a} + Y_{\psi}s\psi + Y_{\overline{\psi}}s\overline{\psi}\,,\end{aligned} \tag{218}$$

where the BRST transformations are given as in Section 2.3.

All ingredients of the QCD Lagrangian can be interpreted as D-dimensional quantities without any changes in the algebraic relations. The D-dimensional version of $\mathcal{L}_{\text{inv}}$ is still fully gauge-invariant, and the full BRST-invariant classical Lagrangian $\mathcal{L}_{\text{cl}}$ is BRST-invariant in D dimensions. Likewise, the Slavnov–Taylor identity (62) is satisfied in D dimensions.

We therefore have

$$\mathcal{S}\left(\int d^{D}x\mathcal{L}_{\text{cl}}\right) = 0\,, \tag{219}$$

for the D-dimensional regularized theory. Now, we can use the quantum action principle in the form of Equation (208), where now, the breaking term $\Delta = 0$. Accordingly, the symmetry of the D-dimensional classical action implies that the regularized Green functions represented by the generating functional Γ_{DReg} satisfy the Slavnov–Taylor identity $\mathcal{S}(\Gamma_{\text{DReg}}) = 0$ at all orders.

This is the precise form of the statement that DReg preserves gauge invariance of QCD manifestly. The analogous statement is also true for QED or other non-chiral gauge theories. One can go one step further and discuss the renormalized level. If counterterms are generated from the classical Lagrangian by the standard procedure of field and parameter renormalization, the bare Lagrangian $\mathcal{L}_{\text{bare}} = \mathcal{L}_{\text{cl}} + \mathcal{L}_{\text{ct}}$ still satisfies the Slavnov–Taylor identity, $\mathcal{S}(\int d^{D}x\mathcal{L}_{\text{bare}}) = 0$. For this reason, even the renormalized, finite functional Γ_{DRen} in the notation of Section 3.1, which is obtained from $\mathcal{L}_{\text{bare}}$, satisfies the Slavnov–Taylor identity without the need for special symmetry-restoring counterterms. The manifest preservation of gauge/BRST invariance at all steps of the construction of QCD dramatically simplifies practical calculations, as well as all-order proofs.

As our second example, we briefly sketch the situation of supersymmetry in regularization by dimensional reduction. As explained in Section 3.5, the dimensional reduction scheme treats vector fields in quasi-four dimensions and should, therefore, be better compatible with supersymmetry. Without going into the details, supersymmetry can also be expressed in terms of a Slavnov–Taylor identity. If a supersymmetric Lagrangian is defined in dimensional reduction as $\mathcal{L}_{\text{SUSY}}^{\text{DRed}}$ and this scheme is defined mathematically consistently, it does not remain supersymmetric. Instead, applying the corresponding Slavnov–Taylor operator yields $\mathcal{S}(\int d^D x \mathcal{L}_{\text{SUSY}}^{\text{DRed}}) = \Delta \neq 0$. The value of Δ for a general supersymmetric gauge theory was provided in Reference [78]. The reason for the nonvanishing value of Δ is that the quasi-four-dimensional space does not permit using Fierz identities. The nonvanishing value of Δ implies that, ultimately, supersymmetry is not preserved by dimensional reduction at all orders.

Nevertheless, dimensional reduction preserves supersymmetry to a very large extent, and the quantum action principle provides a succinct method to check the validity of supersymmetry in concrete cases: The evaluation of concrete Green functions with an insertion of the breaking, $\Delta \cdot \Gamma$, directly determines the potential breaking of the supersymmetric Slavnov–Taylor identity in a concrete sector. This method was used, e.g., to verify that supersymmetry, indeed, is conserved in a variety of important cases, including phenomenologically important 2-loop and 3-loop contributions to the Higgs boson mass prediction in the minimal supersymmetric standard model [102,105].

5. Renormalization in the Context of DReg

In this section, we review the basic renormalization theory in the context of perturbative relativistic quantum field theories, from the point of view of applications of dimensional regularization. Renormalization has both technical and physical aspects. On the most-technical level, renormalization is a procedure to remove ultraviolet divergences and generate finite Green functions, S-matrix elements, and other quantities of interest. It effectively provides a definition of each term in the Gell–Mann–Low formula (67) and may be viewed as a definition of the path integral measure (68). The removal of ultraviolet divergences is not arbitrary, but subject to important physical constraints such as unitarity and causality. In more physical terms, renormalization can be viewed as a reparametrization. This is reflected by the "main theorem of renormalization" (the name was coined in Reference [112], where also a very general proof was given, which essentially relies on the physical causality constraint), which states that all allowed renormalization procedures differ by nothing but reparametrizations. It is also reflected by the customary practical procedure of first regularizing the theory, then introducing counterterms that depend on the regularization and cancel the divergences. These counterterms can be viewed as arising from reparametrizations, or renormalizations, of Lagrangian parameters and fields.

The need for renormalization and the possibility of renormalization to generate a finite theory also reflect further deep physical properties of quantum field theories. The existence of ultraviolet divergences and the resulting need for subtractions and a renormalization procedure result in the possibility of so-called anomalies. These are breakings of symmetries, which are valid in the classical theory, but broken on the quantum level via the regularization and renormalization procedure. Fundamentally, anomalies arise if the unitarity and causality constraints on renormalization are incompatible with the symmetry in question.

The possibility to successfully carry out the renormalization program and its relation to reparametrizations reflects the physical phenomenon of decoupling. Physics at a certain distance and energy scale is insensitive to physics at a much smaller distance and higher energy scales, leading to the important concepts of effective field theories and the renormalization group. Ultra-short distance details influence long-distance physics only via their effect on long-distance parameters. Since any regularization effectively changes the short-distance behavior of the theory in a cutoff-dependent (but unphysical) way, it is not too surprising that the cutoff dependencies, including divergences, can be compensated by reparametrizations such that a finite and regularization-independent limit exists.

In the present section, we provide a brief review of the general theorems governing the previous statements; this discussion has a strong focus on the so-called BPHZ approach to renormalization, and an outcome is that the customary regularization/renormalization procedure is correct. Then, we review the main theorem stating that dimensional regularization may be employed as one such consistent regularization/renormalization framework.

5.1. General Renormalization Theory and Constraints from Unitarity and Causality

Here, we review the basic properties of renormalization as a means to eliminate ultraviolet (UV) divergences and to generate finite relativistic quantum field theories. The discussion is organized along four questions: What are the required properties of any renormalization procedure? Which procedures satisfy these properties? What are the possible differences between different allowed renormalization procedures? How does the usual procedure of regularization and counterterms fit into the fundamental analysis of renormalization?

As we will discuss, all these questions have rigorous and positive answers, first obtained by Bogoliubov/Parasiuk [113] and Hepp [65], with important additional developments by Speer [22,68,114] and Zimmermann [115] and Epstein/Glaser [116]. We refer to lectures by Hepp [117] (contained in [118]) for very detailed and pedagogical explanations and to Reference [119] for an overview.

We begin by explaining the fundamental requirements on any renormalization procedure. A minimal requirement would be that perturbative S-matrix elements become UV finite; a very strong requirement would be the nonperturbative construction of well-defined products of interacting field operators. Following the analysis of the mentioned references, we choose an intermediate approach. In this approach, a *renormalization* is a procedure that constructs all possible time-ordered products of free field operators, or equivalently, a renormalization is a mapping that maps any Feynman diagram to a well-defined and UV finite expression. In detail, the requirement can be efficiently formulated by writing an interaction Lagrangian:

$$\mathcal{L}_{\text{int}}(x) = \sum_i W_i(x) g_{W_i}(x) , \tag{220}$$

where $W_i(x)$ are all local field monomials of interest (including all monomials appearing in the actual Lagrangian of interest, but also possible further composite operators of interest, similar to the discussion of composite operators in Sections 2.3 and 2.4), and where $g_{W_i}(x)$ are number-valued test functions (acting like the sources K_i in Sections 2.3 and 2.4 or in Equation (58) or like localized coupling constants). This Lagrangian generates a perturbative scattering operator $S(g)$, where the argument g denotes the functional dependence on all the g_{W_i}:

$$S(g) = 1 + \sum_{n=1}^{\infty} \frac{i^n}{n!} \int \sum_{i_1 \dots i_n} T_{i_1 \dots i_n}(x_1, \dots, x_n) g_{W_{i_1}}(x_1) \dots g_{W_{i_n}}(x_n) d^4x_1 \dots d^4x_n , \tag{221}$$

where, formally, the appearing T-products would be given by

$$T_{i_1 \dots i_n}(x_1, \dots, x_n) = T(W_{i_1}(x_1) \dots W_{i_n}(x_n)) . \tag{222}$$

However, the expressions in Equation (222) are, in general, ill-defined if $n > 1$ and several of the x_i coincide. Hence, a *renormalization* is a construction of the T-products and, thus, of Equation (221), which satisfies the following properties, adapted from References [112,116]:

Initial conditions:

$$S(0) = 1 , \tag{223a}$$

$$T_i(x) = W_i(x) . \tag{223b}$$

Unitarity:

$$S(g)^{\dagger}S(g) = S(g)S(g)^{\dagger} = 1 \tag{224}$$

for all Hermitian $W_i g_{W_i}$. Here, $S(g)^{\dagger}$ must be written in terms of anti-T-products $\bar{T}_{i_i \ldots i_n}$, which also must be constructed.

Translational invariance:

$$U(1,a)S(g)U(1,a)^{\dagger} = S(g_a), \tag{225}$$

where $U(1,a)$ is the representation of translations on the respective free Fock space and $g_a(x) = g(x-a)$.

Causality:

$$S(g+h) = S(g)S(h) \qquad \text{if } \operatorname{supp}(g) \gtrsim \operatorname{supp}(h), \tag{226}$$

where $\operatorname{supp}(g) \gtrsim \operatorname{supp}(h)$ means that all points in the support of h are outside the support of g and its future light cone, such that the points in $\operatorname{supp}(h)$ cannot be causally influenced by the points in $\operatorname{supp}(g)$.

Via the expansion (221), these requirements translate into constraints on the T-products and $\bar{T}$-products. For instance, the causality requirement is particularly powerful [112] and translates into the relation:

$$\begin{aligned} T_{i_1 \ldots i_n}(x_1, \ldots, x_n) = & T_{i_{j_1} \ldots i_{j_m}}(x_{j_1}, \ldots, x_{j_m}) T_{i_{j_{m+1}} \ldots i_{j_n}}(x_{j_{m+1}}, \ldots, x_{j_n}) \\ & \text{if } \{x_{j_1}, \ldots, x_{j_m}\} \gtrsim \{x_{j_{m+1}}, \ldots, x_{j_n}\} \\ & \text{and } \{j_1, \ldots, j_n\} = \{1, \ldots, n\} \end{aligned} \tag{227}$$

for T-products.

A construction fulfilling all these constraints, thus, amounts to a construction of all T-products of possible field monomials W_i and, thus, of all terms appearing in the Gell–Mann–Low formula (67) and ultimately of Feynman diagrams and Green functions, including Green functions of composite operators. Similar sets of requirements can also be found in the Bogoliubov/Shirkov textbook [120] and, for Feynman diagrams, in Hepp's lectures [117].

Let us briefly comment on the central role of unitarity and causality. Both requirements allow expressing T-products with a certain number of operator factors in terms of T-products (or $\bar{T}$-products) with fewer factors, such as in Equation (227). Hence, higher-order T-products and, thus, the entire renormalization procedure are not arbitrary, but largely fixed. The only ambiguity arises when all arguments are equal, $x_{i_1} = \ldots = x_{i_n}$, in which case, causality and unitarity do not imply a relation to lower-order T-products.

This clarifies that renormalization is not unique and there can be different *renormalization schemes* with different choices to fix these ambiguities. However, it also gives an indication that the ambiguities affect only local terms, such that different schemes differ only by reparametrizations of local terms in the Lagrangian (220). Further, it is in line with the fact that UV divergences are local in position-space and can be canceled (in the presence of a regularization) by adding local counterterms to the Lagrangian.

The local nature of the ambiguities and possible scheme differences can be formulated as a rigorous theorem: The statement is that any two constructions satisfying all requirements listed above differ only in a finite reparametrization (often called *finite renormalization* in the original literature); conversely, if an allowed renormalization is changed by a finite reparametrization, another allowed renormalization is obtained. In our formulation, two different renormalizations may be expressed as $S_T(g)$ and $S_{T'}(G)$, where T and T' denote the two different constructions of T-products, and g and G represent two different sets of the prefactors g_{W_i} in the Lagrangian (220). A finite reparametrization may be written as [112]

$$G_{W_i}(x) = g_{W_i}(x) + \sum_{n=1}^{\infty} G_{W_i,n}(g, Dg)(x), \tag{228}$$

which is a reparametrization of the couplings expressed in terms of $G_{W_i,n}$, which are local functions of all $g_{W_j}(x)$ and their derivatives. The index n denotes the order in perturbation theory. On this level, the statement is that, if both $S_T(g)$ and $S_{T'}(G)$ are allowed renormalizations, then they can be related as

$$S_T(g) = S_{T'}(G) \tag{229}$$

with a suitable finite reparametrization of the form (228), and conversely, if $S_{T'}(G)$ is allowed, then any finite reparametrization of the form (228) effectively defines another allowed renormalization via requiring (229). Reference [112] gave a very general proof based directly on the causality requirement of renormalizations, and Reference [117] gave a proof on the level of Feynman diagrams.

Since reparametrizations do not change the physical content of a theory, this also shows that any two allowed renormalizations are equivalent, i.e., describe the same physics.

Now, we turn to the question about which renormalization procedures exist and how they are related to the counterterm approach often used in practical computations, giving a brief survey of the approaches and results. Historically, the BPH theorem constitutes the first rigorous proof that all the above properties can be established [65,113]. These references used a recursive, so-called R-operation and an intermediate regularization. Though successful, Hepp [117] assessed the approach as "hideously" complicated and noted that a cleaner approach is provided by analytic regularization [62,68]. Working on the level of Feynman diagrams, the idea of analytic regularization is to replace the propagator denominator of any internal line with index k as

$$\frac{1}{\ell_k^2 - m_k^2 + i\varepsilon} \to \frac{1}{(\ell_k^2 - m_k^2 + i\varepsilon)^{\lambda_k}} \tag{230}$$

with complex parameters λ_k. Similar to DReg, there is a domain for λ_k where all integrals are well defined, and analytic continuation leads to poles at the physical value $\lambda_k = 1$. It is then possible to define the renormalized expressions via Laurent expansion in $(\lambda_k - 1)$ and keeping only the zeroth-order term.

In this approach, the finiteness of the construction, as well as the validity of all required properties including causality and unitarity are comparatively easy to prove [117]. The equivalence to the counterterm method was at first only established indirectly by using the equivalence to BPH, but later also directly [68]. A drawback of analytic regularization is that the relation to the Lagrangian is obscured. In contrast to, e.g., DReg (see Section 3.4), the regularization cannot be expressed in terms of a regularized Lagrangian.

Though technically more complicated, the BPH approach and the BPH theorem are very instructive, most importantly since they establish the connection with the customary procedure of regularization and counterterms. In this approach, first, every Feynman diagram is regularized, e.g., using the Pauli–Villars prescription. Then, the renormalization procedure is carried out via the so-called recursive R-operation. For any 1PI graph G, a subrenormalized amplitude is defined by

$$\overline{\mathcal{R}}_G = G + \sum_{H_1 \ldots H_s} G/_{H_1 \cup \ldots \cup H_s} \cdot C(H_1) \ldots C(H_s), \tag{231}$$

where the sum runs over all possible sets of disjoint 1PI subgraphs H_i of G (excluding G itself). The object in the sum denotes the amplitude for the graph, where all the disjoint subgraphs $H_1 \dots H_s$ are shrunk to points and replaced by the counterterms $C(H_1) \dots C(H_s)$. The fully renormalized result and the counterterms are defined as

$$\mathcal{R}_G = \overline{\mathcal{R}}_G + C(G), \tag{232a}$$

$$C(G) = -T\overline{\mathcal{R}}_G, \tag{232b}$$

where T denotes the operation to extract the divergent part. In the BPH approach, T is defined via a Taylor expansion in external momenta of a graph and, therefore, by construction, a polynomial in momentum space.

The BPH theorem [65,113] states that the renormalized graphs $\mathcal{R}_G$ are finite (in the sense of distributions in momentum-space) and that all required properties are valid. The difficult part of the proof is the proof of finiteness. The big advantage of the R-operation is its relationship to the usual counterterm approach. Indeed, it is easy to see that the formula (231) combinatorically corresponds to the prescription to add to G all possible counterterm Feynman diagrams with all possible insertions of counterterm vertices; furthermore, the counterterms are local in position space and, therefore, can be obtained from a local counterterm Lagrangian. For a detailed discussion of the R-operation and a full proof of its relationship to counterterm diagrams and counterterm Lagrangians, we also refer to the monograph [63], Chapter 5.7.

Since both the BPH procedure and analytic regularization constitute allowed renormalizations, they must be physically equivalent in the sense defined above, i.e., they differ only by reparametrizations/finite renormalization. This equivalence has also been directly established in References [68,117], where it was also shown that the required finite renormalization only involves counterterms whose power-counting degree is bounded by the superficial degree of divergence of the original Feynman diagrams.[13]

A further instructive and important renormalization procedure was developed by Zimmermann [115], leading to the notion of BPHZ renormalization. Its main virtue is that it completely eliminates the need for any regularization, but directly constructs finite momentum-space loop integrals. Its technical tool is the famous forest formula, which is a direct solution of the recursive R-operation. It allows constructing loop integrals via repeated applications of Taylor subtractions on the integrand level. A technical obstacle is that care must be taken to avoid ambiguities from different loop momentum routings in case the same subdiagram is inserted into different higher-order diagrams. While the proof of the finiteness of the construction is highly nontrivial, the proof of equivalence to the BPH approach is rather straightforward if an intermediate regularization is employed.

Already, Reference [65] on the BPH theorem and References [62,68,117] on analytic regularization made essential use of the α-parametrization (see Section 3.2.4) in their proofs. The idea of using the α-parametrization was combined with the forest formula in References [66,67,114] to strongly simplify the finiteness proof. These references applied subtractions via Taylor expansions with respect to the αs such that directly finite α integrals were obtained.

5.2. Theorem on Divergences and Renormalization in DReg

5.2.1. Statement of the Theorem

Here, we discuss the central theorem of dimensional regularization, most rigorously established as Theorem 1 in the paper by Breitenlohner/Maison, Reference [4]. In essence, it implies the following: the renormalization of relativistic quantum field theories can be performed using DReg as an intermediate regularization, the renormalized answer is correct and equivalent to the results from other consistent schemes discussed in the previous subsection, and the required subtractions can be implemented as counterterm Lagrangians.

13 Such a renormalization was called "minimal" in Reference [117], but we stress that this is a different notion of minimality than, e.g., in the so-called minimal subtraction prescription within DReg.

In more detail, it can be formulated as follows. Let G be a 1PI Feynman graph (in Reference [4], a theory without massless particles is required; References [23,24] considered the case with massless particles). The corresponding regularized Feynman integral $\mathcal{T}_G$ is defined as discussed in Section 3, making use of the consistently constructed formally D-dimensional covariants and D-dimensional integrals. Reference [4] specifically employed the α-parametrization introduced in Section 3.2.4.

Then, it is possible to apply a subtraction algorithm to the graph that defines first a subrenormalized Feynman integral $\overline{\mathcal{T}}_G$ and, finally, a fully renormalized Feynman integral $\mathcal{R}_G$. Assuming four-dimensional quantum field theory and writing $D = 4 - 2\epsilon$, these objects have the following properties:

- The regularized, but not-yet-renormalized amplitude $\mathcal{T}_G$ is a meromorphic function of D or, equivalently, of ϵ.
- The subrenormalized amplitude $\overline{\mathcal{T}}_G$ may have singularities in ϵ, which are poles of the form:

$$\frac{1}{\epsilon}P_G^{(1)} + \ldots + \frac{1}{\epsilon^{L_G}}P_G^{(L_G)}, \tag{233}$$

  where L_G is the number of closed loops in the graph G. The coefficients $P_G^{(k)}$ are polynomials in the external momenta and the masses appearing in G (corresponding to local terms in position-space). The degree of all these polynomials is bounded by the superficial power-counting degree of the graph $\omega_G = 4L_G - 2I_G + r_G$ with I_G the number of internal lines in G and r_G the power-counting degree of the numerator.
- $\mathcal{R}_G$ is finite, i.e., it is an analytic function of ϵ in a region around $\epsilon = 0$.

The theorem provides several crucial additional details:

- The subtraction is organized according to a forest formula, which is equivalent to Bogoliubov's recursive R-operation (we also refer to the monograph [63] for a detailed explanation). For this reason, the subtraction algorithm is equivalent to adding counterterm Feynman diagrams.
- The subtractions corresponding to subgraphs H of G, called C_H, are given by $\overline{\mathcal{T}}_H$ with analogous properties to $\overline{\mathcal{T}}_G$ as explained above.
- The subtractions corresponding to a subgraph H are independent of the surrounding graph G; they really only depend on H itself (and, of course, its subgraphs).
- The renormalized results for all graphs $\mathcal{R}_G$ are equivalent to the results obtained in the BPHZ framework (before Reference [4], this point had been established also in Reference [22]). This means they differ from the BPHZ results at most by finite, local counterterms at each order, in line with the general theorem discussed around Equations (228) and (229).

The previous, rather technical details have very important consequences for practical calculations and physical interpretations:

- The combinations of all subtractions of all graphs can be written as a counterterm Lagrangian, which is local and contains only terms of dimensionalities limited by the power-counting of the original graphs.
- The renormalized amplitudes constructed in DReg provide a finite quantum field theory, which is consistent with unitarity and causality in the sense analyzed by References [65,113,115,116,120].

We provide even further details:

- Initially, all propagators in the integrals are defined via the $+i\varepsilon$ prescription in momentum-space (which corresponds to time-ordering in position-space) with $\varepsilon > 0$. As long as $\varepsilon > 0$, the dependence of $\mathcal{R}_G$ on external momenta and masses is infinitely differentiable, i.e., of the C^∞ type. After the limit $\varepsilon \to 0$ has been taken, the dependencies take the character of tempered distributions. In this regard, DReg behaves identically to, e.g., BPHZ [65].

- The setup of the subtractions requires that all $1/\epsilon$ poles are subtracted, even if the coefficients happen to be evanescent in the sense defined in Section 3.4. In the coefficients $P_G^{(k)}$ in Equation (233), a four-dimensional limit is not permitted during the subtraction procedure. For the counterterm Lagrangian, this implies that evanescent operators (operators that have no four-dimensional counterpart since they would vanish either in view of Fierz identities or γ_5 identities or because of contractions with $\hat{g}^{\mu\nu}$) must be included in case they are needed to cancel $1/\epsilon$ poles.

5.2.2. Overview of the Proof

The full proof of the theorem explained above requires many ingredients, which need to be analyzed in detail. Most of them are largely independent of the regularization scheme, but related to Feynman graph theory, relationships between graphs and subgraphs, and the structural properties of the α-parametrization. Several key ideas for the proof are common to proofs for BPHZ renormalization. The specific aspects of DReg enter in a very localized form.

Here, we first list the most-important ingredients of the proof:

- The α-parametrized integral can be decomposed into sectors.
- A particularly elegant forest formula holds for each sector of the α-parametrization.
- In each sector, clever variable substitutions can be made, which lead to an explicit general formula for the integral.
- There is a general relationship between the integrand for a certain graph and the integrands for corresponding subgraphs and reduced graphs.
- There are a few simple observations for typical integrals and functions encapsulating the $1/\epsilon$ poles.

The following subsections will illustrate each of these ingredients with the help of suitable examples and will motivate the general statements, which can all be found in Reference [4]. A further subsection will sketch the essential steps of the proof by induction.

5.2.3. Ingredient 1: Sectors of the α Integration

In Equation (144), we already considered a simple one-loop integral transformed into Schwinger or α-parametrization. For each internal line of the diagram, there is one α_l parameter, and all α_l are integrated in the range from 0 to ∞. It is easy to compute one integral explicitly, and the second integral could be computed as well. For the general proof of renormalization, we neither want nor need an explicit computation of all loop integrals. We rather need to transform all integrals into a uniform structure from which we are able to read off the required properties. It turns out that decomposing the α integrations into sectors is extremely helpful in this regard.

The strategy of similar sector decompositions of the α integrations has been employed also in the important proof of the BPH theorem in Reference [65] and in simplified proofs of the BPHZ theorem in References [66,67] and is the basis of modern numerical evaluations of multiloop integrals [121,122]. For the integral (144), the sector decomposition is very simple:

$$\int_0^\infty d\alpha_1 d\alpha_2 = \int_{\text{sector 1}} d\alpha_1 d\alpha_2 + \int_{\text{sector 2}} d\alpha_1 d\alpha_2 \,, \tag{234a}$$

where the two sectors are defined as

$$\text{sector 1} = \{\alpha_1 \le \alpha_2\}\,, \tag{234b}$$
$$\text{sector 2} = \{\alpha_2 \le \alpha_1\}\,. \tag{234c}$$

Let us describe the sector decomposition used for the proof in Reference [4] with the following six-loop example diagram:

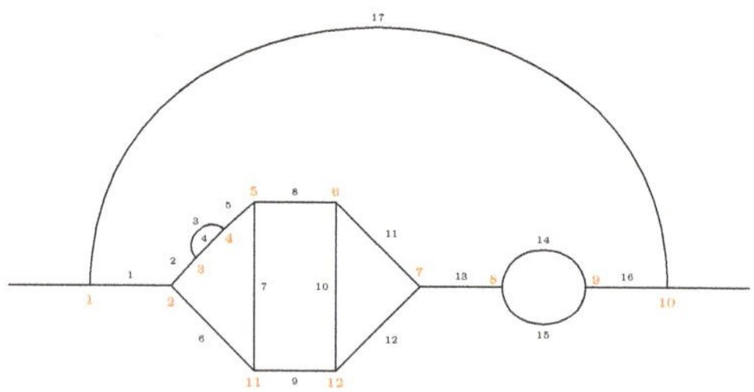

(235)

with line labels and vertex labels as indicated. One particular sector is constructed by the following algorithm. First, we chose one particular one-loop subdiagram. As an example, we chose the diagram consisting of the lines 14, 15 and call it H_1. Next, we chose either a second, disjoint 1-loop subdiagram, or a 2-loop subdiagram, which contains H_1. Let us choose the diagram consisting of Lines 3, 4 and call it H_2. Next, we chose a subdiagram H_3 such that H_3 either contains H_1 and/or H_2 or is disjoint and such that, overall, the union of $H_{1,2,3}$ contains three loops. We might choose H_3 as the two-loop diagram with Lines 2, 3, 4, 5, 6, 7. We continued this way until we reached the six-loop diagram $H_6 \equiv G$ itself. An example choice of subgraphs $C = \{H_1, H_2, \ldots, H_5, H_6\}$ is illustrated in the diagram of Figure 1.

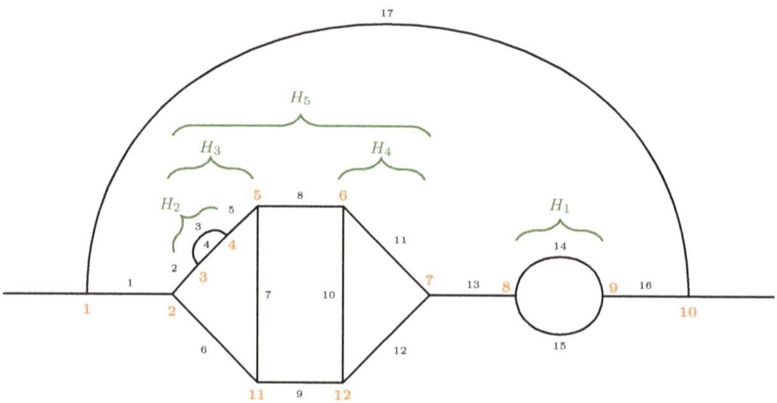

Figure 1. The same diagram as in Equation (235), with additional indications of subdiagrams H_i $(i = 1 \ldots 6)$, which define an example maximal forest.

In this way, we can generally construct what is called a *maximal forest*. In general, the definition of a forest is a set of 1PI subgraphs of G, which are non-overlapping, i.e., either disjoint or nested. A maximal forest is thus a maximal set of 1PI subgraphs that are non-overlapping. The above construction illustrates how one can construct all such maximal forests, and it illustrates that each maximal forest contains as many elements as there are loops in G.

The example also illustrates that each subgraph H_i in a maximal forest contains at least one line that is specific to it, i.e., that is not contained in any smaller subgraphs of the maximal forest. We may define a mapping, called "labelling" in Reference [4], of the form

$$H_i \mapsto \sigma(H_i) = \text{one of the lines specific to subgraph } H_i. \tag{236}$$

In the example, we can choose

$$\sigma(H_1) = 14 \qquad \sigma(H_2) = 3 \qquad \sigma(H_3) = 7 \tag{237a}$$
$$\sigma(H_4) = 11 \qquad \sigma(H_5) = 8 \qquad \sigma(H_6) = 16. \tag{237b}$$

The labeled lines are illustrated in blue color in the diagram of Figure 2.

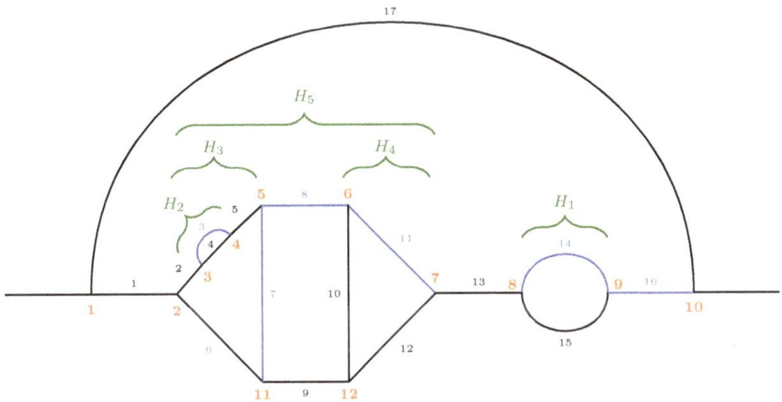

Figure 2. The same diagram as in Figure 1, with additional indications of the labeled lines $\sigma(H_i)$ for each subgraph, according to Equation (237).

For any such choice of a maximal forest together with a labeling for specific lines, $(\mathcal{C}, \supset)$, we define an integration sector for the α_l variables in the following way: in each subgraph H_i, the α for the specific labeled line is the largest, i.e.,

$$\alpha_l \leq \alpha_{\sigma(H_i)} \;\forall\, l \in H_i\,. \tag{238}$$

For the example, the integration sector defined by $(\mathcal{C}, \supset)$ is

$$\alpha_{15} \leq \alpha_{14} \qquad \alpha_4 \leq \alpha_3 \qquad \alpha_{2,3,5,6} \leq \alpha_7 \tag{239a}$$
$$\alpha_{10,12} \leq \alpha_{11} \qquad \alpha_{7,11,9} \leq \alpha_8 \qquad \alpha_{1,14,8,13,17} \leq \alpha_{16}\,. \tag{239b}$$

Note that this does not imply a fixed ordering of all the α_l.

It is elementary to prove a variety of useful properties of maximal forests and labelings. In particular, this way of defining sectors leads to a partitioning of the entire α integration region of any Feynman graph loop integral:

$$\int_0^\infty d\alpha_1 \dots d\alpha_I = \sum_{(\mathcal{C},\supset)} \int_{(\mathcal{C},\supset)} d\alpha_1 \dots d\alpha_I\,. \tag{240}$$

Using the notation $\mathcal{T}_G$ for the regularized amplitude of the graph G, we can therefore write

$$\mathcal{T}_G = \sum_{(\mathcal{C},\supset)} \mathcal{T}_{G,(\mathcal{C},\supset)}\,, \tag{241}$$

with an obvious meaning and where the sum extends over all maximal forests of G and all possible labelings $(\mathcal{C}, \supset)$. This construction of sectors is the essential content of Lemma 3 in Reference [4].

5.2.4. Ingredient 2: Forest Formula after Decomposition into Sectors

In the all-order investigation of renormalization, the graphical language of Feynman diagrams with counterterms has to be formalized in terms of subtractions of divergent integrals. In the historical development of the rigorous BPHZ renormalization, this formalization was first performed via Bogoliubov's recursive definition of the so-called R-operation. This recursive definition was later rewritten into Zimmermann's forest formula [115]. The sector decomposition described above permits a very elegant and powerful alternative version of the forest formula, which simplifies the proof. Such simplified forest formulas were also discussed and applied in the context of BPHZ, e.g., in References [66,114,123].

To explain these relations, we begin with the recursive R-operation, defined in Equations (231) and (232). We recall the main equation, the definition of a subrenormalized amplitude:

$$\overline{\mathcal{R}}_G = G + \sum_{H_1 \dots H_s} G/_{H_1 \cup \dots \cup H_s} \cdot C(H_1) \dots C(H_s)\,, \tag{242}$$

where the counterterms are defined as $C(H) = -T\overline{\mathcal{R}}_H$.

This R-operation is recursive because the definition of the subrenormalized amplitude depends on lower-order counterterms, which in turn are defined via lower-order subrenormalized amplitudes. One may work out the recursion and obtain a direct, non-recursive formula. To illustrate this, consider the case where the full graph G has one 2-loop subgraph γ_2, which in turn has a 1-loop subgraph γ_1. Then, one term in $\overline{\mathcal{R}}_G$ is given by

$$\begin{aligned}\overline{\mathcal{R}}_G &= \dots + G/_{\gamma_2} \cdot C(\gamma_2)\\ &= \dots + G/_{\gamma_2} \cdot [-T\overline{\mathcal{R}}(\gamma_2)]\\ &= \dots + G/_{\gamma_2} \cdot [-T(\gamma_2 + \gamma_2/_{\gamma_1} \cdot C(\gamma_1) + \dots)]\\ &= \dots + G/_{\gamma_2} \cdot [-T\gamma_2 + T(\gamma_2/_{\gamma_1} \cdot T\gamma_1) + \dots]\,.\end{aligned} \tag{243}$$

Hence, working out the recursion leads to subtraction operators T acting on unrenormalized (potentially multiloop) graphs like γ_2 and to iterated subtractions. If we introduce a new notation $T_\gamma \cdot G \equiv G/_\gamma \cdot T\gamma$ for the operation "replace γ within G by $T\gamma$", where products are defined as, e.g., $T_{\gamma_2} \cdot T_{\gamma_1} \cdot G = G/_{\gamma_2} \cdot T(T_{\gamma_1} \cdot \gamma_2)$, then we can rewrite the above terms as

$$\dots + G - T_{\gamma_2} \cdot G + T_{\gamma_2} \cdot T_{\gamma_1} \cdot G\,. \tag{244}$$

We note that both the subgraph $\{\gamma_2\}$, as well as the chain of subgraphs $\{\gamma_1, \gamma_2\}$ constitute forests in the sense defined above.

In general, if γ_1 and γ_2 are subgraphs and elements of a forest of G, we define

$$\gamma_2 \supsetneq \gamma_1: \qquad T_{\gamma_2} \cdot T_{\gamma_1} \cdot G = G/_{\gamma_2} \cdot T(T_{\gamma_1} \cdot \gamma_2) \tag{245a}$$

$$\gamma_1, \gamma_2 \text{ disjoint:} \qquad T_{\gamma_2} \cdot T_{\gamma_1} \cdot G = G/_{\gamma_1 \cup \gamma_2}(T\gamma_1)(T\gamma_2) \tag{245b}$$

while the product $T_{\gamma_2} \cdot T_{\gamma_1}$ is undefined for the case when γ_2 is subgraph of γ_1. Working out the recursion formula, in general, leads to the following forest formula [115]:

$$\mathcal{R}(G) = \sum_{\substack{\mathcal{F}=\text{forest}\\ \text{of } G}} \prod_{\gamma_i \in \mathcal{F}} (-T_{\gamma_i}) \cdot G\,, \tag{246}$$

where the forests may contain the full graph G and where also the empty set is an allowed forest $\mathcal{F} = \emptyset$. The formula for $\overline{\mathcal{R}}(G)$ is similar, but the forests may not contain the full graph G. The T_{γ_i}-operators are by definition always ordered as in Equations (244) and (245) according to nesting. Simply put: operators with bigger subgraphs act on the left, and operators with subgraphs contained in the bigger subgraphs on the right. The forest formula can be easily proven by noting that every forest that does not contain G itself has certain disjoint maximal elements $M_1 \dots M_s$ and can be partitioned into forests of the $M_1 \dots M_s$.

Based on this, the equivalence to the recursive formula can be established by induction over the number of loops.

Now, we turn to the announced elegant simplification of the forest formula due to the sector decomposition. We need to know one additional statement about sectors relevant for combinations like

$$T_\gamma \cdot G = G/_\gamma \cdot T(\gamma) = \sum_{\substack{(\mathcal{C}_1,\sigma_1)\\ \text{for } G/_\gamma}} \sum_{\substack{(\mathcal{C}_2,\sigma_2)\\ \text{for } \gamma}} (G/_\gamma)_{(\mathcal{C}_1,\sigma_1)} \cdot T(\gamma_{(\mathcal{C}_2,\sigma_2)})\,. \tag{247}$$

The statement is that there is a one-to-one correspondence between such combinations for sectors $(\mathcal{C}_1,\sigma_1)$, $(\mathcal{C}_2,\sigma_2)$ for the graphs $G/_\gamma$ and γ and sectors $(\mathcal{C},\sigma)$ for the full graph with the constraint that $\gamma \in \mathcal{C}$. Then, we can split the forest formula into sectors as follows:

$$\mathcal{R}(G) = \sum_{\substack{\mathcal{F}=\text{forest}\\ \text{of } G}} \sum_{\substack{(\mathcal{C},\sigma) \text{ for } G \text{ which}\\ \text{contain all } \gamma\in\mathcal{F}}} \prod_{\gamma_i\in\mathcal{F}} (-T_{\gamma_i}|_{\text{subsector}}) \cdot G|_{\text{subsector}}\,, \tag{248}$$

where it is used that every sector $(\mathcal{C},\sigma)$ with the given constraint generates appropriate subsectors for all subtraction operators T_{γ_i} and the remaining reduced graph and that all possible subsectors are generated in this way. Abbreviating slightly, we can then rearrange as

$$\begin{aligned} \mathcal{R}(G) &= \sum_{\mathcal{F}} \sum_{\mathcal{C}\supseteq\mathcal{F}} \prod_{\gamma_i\in\mathcal{F}} (-T_{\gamma_i}) \cdot G \\ &= \sum_{\mathcal{C}} \sum_{\mathcal{F}\subseteq\mathcal{C}} \prod_{\gamma_i\in\mathcal{F}} (-T_{\gamma_i}) \cdot G \\ &= \sum_{\mathcal{C}} \prod_{\gamma_i\in\mathcal{C}} (1-T_{\gamma_i}) \cdot G\,. \end{aligned} \tag{249}$$

The last step used that the sum over all possible forests $\mathcal{F}$, which are contained in $\mathcal{C}$, effectively generates the power set of $\mathcal{C}$, i.e., the set of all possible subsets of $\mathcal{C}$. This simply leads to the last line, which contains only a summation over all maximal forests $\mathcal{C}$ and the factors $(1-T_{\gamma_i})$. In this way, the forest formula becomes

$$\mathcal{R}(G) = \sum_{(\mathcal{C},\sigma)} \mathcal{R}(G)_{(\mathcal{C},\sigma)}\,, \tag{250a}$$

$$\mathcal{R}(G)_{(\mathcal{C},\sigma)} = \prod_{\gamma_i\in\mathcal{C}} (1-T_{\gamma_i}) \cdot G|_{(\mathcal{C},\sigma)}\,. \tag{250b}$$

The ordering of the $(1-T_{\gamma_i})$-operators is as in the original forest formula, according to the nesting of subgraphs.

This represents an important improvement. The operators $(1-T_{\gamma_i})$ have the effect of replacing an object by the one without the subdivergences from the subgraph γ_i (in the appropriate sector). Intuitively, every such operator improves the finiteness. On a more technical level, consider what any specific T_γ for a multiloop subgraph γ acts on. In the original forest formula, there are terms such as $T_\gamma \cdot G$, which lead to $G/_\gamma \cdot T(\gamma)$. The $T(\gamma)$ is the divergence of the unrenormalized multiloop graph γ, which is typically a very complicated expression, non-polynomial in momentum-space, or non-local in position space. In contrast, in the forest formula modified for sectors, any such multiloop T_γ only acts on expressions where all subdivergences corresponding to subgraphs of γ have already been subtracted:

$$T_\gamma \prod_{\gamma_i\in\mathcal{C},\gamma_i\subsetneq\gamma} (1-T_{\gamma_i}) \cdot G_{(\mathcal{C},\sigma)} = T\left(\prod_{\gamma_i\in\mathcal{C},\gamma_i\subsetneq\gamma} (1-T_{\gamma_i}) \cdot \gamma \right) \cdot G_{(\mathcal{C},\sigma)} \tag{251}$$

Hence, here, the left-most T actually acts on the fully subrenormalized expression $\overline{\mathcal{R}}(\gamma)$ in the appropriate subsector, which can be hoped to have simpler, polynomial/local divergences. These properties of the forest formula help in setting up an inductive proof of renormalization.

5.2.5. Ingredient 3: Sector Variables and Formula for the Integral

Introducing sectors into the α integrations required for Feynman graph integrals has further important advantages. Besides yielding the simpler forest formula, the sectors allow rewriting the actual integrals such that the power counting and the structure of divergences are isolated in a quite transparent way. Here, we illustrate this in a very simple case; then we provide the general result and give comments.

Let us focus on the integral (144), $(4\pi)^{-D/2} i^{1-D/2} \int_0^\infty d\alpha_1 d\alpha_2 \mathcal{U}^{-D/2} e^{iW}$, and consider the sector $\alpha_1 \leq \alpha_2$. In this sector, we introduce sector-specific variables: the largest α in the sector is replaced by a new variable t^2; the other α is rewritten as $t^2\beta$ in terms of a scaling variable β, which runs from 0 to 1. In total, we carry out the following substitution of variables and the integration measure in the sector:

$$\alpha_2 = t^2\,, \tag{252a}$$

$$\alpha_1 = t^2\beta\,, \tag{252b}$$

$$\int_{0\leq\alpha_1\leq\alpha_2} d\alpha_1 d\alpha_2 = 2\int_0^\infty dt t^{(2I-1)} \int_0^1 d\beta\,, \tag{252c}$$

where $I = 2$ is the number of internal lines. The integral (144) depends on two functions, the Symanzik polynomial $\mathcal{U}$ and the exponent W given in Equation (147). After the variable substitution, the Symanzik polynomial takes the value:

$$\mathcal{U} = M = t^2(1+\beta)\,, \tag{253}$$

and we observe that we can factor out the variable t^2. This is no accident. As already mentioned in Section 3.2.4, the behavior of $\mathcal{U}$ if some αs vanish reflects the ultraviolet behavior of the original Feynman integral. If all αs simultaneously vanish $\propto t^2$, $\mathcal{U}$ generally behaves as t^{2L}, where L is the number of loops in the graph. We can exhibit this behavior by defining a new function $\tilde{d}$:

$$\mathcal{U} = t^2\tilde{d}\,, \qquad \tilde{d} = 1+\beta \geq 1\,. \tag{254}$$

The indicated inequality provides a very important lower bound on the function $\tilde{d}$.

A second observation is that we can essentially eliminate the t-variable from the exponent W by rescaling the physical variables p, $u_{1,2}$, and m as

$$\tilde{p} = t\,p\,, \tag{255a}$$

$$\tilde{m} = t\,m\,, \tag{255b}$$

$$\tilde{u}_{1,2} = t^{-1}\,u_{1,2}\,. \tag{255c}$$

The rescaled variables are dimensionless. In terms of these variables, we can write the exponent as

$$W = \frac{\tilde{p}^2\beta - \beta\tilde{u}_2\cdot\tilde{p} + \tilde{u}_1\cdot\tilde{p} - \frac{1}{4}(\tilde{u}_1+\tilde{u}_2)^2}{(1+\beta)} + (it^2\varepsilon - \tilde{m}^2)(1+\beta)\,, \tag{256}$$

where, indeed, t does not appear explicitly, except in the product $t^2\varepsilon$.

Using all these ingredients, we can rewrite the α integral (144) in the considered sector as

$$(-i)^2(4\pi)^{-D/2} i^{1-D/2} 2 \int_0^\infty dt t^{-D+2I-1} \int_0^1 d\beta \tilde{d}^{-D/2} e^{iW} \,, \tag{257}$$

where we record the following observations:

- The power-like behavior of $\int dt t^{-D+2I-1}$ corresponds to the superficial ultraviolet power counting of the original loop integral (144), which behaves like $\int d^D k k^{-2I}$.
- The remaining integrand $\tilde{d}^{-D/2} e^{iW}$ has essentially no explicit dependence on t at all; it only depends on t via the rescaled variables (255) and via $t^2\varepsilon$.
- If $\varepsilon > 0$ in the $+i\varepsilon$ prescription, e^{iW} decreases exponentially for large t, and the full dependence of the integrand on the rescaled variables (255) and on β is of the C^∞-type. The result of the β integration is still C^∞ in the rescaled variables.

We need a second example to shape our understanding of the general case. Let us consider again the six-loop diagram of Section 5.2.3 and fix the same sector $(\mathcal{C}, \sigma)$ discussed there; see Equation (239). Which variable substitutions analogous to Equations (252) and (255) should we now choose? The sector is defined by a maximal forest with six subgraphs, each subgraph containing one specific labeled line, and for each subgraph, there is an inequality stating that the labeled α is the largest. The idea, generalizing the one-loop case, is to introduce one t_i-variable for each subgraph H_i and to define the labeled αs in terms of these t_i-variables. The $t_6 \equiv t_G$-variable corresponding to the full graph runs from 0 to ∞, and all the other t_i run from 0 to 1. Then, all inequalities for the labeled αs are implemented by the following scheme:

subgraph:	labelled α substitution:	rewrite	
H_1	$\alpha_{14} = t_1^2 t_6^2$	$t_1^2 \xi_1^2$	(258a)
H_2	$\alpha_3 = t_2^2 t_3^2 t_5^2 t_6^2$	$t_2^2 \xi_2^2$	(258b)
H_3	$\alpha_7 = t_3^2 t_5^2 t_6^2$	$t_3^2 \xi_3^2$	(258c)
H_4	$\alpha_{11} = t_4^2 t_5^2 t_6^2$	$t_4^2 \xi_4^2$	(258d)
H_5	$\alpha_8 = t_5^2 t_6^2$	$t_5^2 \xi_5^2$	(258e)
H_6	$\alpha_{16} = t_6^2$	$t_6^2 \xi_6^2$	(258f)

where also abbreviation variables ξ_i were introduced; they are products of all the "other t_i", as appropriate. In the next step, we introduce β_k-variables for all the remaining, non-labeled, αs, where the β_k all run from 0 to 1. We remark that $t_6 \equiv t_G$ is dimensionful, while all other t_i and β variables are dimensionless. In addition, we introduce two further useful notations, illustrated in the graph in Figure 3. First, for each subgraph in $\mathcal{C}$, we define a reduced subgraph $\bar{H}_i = H_i/_{\mathcal{M}(H_i)}$, where $\mathcal{M}(H_i)$ is the set of maximal elements in $\mathcal{C}$, which are properly contained in H_i. The lines in $\bar{H}_i$ are the lines specific to H_i, i.e., the lines contained in H_i, but in no smaller subgraph in $\mathcal{C}$. Clearly, the full graph is partitioned into $\bar{H}_i$, i.e., every line is in one unique $\bar{H}_i$. Second, we denote by $\underline{q}_{\bar{H}_i}$ a set of independent external momenta of $\bar{H}_i$, where we, in principle, allow nonzero incoming momenta into all vertices of the graph (the graph is drawn as if it has only two external momenta, but the renormalization procedure becomes more systematic if every graph is generalized to allow arbitrary incoming momenta into all vertices). This leads to the following scheme:

red. subgraph:	αs	indep. ext. momenta	
$\tilde{H}_1 = H_1$	$\{\alpha_{15}, \alpha_{14}\} = \{\beta_{15}, 1\} \times t_1^2 \zeta_1^2$	p_8	(259a)
$\tilde{H}_2 = H_2$	$\{\alpha_4, \alpha_3\} = \{\beta_4, 1\} \times t_2^2 \zeta_2^2$	p_3	(259b)
$\tilde{H}_3 = H_3/_{H_2}$	$\{\alpha_{2,5,6}, \alpha_7\} = \{\beta_{2,5,6}, 1\} \times t_3^2 \zeta_3^2$	p_2, p_5, p_{11}	(259c)
$\tilde{H}_4 = H_4$	$\{\alpha_{10,12}, \alpha_{11}\} = \{\beta_{10,12}, 1\} \times t_4^2 \zeta_4^2$	p_6, p_7	(259d)
$\tilde{H}_5 = H_5/_{H_3 \cup H_4}$	$\{\alpha_9, \alpha_8\} = \{\beta_9, 1\} \times t_5^2 \zeta_5^2$	$p_6 + p_7 + p_{12}$	(259e)
$\tilde{H}_6 = H_6/_{H_5 \cup H_1}$	$\{\alpha_{1,13,17}, \alpha_{16}\} = \{\beta_{1,13,17}, 1\} \times t_6^2 \zeta_6^2$	$p_1, p_8 + p_9, p_{10}$	(259f)

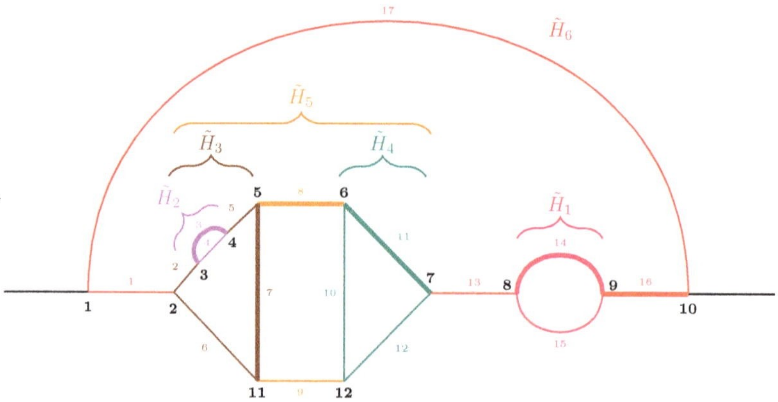

Figure 3. Illustration of sectors and sector variables t and β in Equations (258) and (259). The example is the 6-loop diagram and its subdiagrams already used in Equation (235) and Figures 1 and 2. Here, we chose six different colors for the reduced subdiagrams $\tilde{H}_i$ ($i = 1 \ldots 6$) into which the diagram can be partitioned.

The reduced subgraphs $\tilde{H}_i$ are formed solely by the lines proper to them, and the lines shared amongst the H_i are shrunk to a point. The subgraphs H_1, H_2, H_4 are identical to the reduced ones and, hence, take the obvious form as depicted in Figure 3. In the case of H_3, H_5, H_6, the reduced subgraphs are obtained by shrinking different subgraphs to a point. Let us illustrate this by specifying the form of these reduced subgraphs as follows:

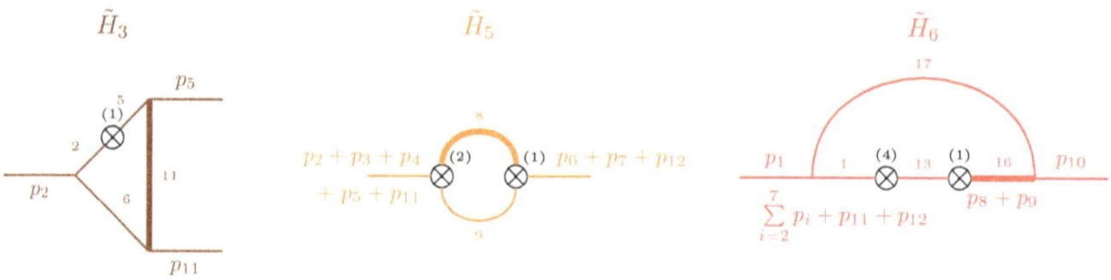

(260)

The crossed dots of order (n) denote the counterterm insertion due to shrinking the respective n-loop subgraph to a point. Since we assumed that, to each vertex V_i, there is associated an entering momentum p_i, shrinking a subgraph comprised of vertices $V_{i_1}, \ldots, V_{i_k}$ leads to a combination of incoming momenta $p_{i_1} + \cdots + p_{i_k}$ for that counterterm vertex, as indicated in the graphs. In choosing independent momenta, we can make use of momentum conservation. For a reduced subgraph with n vertices, it is sufficient to specify $n - 1$ incident momenta to the vertices. They uniquely characterize the momenta of a given reduced subgraph. What is more, all momenta of the graph can be reconstructed by linear combinations of these independent external momenta. A specific choice is given in Equation (259). Clearly, in this way, all inequalities of the sector (239) are implemented, and the combination of all the selected independent incoming momenta of the $\bar{H}_i$ span all independent incoming momenta of the full graph and can be used as independent variables in the result. The variable substitution leads to the following replacement of the integration measure, analogous to Equation (252):

$$\int_{(\mathcal{C},\sigma)} d\alpha_1 \ldots d\alpha_{17} = 2^L \int_{\substack{t_6=0\ldots\infty \\ t_{1\ldots5}=0\ldots1}} \prod_{i=1}^{6} dt_i t_i^{(2I_{H_i}-1)} \int_0^1 \prod_k d\beta_k , \tag{261}$$

where I_{H_i} is the number of internal lines in H_i.

This example provides us with sufficient information to construct the general result for the integral representation of a general 1PI graph G in a specific sector $(\mathcal{C},\sigma)$. As in the example, the sector defines a chain of subgraphs H_i (as many as there are loops; one of them is equal to the full graph G). The sector also defines a particular replacement of all αs in terms of t_i and β_k; for each subgraph H_i, it is also useful to define the variable ξ_i for the product of all the "other t_i". All lines of G are partitioned into lines of the reduced graphs $\bar{H}_i$, and for each $\bar{H}_i$, one can choose a set of independent incoming momenta $\underline{q}_{H_i}$, which in total span all incoming momenta of the full graph. Since each line carries one mass variable and one u-variable, we can also partition these variables into sets of masses m_{H_i} and sets of us, u_{H_i}, corresponding to the respective $\bar{H}_i$.

With these variables, we can rescale physical quantities, generalizing Equation (255) as

$$\tilde{\underline{q}}_{H_i} = t_i \xi_i \underline{q}_{H_i} , \tag{262a}$$

$$\tilde{m}_{H_i} = t_i \xi_i m_{H_i} , \tag{262b}$$

$$\tilde{u}_{H_i} = (t_i \xi_i)^{-1} u_{H_i} . \tag{262c}$$

We allow the integral to contain a numerator expressed as a derivative with respect to u-variables as in Equations (145) and (146a), but we assume that the derivative operator Z in the numerator is a product of Z_{H_i}, where each Z_{H_i} only depends on variables specific to $\bar{H}_i$. This is always the case in actual Feynman diagrams. For simplicity, we follow Reference [4] and assumed that all Z_{H_i} are homogeneous polynomials in the variables $\partial/\partial u_{H_i}$ and m_{H_i} of some degree $r_{\bar{H}_i}$. Then, we can write

$$\tilde{Z}_{H_i} = (t_i \xi_i)^{r_{\bar{H}_i}} Z_{H_i} \tag{263}$$

where $\tilde{Z}_{H_i}$ is the same homogeneous polynomial expressed with $\partial/\partial \tilde{u}_{H_i}$ and $\tilde{m}_{H_i}$. Writing $D = 4 - 2\epsilon$, we can finally define a power-counting degree of each reduced subgraph $\bar{H}_i$ and the complete (sub)graphs H_i as

$$\omega_{\bar{H}_i} = 4L_{\bar{H}_i} - 2I_{\bar{H}_i} + r_{\bar{H}_i} , \tag{264a}$$

$$\omega_{H_i} = \sum_{\substack{H' \subseteq H_i \\ H' \in \mathcal{C}}} \omega_{\bar{H}'} . \tag{264b}$$

This clearly corresponds to the superficial power-counting degree of the original momentum integral.

With these building blocks, we can formulate the general result for the integral specified in Equations (145) and (146a). Decomposing the integral into sectors as in Equation (241),

$$\mathcal{T}_G = \sum_{(\mathcal{C},\sigma)} \mathcal{T}_{G,(\mathcal{C},\sigma)} , \tag{265}$$

and setting again $D = 4 - 2\epsilon$, the result for each sector can be written as

$$\mathcal{T}_{G,(C,\sigma)} = c_D^L 2^L \int_{\substack{t_L=0\ldots\infty \\ t_{1\ldots L-1}=0\ldots 1}} \prod_{i=1}^{L} \frac{dt_i}{t_i} (t_i \xi_i)^{-\omega_{\bar{H}_i}+2\epsilon} \tilde{Z}_{H_i} \times \int_0^1 \prod_k d\beta_k \tilde{d}_G^{-D/2} e^{iW_G} \Bigg|_{u=0} . \tag{266}$$

The properties of the appearing objects are:

- All variables t_i, ξ_i, and β_k, the rescaled physical variables $\tilde{\underline{q}}_{H_i}$, $\tilde{m}_{H_i}$, and $\tilde{u}_{H_i}$, and the power-counting degrees $\omega_{\bar{H}_i}$ are defined above.
- The explicit powers of t_i correspond to the original superficial power-counting degrees of the momentum integrals over the subdiagrams H_i. For each t_i integral, a factor $(t_i \xi_i)^{2\epsilon}$ was split off, which may be viewed as the essence of the D-dimensional integration measure.
- The remaining integrand $\tilde{d}_G^{-D/2} e^{iW_G}$ has no explicit dependence on t_L at all. It depends on t_L only via the rescaled physical variables. The other t_i with $i = 1 \ldots L-1$ typically appear explicitly, however.
- The function $\tilde{d}_G$ is a rescaled Symanzik polynomial, which satisfies $\tilde{d}_G \geq 1$ in the integration region.
- For $\varepsilon > 0$ in the $+i\varepsilon$ prescription, the function e^{iW_G} is exponentially decreasing for large t_L.
- The product $\tilde{d}_G^{-D/2} e^{iW_G}$, therefore, is analytic in ϵ and C^∞ in t_i, β_k, and the rescaled physical variables $\tilde{\underline{q}}_{H_i}$, $\tilde{m}_{H_i}$, and $\tilde{u}_{H_i}$.

This statement is the starting point for the inductive proof of renormalization in DReg given in Reference [4], and it is a direct consequence of Lemma 4 of that Reference.

5.2.6. Ingredient 4: Integrand Relation between Graphs and Subgraphs

An important step in the proof is the application of subtraction operators T_H to a graph G. In order to analyze this operation, relationships between the original graph G, the reduced graph $G/_H$, and the subgraph H are needed. These relationships are again essentially independent of D-dimensional treatments. They rely on detailed analysis of the graphs themselves and the relationships between graphs and the α-parametrizations.

The required theory involves incidence matrices and graph theoretical representations of the Symanzik polynomial $\mathcal{U}$, or $\tilde{d}_G$, and the exponent W_G. Although the theory is very elegant and not too difficult, we do not develop it here. Hence, we only list several important statements without proof. For the proofs, we refer to Reference [4] and the references therein. Further discussions were given, e.g., in References [51,69,124]

Consider the Symanzik polynomial $\mathcal{U}_G$ for a graph G, and let H be a subgraph of G. $\mathcal{U}$ is a homogeneous polynomial in all αs of degree L. Consider the case where all αs corresponding to the subgraph H are rescaled by a factor ρ, while all other αs remain fixed. Then, for small ρ, we have

$$\mathcal{U}_G(\alpha\text{s in } H \text{ rescaled by } \rho) = \underbrace{\mathcal{U}_{G/_H}}_{\rho\text{-independent}} \underbrace{\mathcal{U}_H}_{\propto \rho^{L_H}} + \mathcal{O}(\rho^{L_H+1}) , \tag{267}$$

i.e., at the lowest nonvanishing order, the Symanzik polynomial factorizes into the two individual Symanzik polynomials for the reduced graph and the subgraph. If G and H are part of an integration sector as defined above, then variables t_G, t_H (and possibly further, t_i), and β_k exist, and rescaled Symanzik polynomials $\tilde{d}$ can be defined for each of these graphs.

In this case, $\tilde{d}_G$ for the full graph cannot depend on t_G, but it can depend on t_H, while $\tilde{d}_{G/_H}$ and $\tilde{d}_H$ can neither depend on t_G nor on t_H. Their relationship is the factorization:

$$\tilde{d}_G|_{t_H=0} = \tilde{d}_{G/_H}\tilde{d}_H\,. \tag{268}$$

A similar relationship can be established for the exponent W_G appearing in the general result of the integral (266). Defining W_H and $W_{G/_H}$ using the same variable transformations for the sector $(\mathcal{C},\sigma)$, suitably adapted to the subgraph and reduced graph, the relationship is

$$W_G|_{t_H=0} = W_{G/_H} + W_H|_{t_H=0}\,, \tag{269}$$

if all these quantities are expressed in terms of rescaled variables $\underline{\tilde{q}}$, $\tilde{u}$, and $\tilde{m}$. This property can be established in an elementary way once the exponents are constructed via incidence matrices.

For the same conditions, a further, more intricate property can also be established and is important. It is the following property involving derivatives:

$$\frac{d}{dt_H}\tilde{d}_G^{-D/2}e^{iW_G}\bigg|_{t_H=0} = \zeta_H U_H\left[\frac{d}{dt_H}\tilde{d}_H^{-D/2}e^{iW_H}\right]_{t_H=0}\cdot \tilde{d}_{G/_H}^{-D/2}e^{iW_{G/_H}}\,. \tag{270}$$

Here, $U_H[X]$ denotes an insertion operator that effectively inserts its argument X as a vertex into a bigger graph. To achieve this insertion, the external momenta of the argument X must become internal momenta of the bigger graph, in this case of $G/_H$. Technically, U_H acts by shifting in its argument the momentum variables $\underline{\tilde{q}}_H$ by terms involving derivatives with respect to u-variables for the bigger graph $G/_H$.

This is a statement of pivotal importance for the full proof of the theorem stated in Section 5.2.1 since it allows relating the divergences of a full graph to divergences of counterterm graphs and, thus, allows making manifest the cancellation of subdivergences. It is essentially the content of Lemma 5 of Reference [4].

5.2.7. Ingredient 5: Simple Integrals and Non-Analytic Functions of $D-4$

Now, we discuss several simple integrals and special functions that arise in DReg due to the D dimensionality of spacetime. They encapsulate how the regularization acts, how divergences arise as $1/(D-4)$ poles, and how divergences cancel by adding suitable counterterms. We set again $D = 4-2\epsilon$.

First, we discuss a simple type of integral, defined as

$$f(z) = \int_0^\infty dt t^{z-1} g(t)\,, \tag{271}$$

where z is a complex variable and $g(t)$ is a C^∞ function, which either decreases exponentially for $t\to\infty$ or which involves the step function $\theta(1-t)$ cutting off the integral at $t=1$. This simple integral appears in the general result (266), but also in the one-loop example (257). In all these cases, the t-integration involves one factor, which is of the form $t^{n-1+2\epsilon}$, where n is an integer. This corresponds to the above form for $z = n+2\epsilon$. This factor is nonanalytic in t around $t=0$. The remaining t-dependences in Equations (266) and (257) are complicated, but are C^∞ functions in t, which indeed fulfil the requirements on $g(t)$ listed above. In the case of the t_L integration, the remaining integrand exponentially decreases, and in the case of all other t_i integrations, the integration stops at $t_i = 1$.

The above function $f(z)$ is a generalization of the Γ-function, where $g(t) = e^{-t}$. The Γ-function is known to have simple poles at $z=0, z=-1, z=-2, \ldots$. It is easy to see that the same is true for the more general $f(z)$. Clearly, when for $\mathrm{Re}(z) > 0$, the integral defining $f(z)$ converges and defines an analytic function. To study negative $\mathrm{Re}(t)$, we can add to and subtract from $g(t)$ a Taylor polynomial $\sum g^{(k)}(0)t^k/k!$, where $g^{(k)}$ denotes the k-th derivative. Integrating this polynomial from 0 to 1, we obtain

$$f(z) = \int_0^\infty dt t^{z-1} \left[g(t) - \theta(1-t) \sum_{k=0}^{n} \frac{g^{(k)}(0)}{k!} t^k \right] + \sum_{k=0}^{n} \frac{g^{(k)}(0)}{k!} \frac{1}{z+k} . \tag{272}$$

For any non-negative integer n and for $\mathrm{Re}(z) > 0$, the value and convergence properties of the integral are not changed. However, the square bracket behaves like t^{n+1} for small t; hence, the integral now converges even for negative z, as long as $\mathrm{Re}(z) > -n-1$. Hence, this formula represents an analytic continuation of $f(z)$ onto the entire complex z plane. It makes also manifest that this analytically continued $f(z)$ has single poles at $z = -0, z = -1, z = -2, \ldots$.

We can rewrite the result in the form of an integration rule for the typical t-integrals appearing in DReg by replacing $z = -n + 2\epsilon$ with integer non-negative n and $\epsilon \approx 0$. We then have the rule:

$$\int_0^\infty dt t^{-n-1+2\epsilon} g(t) = \frac{1}{n!} \left(\frac{d}{dt} \right)^n g(t) \Bigg|_{t=0} \frac{1}{2\epsilon} + \text{regular expression} , \tag{273}$$

where the form of the regular expression can be read off from Equation (272). The t-integrals in the general formula (266) are to be analytically continued in this way. Hence, this rule immediately shows that any t-integration can only lead to single $1/\epsilon$-poles and not to more complicated divergences as $\epsilon \to 0$.

Next, we consider two special simple classes of nonanalytic functions of t. They are defined as the two kinds of sets (for integer K, L):

$$K < L: \quad J_K^L = \left\{ f(t,\epsilon) = \frac{c_1 t^{2\epsilon} + \ldots + c_L t^{2L\epsilon}}{\epsilon^K} = \text{finite for } \epsilon \to 0 \right\} , \tag{274a}$$

$$K \leq L: \quad \tilde{J}_K^L = \left\{ f(t,\epsilon) = \frac{c_0 + c_1 t^{2\epsilon} + \ldots + c_L t^{2L\epsilon}}{\epsilon^K} = \text{finite for } \epsilon \to 0 \right\} . \tag{274b}$$

In the definitions of the sets, the lower index K refers to the ϵ-power in the denominator, and the upper index L can be thought of as the loop number at which the functions become of interest. The coefficients c_i are arbitrary except for the constraint that the defined functions are finite for $\epsilon \to 0$.

Let us illustrate how such functions can appear by considering a 2-loop diagram G with a 1-loop subdiagram H. We imagine a calculation not only of the diagrams themselves, but of the entire renormalization procedure, taking into account suitable counterterm diagrams canceling subdivergences. In the imagined calculations, we use the general formula (266). If the one-loop diagram H is computed in isolation, it involves one t_1-integral whose essential nonanalytic part is simply

$$t_1^{2\epsilon} \in J_0^1 , \tag{275}$$

which is an element of the set J_0^1 and which may be attributed to the D-dimensional measure. The result of the t_1-integration via the rule (273) then leads particularly to a $1/(2\epsilon)$ pole, and a counterterm for diagram H can be defined that cancels this divergence. In the 2-loop calculation of G, the 1-loop diagram H appears as a subdiagram with corresponding t_1 integration. Here, the t_1 variable is accompanied by ξ_1, which is here simply $\xi_1 = t_2$. After the t_1 integration, the nonanalytic factor $\xi_1^{2\epsilon}$ remains and combines with the $1/(2\epsilon)$ pole. In the corresponding counterterm diagram, where the subdiagram H is replaced by the counterterm canceling its $1/(2\epsilon)$ pole, there is no t_1 integration and no appearance of the variable ξ_1. Therefore, after the t_1 integration and after combining with the counterterm diagram canceling the subdivergence, a combined function:

$$\frac{t_2^{2\epsilon}-1}{2\epsilon} \in \tilde{J}_1^1 \tag{276}$$

appears. The finiteness of functions in the set $\tilde{J}_1^1$ reflects the successful cancellation of the subdivergence. Proceeding with the computation of the two-loop diagram G, this function is combined with the measure factor, such that the interesting nonanalytic part of the t_2-integrand is

$$t_2^{2\epsilon}\frac{t_2^{2\epsilon}-1}{2\epsilon} \in J_1^2. \tag{277}$$

This example illustrates the general idea: Functions in J_K^L are the functions that actually appear as the nonanalytic factors in the t_L integrations at the L-loop level during the renormalization procedure. After carrying out a t_L integral and after combining with the suitable counterterm contribution, a function in the set $\tilde{J}_{K+1}^L$ appears. At the next loop level, the integrand needs to be prepared by suitable rearrangements and combined with the measure factor $t_{L+1}^{2\epsilon}$ to produce a function of the set J_{K+1}^{L+1}, and so on.

For this reason, it is helpful to study the properties of functions in these sets on their own, before tackling the actual loop integrations. Some particularly useful properties are as follows:

(i) Any function $f \in J_K^L$ has the limit $f(t,0) = \text{const} \times (\ln t)^K$.

(ii) For a function $f \in J_K^L$, the integral $\int_1^t \frac{dt'}{t'} f(t',\epsilon)$ produces an element of the next set $\tilde{J}_{K+1}^L$.

(iii) The converse is also true, i.e., every element of $\tilde{J}_{K+1}^L$ can be written in terms of such an integral.

(iv) A function $f \in J_K^L$ where the first argument is a product can be factorized as $f(\xi t,\epsilon) = \sum_j f_{1j}(\xi,\epsilon) f_{2j}(t,\epsilon)$, where all functions on the right-hand side are elements of $f_{nj} \in J_{K_{nj}}^L$, where $K_{1j} + K_{2j} = K$. This property is obviously important to prepare higher-loop integrands such that t integrals act on isolated functions depending only on t, not on ξ.

(v) There is a simple product rule $f_{K_1}^{L_1} f_{K_2}^{L_2} \in J_{K_1+K_2}^{L_1+L_2}$ for functions $f_{K_i}^{L_i} \in J_{K_i}^{L_i}$. This property is also important on the multiloop level in case a multiloop diagram contains two disjoint divergent subdiagrams.

The properties can all be proven using elementary integration tricks and l'Hopital's rule for limits. Such properties of these functions are the content of Lemma 2 of Reference [4].

5.2.8. Sketch of Proof by Induction

All explained ingredients are important in the full proof of the central Theorem 1 in Reference [4] and stated in Section 5.2.1. Here, we give a sketch of this proof. The proof applies the α parametrization of integrals decomposed into sectors as in Equation (241). The renormalization procedure is then expressed in terms of the forest formula (250). This formula provides the basis for an inductive proof, where a graph G and a sector are fixed, and then, all factors $(1 - T_{H_i})$ in the forest formula are successively applied in the correct ordering. The base case of the induction is provided by the general formula (266). The induction step needs to carry out the actual integration over one t variable and some β variables corresponding to the next $(1 - T_{H_i})$ factor. The step uses the properties of the special functions of ϵ defined in Section 5.2.7, and the relationships between the graph, subgraph, and reduced graph described in Section 5.2.6.

Obtaining the precise form of the induction hypothesis is highly nontrivial, but it can be motivated using all the developed insight. It can be formulated as follows. Consider a 1PI graph G and a sector $(\mathcal{C},\sigma)$. All the following quantities are specific to this sector, but for brevity, we will omit all indices denoting this dependence. The graph has L_G loops, and the sector contains L_G subgraphs $H_1,\ldots,H_{L_G}$. Without loss of generality, we assume the labeling such that the subgraphs are already ordered according to their allowed appearance in the forest formula (250), such that if $H_j \supseteq H_i$, then also $j \geq i$ (the ordering is not unique). Then, after evaluating $L \leq L_G$ factors in the forest formula, we obtain the expression (suppressing the dependence on the sector $(\mathcal{C},\sigma)$):

$$\mathcal{R}_X(G) \equiv (1-T_{H_L})\cdot\ldots\cdot(1-T_{H_2})\cdot(1-T_{H_1})\cdot G\,. \tag{278}$$

This represents a partially renormalized graph where L loops and L subgraphs have already been treated in previous induction steps. Section 5.2.4 gave arguments that this expression should have simple divergence properties when acted upon by further T_{H_i} operators. Despite this, the partially renormalized expression on its own clearly can have a very complicated analytical structure and can still have non-polynomial divergences, which the proof needs to deal with. The label X denotes the set of all subgraphs that have already been treated, and we also define X_0 as the subset of X, which contains only maximal subgraphs, i.e.,

$$X = \{H_1,\ldots,H_L\}\,, \qquad\qquad X_0 = \{M_1,\ldots,M_S,H_L\}\,, \tag{279}$$

where it is used that H_L itself is necessarily a maximal subgraph in X and where names have been given to all other elements of X_0.

The induction hypothesis states that, after evaluating all t_i and β_k integrals corresponding to lines in the already treated graphs in X, we obtain

$$\begin{aligned}\mathcal{R}_X(G) &= \text{sum of terms like}\\ &\qquad \int_{\geq L+1}\prod_{M\in X_0}\xi_M^{-\omega_M}\tilde{f}_M(\xi_M,\epsilon)g_{G,X}\Bigg|_{\tilde{u}=0}\,,\end{aligned} \tag{280}$$

where the integration factors for the remaining integrals are abbreviated as

$$\int_{\geq L+1} = c_D^{L_G-L}\int\prod_{i=L+1}^{L_G}\frac{dt_i}{t_i}(t_i\xi_i)^{-\omega_{\bar{H}_i}+2\epsilon}\int\prod_{k\in G/X_0}d\beta_k\tilde{Z}_{H_i}\,. \tag{281}$$

Here, the integration boundaries of the t_i and β_k integrals are as in Equation (266), and the notation $k\in G/X_0$ corresponds to all indices k corresponding to any line outside the already treated graphs in the set X. In the product over the maximal subgraphs M (which includes the case $M=H_L$), each M is equal to one particular $H_{j(M)}$, and for simplicity, we identified the indices $\xi_M \equiv \xi_{j(M)}$.

We provide the following comments on the induction hypothesis:

- The "sum of terms like" refers to the expression in the integrand, which really is of the form $\sum_a\prod_M\tilde{f}_{M,a}g_{G,X,a}$. Since the proof can be carried out for each such term, we drop the index a and this summation.
- The integration variables t_i and β_k and the $\tilde{u}_k$ variable for the already treated graphs do not exist anymore, since they have been integrated over/set to zero. Hence, the only appearing t_i, β_k, and $\tilde{u}_k$ are the ones for $i=L+1,\ldots,L_G$ and for $k\in G/X_0$.
- The sets of physical variables $\tilde{q}_{\underline{H}_i}$, $\tilde{m}_{H_i}$ and the remaining $\tilde{u}_{H_i}$ (for $H_i\notin X$) are rescaled only by the remaining t_is, i.e., Equation (262) applies in a modified form where, on the right-hand side, $t_i = 1\,\forall i\leq L$ and where the $\tilde{u}_{H_i}$ for $i\leq L$ do not exist.
- The particularly nontrivial and interesting part of the statement is the integrand in Equation (280). It displays the analytic structure of the partially renormalized graph

and the result of all the evaluated t_i and β_k integrals. The result is a product of functions $\tilde{f}_M$, which are nonanalytic in the remaining t_i, and the function $g_{G,X}$.

- Each function $\tilde{f}_M$ is an element of a set $\tilde{J}^L_K$ with $K \leq L$. These functions are thus nonanalytic in the remaining t_i, but have a finite limit for $\epsilon \to 0$, reflecting the successful subtraction of subdivergences. The functional form of each $\tilde{f}_M$ is further specific to the chain of subgraphs $X_M = \{H' \subseteq M, H' \in \mathcal{C}\}$ and does not depend on any details of graphs or a part of graphs outside M. Only the argument ξ_M has a dependence on t_i variables corresponding to bigger graphs.
- The function $g_{G,X}$ carries the complicated dependence on all physical variables and all other t_i and β_k variables. $g_{G,X}$ is C^∞ in all these remaining integration variables and all the physical variables $\tilde{\underline{q}}_{H_i}$, $\tilde{m}_{H_i}$, and $\tilde{u}_{H_i}$ rescaled as defined above. It is analytic in ϵ, again reflecting the cancellation of subdivergences, and it has no explicit dependence on t_{L_G} corresponding to the full graph G (except for the product $t^2_{L_G}\epsilon$, similar to Equation (256)). Its functional form is specific to the full graph G and the treated graphs $H_i \in X$.

The induction base case is the one where $L = 0$ and no subgraph has been treated yet. In this case, the sets X and X_0 are empty, and $\mathcal{R}_X(G)$ simply refers to the unrenormalized result $\mathcal{T}_G$. The form of the unrenormalized result is given in Equation (266), and it directly confirms the induction hypothesis (280) with $g_{G,\emptyset} = 2^{L_G}\tilde{d}^{-D/2}e^{iW_G}$.

For a sketch of the induction step, we assumed $L \geq 1$ and assumed the partial renormalization was carried out up to loop number $L-1$ and that the induction hypothesis holds at loop number $L-1$. It is then useful to introduce the notation for the previously treated subgraphs and previously treated maximal subgraphs. We write

$$X' = \{H_1, \dots, H_{L-1}\} \qquad\qquad X'_0 = \{m_1, \dots, m_s\} \cup \{M_1, \dots, M_S\}, \tag{282}$$

and we keep the definitions of Equation (279) such that $X = X' \cup \{H_L\}$ and such that the subgraphs m_i are the maximal subgraphs of H_L. The remaining subgraphs are H_L, as well as H_i with $i \geq L+1$; the lines and β_k are the ones with $k \in G/_{X'_0}$ or, equivalently, the ones with $k \in G/_{X_0}$ or with $k \in \bar{H}_L$. The induction hypothesis for loop number $L-1$ can, therefore, be cast into the form:

$$\begin{aligned}\mathcal{R}_{X'}(G) &= \text{sum of terms like} \\ &\int_{\geq L+1} \prod_{M \in X_0 \backslash \{H_L\}} \xi_M^{-\omega_M} \tilde{f}_M(\xi_M, \epsilon) \\ &\times c_D \int \frac{dt_L}{t_L} (t_L \xi_L)^{-\omega_{\bar{H}_L} + 2\epsilon} \tilde{Z}_{H_L} \int \prod_{k \in \bar{H}_L} d\beta_k \\ &\times \prod_{m_i} \xi_{m_i}^{-\omega_{m_i}} \tilde{f}_{m_i}(\xi_{m_i}, \epsilon) g_{G,X'} \Bigg|_{\tilde{u}=0} .\end{aligned} \tag{283}$$

In this way of writing, the role of the graph H_L, which is to be treated next, is exhibited, while the factors in the first line contain the same integration factors and almost the same $\tilde{f}_M$ factors as Equation (280). The physical variables appearing here inside $\tilde{Z}_{H_i}$ and $g_{G,X'}$ are rescaled with all t_i for $i \leq L$, and all comments made for the induction hypothesis apply with suitable modifications.

In the induction step, we need to assume the validity of Equation (283) and carry out the next step, construct $\mathcal{R}_X(G)$, and prove that it takes the form (280) with all listed properties. The construction involves the evaluation of all integrals in the last two lines of Equation (283). It also involves the application of the next subtraction operator $(1 - T_{H_L})$, which also only affects the last two lines of Equation (283), in particular because the integration factors $\int_{L+1}$ stay unchanged if the subgraph H_L is replaced by its counterterm.

We begin with several immediate simplifications of the factors in the last two lines of Equation (283). First, we observe that all the ξ_{m_i} in the last line are equal to each other, and they are equal to $\xi_{m_i} = t_L\xi_L$. The reason is that the ξ_{m_i} are the products of t_j for all subgraphs in C that contain m_i and that the m_i are maximal subgraphs of H_L. One consequence is that the $\omega_{\bar{H}_L}$- and ω_{m_i}-dependent terms combine simply to $(t_L\xi_L)^{\omega_{H_L}}$. A less trivial consequence is that all nonanalytic functions and the measure factor for t_L can be combined as

$$f_{H_L}(t_L\xi_L, \epsilon) = (t_L\xi_L)^{2\epsilon} \prod_{m_i} \tilde{f}_{m_i}(t_L\xi_L, \epsilon), \tag{284}$$

which is an element of the set J_K^L for some $K < L$ thanks to the properties of the functions discussed in Section 5.2.7. Second, after the β_k integrations and after applying the derivative operator $\tilde{Z}_{H_L}$ and setting $\tilde{u}_{H_L} = 0$, we obtain

$$\bar{g}_{G,X'} = \tilde{Z}_{H_L} \int \prod_{k \in \bar{H}_L} d\beta_k g_{G,X'} \Bigg|_{\tilde{u}_{H_L}=0}. \tag{285}$$

This function is still C^∞ in the remaining variables and analytic in ϵ. Hence, the last two lines of Equation (283) can be written as

$$c_D \int \frac{dt_L}{t_L} (t_L\xi_L)^{-\omega_{H_L}} f_{H_L}(t_L\xi_L, \epsilon) \bar{g}_{G,X'} . \tag{286}$$

The more difficult part of the induction step is the evaluation of the t_L integral and the application of the $(1 - T_{H_L})$ subtraction operator. Two cases need to be distinguished. The first case is when the next step is the final step of renormalization, i.e., when $L = L_G$ and $H_L = G$. The second case is when $L < L_G$ and H_L is still a proper subgraph of the full graph G.

To sketch the first case with $L = L_G$ and $H_L = G$, we note that, in this case, the second line of Equation (283) is just the factor one, since there are no remaining integrations and there are no other maximal subgraphs M. Likewise, the remaining $\xi_L = 1$, and from the induction hypothesis, we know that the variable $t_L = t_{L_G}$ does not appear explicitly in $\bar{g}_{G,X'}$; this variable only enters via rescaled physical variables $\tilde{q}_{H_i}$ and $\tilde{m}_{H_i}$, i.e., via the products of t_L and physical momenta and masses. Plugging in the general form of the function f_{H_L} yields a sum of terms as

$$\sum_n \frac{c_n t_L^{2n\epsilon}}{\epsilon^K} \bar{g}_{G,X'} , \tag{287}$$

which need to be integrated over t_L with the measure $\int dt_L t_L^{-\omega_{H_L}-1}$. This integral is performed via the general rule (273). This rule leads to a regular expression and a singular term. The regular expression can be shown to be analytic in ϵ and C^∞ in all other variables. The singular term contains poles in ϵ and takes the form:

$$\sum_n \frac{c_n}{2n\epsilon^{K+1}} \frac{1}{\omega_{H_L}!} \left[\left(\frac{d}{dt_L} \right)^{\omega_{H_L}} \bar{g}_{G,X'} \right]_{t_L=0}. \tag{288}$$

This singular term can be shown to have all desirable properties. The poles in ϵ are at most of degree $1/\epsilon^{L_G}$. The coefficients are polynomials in the physical variables, masses, and momenta, of degree ω_{H_L}. Here, and in Reference [4], the factor of the dimensional regularization scale $\mu^{2\epsilon}$ is omitted from the definition of renormalized amplitudes. If this factor is included, it is also possible to prove that the divergent polynomial is independent of μ. It is, therefore, possible to define the subtraction operator $(1 - T_G)$ for this sector such that it subtracts this polynomial divergence; the resulting finite remainder satisfies all properties listed after the induction hypothesis (280). It is further possible to define the full divergent part of the full diagram, $T\mathcal{R}_G$, as the sum of all these singular terms arising in this way in all sectors. This object has all properties required for a possible contribution to a counterterm Lagrangian: in position-space, it is local; it has the correct power-counting degree; its value depends only on the graph G and not on its embedding into bigger graphs.

Finally, we also sketch the remaining induction step for the case $L < L_G$ and $H_L \neq G$. Here, evaluating the t_L integral and applying the subtraction operator $(1 - T_{H_L})$ to Equation (286) lead to three terms: the regular expression from the t_L integration, the singular expression from the t_L integration, and the counterterm contribution from T_{H_L}, where T_{H_L} is defined via the full renormalization of the graph H_L in isolation. All terms need to be rearranged by using the properties of the f functions discussed in Section 5.2.7, in particular of the factorization property of these functions. Furthermore, the singular expression of the t_L integration has to be rearranged by using properties such as (270) for the relationships between graphs, subgraphs, and reduced graphs. In these ways, it is possible to show that the combination of all terms acquires the form of the induction hypothesis (280) and that all announced properties are fulfilled.

In this way, all properties announced in Section 5.2.1 are established, except for the equivalence to BPHZ. Illustrating this point requires comparing the structure of the appearing integrals in the DReg and the BPHZ approaches. For this, we refer to the original literature [4,22].

6. Renormalization and Symmetry

In the preceding section, we saw how the renormalization program allows subtracting the divergences from Feynman diagrams. Importantly, the subtraction terms are polynomials in momenta constrained by power counting, and the subtraction is equivalent to adding certain counterterms to the Lagrangian. By choosing a certain renormalization scheme, the remaining ambiguities of finite counterterms can be fixed, and the Lagrangian supplemented by those counterterms defines a finite four-dimensional theory.

In this section, we consider the problem of renormalization in the presence of symmetries, specifically gauge invariance. On the one hand, symmetries put additional restrictions on certain quantities, which allows for simplifications. On the other hand, we also have to ask about the compatibility of symmetries and regularization and whether they can be restored if intermediately broken. Since regularization may, in general, spoil the classical symmetry, we shall require its validity as part of the definition of our theory. The symmetry of interest for us is gauge invariance promoted to BRST invariance as described in Section 2.3. On the level of Green functions, this symmetry is implemented by the Slavnov-Taylor identity as described in Section 2.5. In a more compact notation (cf. Equation (92)), it can be written as

$$\mathcal{S}(\Gamma_{\text{ren}}) = \int \mathrm{d}^4x \, \frac{\delta\Gamma_{\text{ren}}}{\delta\phi(x)} \frac{\delta\Gamma_{\text{ren}}}{\delta K_\phi(x)} \stackrel{!}{=} 0. \tag{289}$$

Here, we assumed for simplicity that all symmetry transformations, i.e., both linear and nonlinear, are coupled to sources K_ϕ. The Slavnov–Taylor identity is the pivotal tool in the proof of the renormalizability of quantized Yang–Mills gauge theories, including the proof that the quantum theory actually is physically sensible.

The first proofs of the renormalizability of non-Abelian gauge theories were given by 't Hooft, Lee, and Zinn-Justin in References [125–130], all employing various versions of Slavnov–Taylor identities. These proofs establish not only the finiteness and validity of the Slavnov–Taylor identity, but also the interpretation of the quantum theory with a unitary and gauge-fixing independent S-matrix defined on a Hilbert space of quantum states with a positive norm. Later, the proofs were generalized by Becchi, Rouet, Stora, and Tyutin (BRST) to the case where nothing is known about the symmetry properties of the employed regularization scheme, establishing the approach of algebraic renormalization [41–44]; see also the reviews by Piguet/Rouet and Piguet/Sorella [47,119]. A particularly satisfactory formulation was achieved with the Kugo/Ojima formalism [45], where the existence of a nilpotent operator Q_B was derived from the Slavnov–Taylor identity. Q_B generates BRST transformations on the level of asymptotic states, and its role on the level of quantum states is similar to the role of the BRST operator s on the classical level; see Equations (46)–(49). It may be used to define the physical Hilbert space as the quotient space:

$$\mathcal{H}_{\text{phys}} = (\ker Q_B)/(\text{im}\, Q_B)\,. \tag{290}$$

Hence, two states are equivalent if they differ by a total Q_B-variation. A single state is called physical if $Q_B|\psi\rangle = 0$, provided it is not some total variation, i.e., $|\psi\rangle \neq Q_B|\chi\rangle$ for some $|\chi\rangle$, in which case it would be equivalent to the zero vector. The fields act Lorentz covariantly on the whole space including unphysical states, and because of the Slavnov–Taylor identity, Q_B commutes with the S-matrix. Hence, the physical S-matrix defined on the physical Hilbert space maps physical states to physical states; it is Lorentz-invariant, unitary, and causal. All these properties can be shown in a very elegant way [45]. We thus see that, if we make sure that the Slavnov–Taylor identity is obeyed after renormalization, we are guaranteed a consistent quantum field theory.

Hence, the logic now is the following. In Section 2, we defined gauge theories that classically satisfy the BRST symmetry. Then, we established dimensional regularization as a framework for treating such theories perturbatively in loop orders. Now, we are in a position to define our renormalized theory with the fundamental Slavnov–Taylor identity intact and study the possible obstructions posed by regularization. To this end, we shall first discuss the counterterm structure for manifestly preserved symmetries during renormalization in Section 6.1. Then, in Section 6.2, we give a brief overview of the field of algebraic renormalization, which is the appropriate setting in which to discuss the breaking and restoration of symmetries. Finally, we discuss how the general analysis of algebraic renormalization can be specialized to the case of dimensional regularization in Section 6.3.

6.1. Counterterms in Symmetry-Preserving Regularization

We first recall the simple case where a symmetry is manifestly preserved at all steps of the calculation. This is the standard case often encountered in textbook discussions and practical calculations using DReg in QED and QCD, for reasons described already in Section 4.3. There, one frequently uses so-called renormalization transformations of the generic form:

$$g \to g + \delta g \tag{291a}$$

$$\phi_i \to \sqrt{Z_{ij}}\,\phi_j, \tag{291b}$$

for coupling constants g and quantum fields ϕ_i with associated parameter and field renormalization constants δg and $\delta Z_{ij} = Z_{ij} - \delta_{ij}$. The renormalization constants are to be understood as power series in loop orders or, equivalently, in the renormalized parameters.

This procedure is applied onto the classical action S_0 and, thereby, defines a bare action S_{bare} (cf. (109)), itself giving rise to the counterterm action:

$$S_{\text{ct}} = S_{\text{bare}} - S_0. \tag{292}$$

The divergent parts of these generated counterterms cancel the UV divergences of loop diagrams, and the finite parts of the counterterms can be used to fulfil certain renormalization conditions, as mentioned in Section 3.1.

In terms of the Slavnov–Taylor identities, the standard case is expressed by the statement

$$\mathcal{S}(\Gamma_{\text{reg}}) = 0, \tag{293}$$

which, as explained in Section 4.3, means that the regularized Green functions already satisfy the Slavnov–Taylor identity. If applicable, similar equations should hold for other identities such as the ones discussed in Section 2.6 (e.g., ghost equation). This is indeed the case in QED and QCD in DReg at all orders. The basis of this statement was explained in Section 4. The manifest symmetry at the regularized level (293) has two implications for the structure of renormalization. First, the possible divergences are restricted by Equation (293), which, in turn, also restricts the structure of counterterms needed to cancel divergences. Second, possible finite counterterms are also restricted by Equation (293), together with the ultimate requirement (289) for the renormalized theory. Both implications can be simultaneously evaluated as follows. Assuming that the theory has been renormalized up to order $\mathcal{O}(\hbar^{n-1})$, we are interested in the $\mathcal{O}(\hbar^n)$-order counterterms $\mathcal{L}^n_{\text{ct}}$ and the $\mathcal{O}(\hbar^n)$ divergences of the regularized theory. The renormalized theory at order $\mathcal{O}(\hbar^n)$ can be written as

$$\Gamma^{(n)}_{\text{ren}} = \Gamma^{(n)}_{\text{reg,fin}} + \Gamma^n_{\text{reg,div}} + S^n_{\text{ct}}. \tag{294}$$

For further analysis, it is customary to introduce the linearized Slavnov–Taylor operator s_Γ, defined by expanding the Slavnov–Taylor operator $\mathcal{S}(\Gamma)$ for both linearly and nonlinearly transforming fields ϕ and Φ, respectively:

$$\mathcal{S}(\Gamma) = \int \mathrm{d}^4x \frac{\delta\Gamma}{\delta K_i(x)} \frac{\delta\Gamma}{\delta\Phi_i(x)} + \int \mathrm{d}^4x\, s\phi_i(x) \frac{\delta\Gamma}{\delta\phi_i(x)}, \tag{295}$$

as follows:

$$\mathcal{S}(\Gamma + \zeta\mathcal{F}) = \mathcal{S}(\Gamma) + \zeta s_\Gamma \mathcal{F} + \mathcal{O}(\zeta^2), \tag{296}$$

for some functional $\mathcal{F}$. Its concrete form is given by

$$s_\Gamma = \int \mathrm{d}x \left(\frac{\delta\Gamma}{\delta K_i(x)} \frac{\delta}{\delta\Phi_i(x)} + \frac{\delta\Gamma}{\delta\Phi_i(x)} \frac{\delta}{\delta K_i(x)} + s\phi_i(x) \frac{\delta}{\delta\phi_i(x)} \right). \tag{297}$$

Of special interest is the case of the classical action Γ_{cl}, for which we define

$$b \equiv s_{\Gamma_{\text{cl}}}, \tag{298}$$

as the linearized Slavnov–Taylor operator based on the classical action. In agreement with the nilpotency of the BRST operator (49), the algebraic structure of the Slavnov–Taylor operator leads to two nilpotency relations:

$$s_\Gamma \mathcal{S}(\Gamma) = 0, \tag{299}$$

$$s_\Gamma s_\Gamma = 0 \quad \text{if} \quad \mathcal{S}(\Gamma) = 0. \tag{300}$$

Substituting the decomposition of Equation (294) into Equations (293) and (289), we first obtain

$$b\, \Gamma^n_{\text{reg,div}} = 0. \tag{301}$$

This establishes the restriction on the possible divergences. Second, we obtain

$$b\, S^n_{\text{ct}} = 0, \tag{302}$$

both for the divergent and the finite parts. The most-general solution of this equation in terms of admissible counterterm actions yields the counterterm structure, which is sufficient to cancel the divergences and required to establish the symmetry. The corresponding calculations were carried out in the original references on the renormalization of Yang–Mills theories cited at the beginning of this section; textbook discussions can be found, e.g., in the textbooks by Zinn-Justin, Weinberg, and Böhm/Denner/Joos [32,34,50].

For most theories of interest including the SM, the outcome is the familiar statement cited in the beginning (cf. Equation (291)) that all counterterms can be obtained by renormalization transformation of the classical action. A second related outcome is then that any two consistent regularization/renormalization prescriptions that both fulfil the symmetry requirement (289) can only differ by a reparametrization of the form (291). This latter result is a stronger statement than the one of Equation (228) because a smaller number of parameters is affected.

6.2. Broken Symmetries and Algebraic Renormalization

Now, we turn to the case of interest for, e.g., chiral gauge theories in which the symmetry is not manifestly preserved by the regularization. This case is characterized by

$$\mathcal{S}(\Gamma_{\text{reg}}) \neq 0, \tag{303}$$

in contrast to Equation (293). Clearly, the required structure of the counterterms is more complicated. Now, the divergences and required divergent counterterms may be non-symmetric and not fulfil Equation (301). In this case, one has to determine them by explicit calculation of the divergences of Green functions instead of reading off their structure from a renormalization transformation such as (291). In this way, the theory can be rendered finite despite the broken symmetry (303).

Even on the finite level, the symmetry breaking (303) might still persist. Finite counterterms then have to be determined such that the fully renormalized theory fulfills the basic requirement (289). In some cases, it can actually be impossible to find such counterterms; the symmetry is then said to be broken by an anomaly. Since we considered the Slavnov–Taylor identity as part of the definition of the theory, an anomalous breaking of the Slavnov–Taylor identity means that the theory is inconsistent and not renormalizable. In cases without an anomaly, it is indeed possible to recover the symmetry by appropriately chosen finite counterterms.

Even though the precise form of the symmetry breaking depends on the regularization, it is possible to study the general case of (303) in a regularization-independent way. This study is the content of algebraic renormalization, pioneered by BRST [41–44]; see also the reviews [47,119]. The main insight of the procedure is that the possible breakings are restricted in two ways. On the one hand, they are restricted by the Slavnov–Taylor identity itself, similar to the possible divergent structures in Equation (301). On the other hand, they are restricted by a regularization-independent version of the quantum action principle.

Those two restrictions taken together provide a regularization-independent analysis of the renormalization of gauge theories. In the following, we shall first sketch the quantum action principle in the BPHZ framework of renormalization, where it was originally established and subsequently used for algebraic analysis, as well as exhibit a connection to the regularized quantum action principle of DReg. The central point is then to review how the aforementioned restrictions can be used to restore the broken symmetry by suitable counterterms provided there are no anomalies.

6.2.1. The Quantum Action Principle in BPHZ

As discussed in Section 5.1, the BPHZ approach to renormalization constituted one of the first full discussions of all-order renormalization, rigorously establishing the possibility to obtain finite Green functions and S-matrix elements in agreement with basic postulates such as causality and unitarity. In this framework, Lowenstein and Lam derived various theorems now summarized as the quantum action principle [106–110]. The theorems are similar to the regularized quantum action principle in DReg discussed in Section 4. The difference is that the theorems discussed here are valid in strictly four dimensions, for the fully renormalized theory.

Furthermore, this form of the quantum action principle is generally valid not only in the BPHZ framework, but in all regularization/renormalization frameworks that are equivalent; hence, it also applies to results obtained using DReg, if the $\text{LIM}_{D\to 4}$ defined in Equation (113) has been taken. The algebraic method is based on this general formulation, and its results hold for all such equivalent frameworks. In BPHZ, finite expressions and the Gell–Mann–Low formula are defined by an iterative operation on momentum-space integrals whereby Taylor series contributions up to some UV subtraction degree are subtracted from the integrands, giving finite integrals by power counting. Further, normal products, i.e., products of fields and their derivatives at the same spacetime point, may be defined as finite parts of certain Wick-ordered insertions into the Green function. One can derive so-called Zimmermann-identities, which linearly relate over-subtracted normal products, i.e., of a higher UV degree than the canonical operator dimension, to minimally subtracted ones. These prove a powerful tool in, e.g., deriving field equations and studying anomalies.

A first version of the quantum action principle can be used to express the relation of some infinitesimal variation of Green functions, or equivalently generating functionals, with the insertion of a normal product. Reference [107] considered differential vertex operations (DVOs), which are insertions of integrated normally ordered local field polynomials into the Gell–Mann–Low formula corresponding to the respective Green function:

$$\Delta \cdot G_{i_1,\ldots,i_n}(x_1,\ldots,x_n) = \langle 0|T \int \mathrm{d}y\, N[P(y)]\phi^{i_1}(x_1)\ldots\phi^{i_n}(x_n)|0\rangle. \tag{304}$$

Then, one can connect the variation of the Green function with respect to some parameter with those DVOs, i.e., taking some infinitesimal variation as $\mathcal{L}_{\text{int}} \to \mathcal{L}_{\text{int}} + \sum_k \varepsilon_k P_k(x)$, it follows:

$$\left.\frac{\partial G^{\varepsilon}}{\partial \varepsilon^k}\right|_{\varepsilon=0} = i\Delta \cdot G. \tag{305}$$

This result is valid for BPHZ-renormalized disconnected, connected, and 1PI Green functions and, therefore, also for the corresponding generating functionals.

It can be used to derive the renormalized QAP for a generic parameter of the theory λ:

$$\frac{\partial \Gamma}{\partial \lambda} = i\Delta_\lambda \cdot G, \tag{306}$$

where $\Delta_\lambda = \int \mathrm{d}x\, N[\frac{\partial \mathcal{L}}{\partial \lambda}]$.

There are several further versions of the quantum action principle with regard to variations of parameters or (external) fields. In particular, References [108–110] established a version of the action principle with respect to variations of dynamical fields (see, e.g., Reference [108], Equation (5.4)). The left-hand side being equal to zero due to the conservation of some current, the resulting relation corresponds to Equation (89) for the more general case of a non-invariant Lagrangian $\delta\mathcal{L} \neq 0$ under some symmetry transformation. It is rigorously established in terms of the generating functional for general Green functions renormalized in the BPHZ framework, and it can be connected to the generating functional of 1PI Green functions via Legendre transformation.

Thus, the finite BPHZ framework is a setting in which the formally derived identities among generating functionals such as the ones described in Sections 4.1 or 2.5 can be given a sensible all-order meaning.

In addition, in any regularization/renormalization procedure in agreement with the basic postulates, there is a way to cancel divergences and to obtain finite Green functions. These may differ from the ones obtained in BPHZ (or any other regularization), but in view of the theorems discussed in Section 5.1, the differences can only amount to local counterterms at each order.

In the following, we summarize important statements of the quantum action principle valid for any such finite Green functions defined via any consistent regularization and subtraction of divergences. The statements can be cast in a variety of forms, similar to Section 4.1. Here, we provide the formulation for the effective action Γ, as reviewed in Reference [47]. First, the equations of motion for the generating functionals can be written as

$$\frac{\delta \Gamma}{\delta \phi_i(x)} - \Delta_i(x) \cdot \Gamma = 0\,. \tag{307}$$

For variations with respect to the parameters, we have

$$\frac{\partial \Gamma}{\partial \lambda} = \int \mathrm{d}x\, \Delta(x) \cdot \Gamma\,. \tag{308}$$

As discussed in Sections 2.4 and 2.5, in the case of nonlinear symmetry transformations, it is useful to couple the composite operators to some external field, say $\rho^a(x)$. Then, one can arrive at the following version of the quantum action principle relevant for such nonlinear symmetry transformations:

$$\frac{\delta\Gamma}{\delta\rho_a(x)}\frac{\delta\Gamma}{\delta\phi_i(x)} = \Delta^{ai}(x)\cdot\Gamma\,. \tag{309}$$

In all previous Equations (307)–(309), the quantities Δ denote insertions of local composite field operators, whose dimensions are bounded by power counting and whose tree-level value is fixed in terms of the classical expression $\Gamma_{\rm cl}$. For example, in the case of Equation (307), Δ_i is a local composite field operator whose dimension is bounded by $(D-d_i)$, where d_i denotes the power-counting dimension of the corresponding field ϕ_i, and

$$\Delta_i = \frac{\delta\Gamma_{cl}}{\delta\phi_i} + \mathcal{O}(\hbar)\,. \tag{310}$$

6.2.2. Comparing Quantum Action Principles in BPHZ and DReg

The quantum action principles discussed in the previous subsection for BPHZ and in Section 4 for DReg are similar, but different. Here, we briefly comment on their relationship. The BPHZ version is valid for any regularization/renormalization procedure, including DReg. However, it is valid for the finite theory, in DReg for the theory after taking $\text{LIM}_{D\to4}$ as defined in Equation (113). The definition of this limit includes setting evanescent quantities (such as the $(D-4)$-dimensional metric $\hat{g}^{\mu\nu}$) to zero. The insertions Δ appearing, e.g., in Equations (309) are always finite, four-dimensional normal product insertions into the finite Green functions.

In contrast, in the DReg case, the counterpart Equation (205) is valid for general $D\neq4$, including evanescent quantities. In addition, if the identity corresponds to a symmetry such as the Slavnov–Taylor identity, which is valid at the tree level and in four dimensions, then the insertion Δ appearing in Equation (205) is purely evanescent.

It may not be immediately obvious how this can be reconciled with the purely four-dimensional case of BPHZ. This is, however, important as we shall be making use of general considerations following from the algebraic framework while working in DReg. In fact, both versions of the quantum action principle are valid and useful. The BPHZ version is useful to establish general existence proofs, which we can rely on also within DReg, but the DReg version is useful for explicit computations since, there, the explicit form of the insertion Δ is known.

The key is provided by the Bonneau identities established in References [73,74]. These identities precisely state that the insertion of an evanescent operator in DReg as in Equation (205) may in the $\text{LIM}_{D\to4}$ be rewritten as an insertion of a finite, four-dimensional operator as in Equation (309). In this way, the BPHZ quantum action principle can also be rederived from the one in DReg.

On the technical level, the Bonneau relationship also provides the coefficients in the expansion of evanescent operator insertions in terms of four-dimensional, finite insertions. They are given by the residue of the simple $1/(D-4)$ pole of the insertion of the evanescent operator into Green functions. The proof is essentially achieved by taking dimensionally renormalized amplitudes $\mathcal{R}_G$ associated with a graph G and comparing the vertex insertions $\hat{g}^{\mu\nu}[\mathcal{O}_{\mu\nu}\cdot\mathcal{R}_G]$, on the one hand, with the vertex insertions with $[\hat{g}^{\mu\nu}\mathcal{O}_{\mu\nu}]\cdot\mathcal{R}_G$, on the other hand.

At the one-loop level, the Bonneau identities are not surprising since evanescent quantities can only contribute in the $\text{LIM}_{D\to4}$ if they hit $1/(D-4)$ poles, which, at the one-loop level, have local coefficients, which may be interpreted as a four-dimensional local operator. However, their validity lies in their all-order nature. We mention here that Bonneau identities can also be used to obtain information on renormalization group equations in the presence of symmetry breakings of the regularization; see, e.g., References [25,28,131].

6.2.3. Algebraic Renormalization and Symmetry Restoration

With the quantum action principle at our disposal, we can now describe the logic of the algebraic renormalization of gauge theories. The starting point is the possible breakings of the Slavnov–Taylor identity (or similar identities) as given by Equation (303) due to the regularization. The quantum action principle provides a useful tool in restricting the structure of the breaking and in determining whether the symmetry can be restored, i.e., whether there are anomalies. For that, we proceed inductively order by order in perturbation theory. The goal is to determine the required finite, symmetry-restoring counterterms S^n_{ct} step by step for each n.

At the lowest order, at the classical level $n = 0$, the Slavnov–Taylor identity is valid by construction. This forms the basis of the inductive procedure. Let us then suppose the theory is renormalized completely; hence, it is finite, and the Slavnov–Taylor identity is fulfilled at some order $n-1$. In addition, at the next order n, the divergences are already canceled by appropriate singular counterterms. Hence, we have

$$\mathcal{S}(\Gamma^{(n),\mathrm{fin}}_{\mathrm{subren}}) = \mathcal{O}(\hbar^n), \tag{311}$$

where we have introduced the notation:

$$\Gamma^{(n),\mathrm{fin}}_{\mathrm{subren}} = \Gamma^{(n)}_{\mathrm{subren}} + S^n_{\mathrm{sct}}, \tag{312}$$

which denotes the effective action finite at order n after subrenormalization and adding the necessary divergent n-loop counterterms. This quantity corresponds to the set of finite Green functions for which the validity of the quantum action principle in BPHZ has been proven, and it can be defined in any other regularization scheme equivalent to BPHZ.

The task is then to study the possible breakings on the RHS of Equation (311), as well as the possible structure of the counterterms. As mentioned before, the breaking is restricted in two ways. First, we may employ the quantum action principle to find

$$\mathcal{S}(\Gamma^{(n),\mathrm{fin}}_{\mathrm{subren}}) = \hbar^n \Delta \cdot \Gamma^{(n),\mathrm{fin}}_{\mathrm{subren}} = \hbar^n \Delta + \mathcal{O}(\hbar^{n+1}). \tag{313}$$

The important point is that Δ is a local polynomial in fields and derivatives, also restricted by power counting. This property was announced in Section 2.5, where the Slavnov–Taylor identity was formally derived from the path integral.

Second, applying the linearized BRST operator $s_{\Gamma_{\mathrm{cl}}} \equiv b$ to Equation (313) using Equation (299) and extracting the $\mathcal{O}(\hbar^n)$ terms, we arrive at a consistency condition (also called the Wess–Zumino consistency condition):

$$b\Delta = 0. \tag{314}$$

Hence, the possible breaking Δ is restricted very similarly (cf. Equation (301)) to the possible divergences Γ_{div} in Section 6.1. Both Γ_{div} in Equation (301) and Δ in Equation (314) are local polynomials restricted by power counting, which are annihilated by b, but Γ_{div} is of ghost number 0, whereas Δ has ghost number 1. Now, one can make a distinction. If Δ is a b-exact term, i.e., if there exists another local polynomial Δ' with

$$\Delta = b\Delta', \tag{315}$$

it is called a trivial element of the cohomology of the BRST operator. In this case, we can supplement the original action with a new n-loop order counterterm:

$$S^n_{\mathrm{fct}} = S^n_{\mathrm{fct,non\text{-}inv}} + S^n_{\mathrm{fct,inv}} = -\Delta' + S^n_{\mathrm{fct,inv}}, \tag{316}$$

where the last term reflects the freedom to add to the action any finite, symmetric counterterm, obeying $b\, S^n_{\text{fct,inv}} = 0$. Hence, we end up with

$$\mathcal{S}(\Gamma^{(n),\text{fin}}_{\text{subren}} + \hbar^n S^n_{\text{fct}}) = \mathcal{S}(\Gamma^{(n),\text{fin}}_{\text{subren}}) + b\,\hbar^n S^n_{\text{fct}} + \mathcal{O}(\hbar^{n+1}) = \mathcal{O}(\hbar^{n+1}), \tag{317}$$

where the last step follows from the induction hypothesis. Compatibility with ghost and gauge fixing equation was shown in Reference [47].

Hence, under the condition (315), we can find a counterterm action $S^n_{\text{fct,non-inv}}$ that defines finite, non-invariant counterterms that repair the symmetry. Furthermore, it is possible to add any number of finite, invariant counterterms to the action as they satisfy $b\, S_{\text{fct,inv}} = 0$ and, hence, do not disturb the STI. These invariant counterterms behave like the finite counterterms discussed in Section 6.1 and can be used to satisfy certain renormalization conditions.

One task of the algebraic renormalization program is, therefore, to determine the most-general solution of the equation $b\Delta = 0$. If all possible solutions are b-exact, then this constitutes a proof that the Slavnov–Taylor identity can be established at all orders in the renormalized theory.

However, if we cannot write the breaking Δ as a b-exact term, the symmetry cannot be repaired. This is an anomaly. In the case of the Slavnov–Taylor identity, such an anomaly is disastrous since it destroys the interpretation of the theory as a sensible quantum theory; see the discussion at the beginning of the present section. Anomalies are thus nontrivial elements of the cohomology of the b-operator, i.e., expressions that are annihilated by b, but are not b-exact.

The previous remarks constitute crucial insights into the BRST formalism [41–44]. The analysis of whether a gauge theory is renormalizable, i.e., whether the Slavnov–Taylor identity can be restored at each order, can be made on a purely classical level, by finding all possible solutions of Equation (314) and checking whether they are all b-exact.

The actual computation can be found in the original references and in the reviews [47,119]. It can be sketched as follows. From the Wess–Zumino consistency condition (314) and the nilpotency of the BRST operator, one can derive a set of equations, the so-called descent equations. Solving these gives a general expression of the possible anomalies of a theory. In the present case of interest for a generic Yang–Mills theory, it can be shown that the consistency condition simplifies to $s\Delta(G,c) = 0$ (see, e.g., [47]) with dependence on the gauge and the ghost field only. Writing Δ as an integrated local product and solving the descent equations lead to the famous Adler–Bell–Jackiw gauge anomaly first discovered in References [6–8] (note the different relative sign of the first term of Equation (318) compared to [47], which comes from a different sign convention in the covariant derivative; see Equation (6)),

$$\Delta = L \times \varepsilon_{\mu\nu\rho\sigma} \text{Tr} \int d^4x\, c_a \partial^\mu \left(-g d^{abc}_A \partial^\nu G^\rho_b G^\sigma_c + g^2 \frac{\mathcal{D}^{abcd}_A}{12} G^\nu_b G^\rho_c G^\sigma_d \right), \tag{318}$$

where L is a coefficient that can be determined from explicit calculations and which depends on the theory inputs. The group symbols are given by

$$d^{abc}_A = \text{Tr}\left(T^a_{\text{adj}} \{ T^b_{\text{adj}}, T^c_{\text{adj}} \} \right), \tag{319}$$

and

$$\mathcal{D}^{abcd}_A = d^{nab}_A f^{ncd} + d^{nac}_A f^{ndb} + d^{nad}_A f^{nbc}, \tag{320}$$

where T^a_{adj} denotes adjoint generators under which ghosts and gauge fields transform; cf. Equation (2). Expression (318) must vanish by itself, i.e., it cannot be absorbed by the counterterms, for the theory to be consistent. In the case of a single left-handed fermion, it can be shown by a one-loop calculation that the anomaly is proportional to

$$\frac{1}{2} d_A^{abc} \operatorname{Tr}(T^a\{T^b, T^c\}), \tag{321}$$

which means that its cancellation depends on an appropriate choice of the matter content of the theory. The famous Adler–Bardeen theorem guarantees that, if the gauge anomaly vanishes at one-loop order, it also vanishes at all orders; cf. [47]. The expression in Equation (321) cannot vanish by itself, but in such a theory with a family of left-handed fermions, their charges may add up to zero, as is the case in the SM. For some gauge groups such as SU(2), the above expression vanishes identically due to the vanishing of some group symbols. Hence, there can be no anomaly.

In summary, we have sketched how algebraic renormalization allows identifying the general structure of the breaking of the Slavnov–Taylor identity. It constitutes a setting in which the restoration of the symmetry can be proven to all orders for trivial elements of the BRST cohomology such as spurious breakings introduced by the BMHV algebra. In the case of non-spurious breakings, e.g., the gauge anomaly, one can derive explicit conditions for its cancellation that a sensible theory must satisfy. Further, nonrenormalization theorems, as in the case of the Adler–Bardeen theorem, can be shown and allow evaluating the gauge anomaly in a simple way. The main technical tool that serves to establish these findings is the general quantum action principle valid in many equivalent subtraction schemes. A key advantage of the algebraic proof is that there is no need for an invariant regularization, which for, e.g., chiral gauge theories does not exist.

6.2.4. Outlook and Further Remarks on Anomalies and Algebraic Renormalization

At this point, we interject with a brief outlook on anomalies and further applications of the techniques of algebraic renormalization. Next to the perturbative chiral gauge anomalies discussed above and discovered in References [6–8], there exist global chiral anomalies [132] and perturbative mixed gauge–gravitational anomalies [133–135]. A chiral gauge model can be renormalized only if all these chiral anomalies cancel, which may be achieved by a proper choice of fermion representations of the chiral model; for example, see Reference [11] and the references therein. Equation (318) is necessary, but not sufficient if gravity and nonperturbative effects are taken into account.

Important theories such as the Standard Model of particle physics are renormalizable. In particular, the electroweak SM was completely treated in algebraic renormalization in Reference [57], establishing the SM as a fully all-order consistent, renormalizable theory. Reference [56] gave a similar proof using the background field gauge (see footnote 3), and Reference [59] gave a similar proof for the supersymmetric SM. These papers complement earlier extensive discussions of the renormalization of the electroweak SM by, e.g., References [60,136]; see also Reference [137].

The validity of the Slavnov–Taylor identity and the techniques of algebraic renormalization can also be used to establish further interesting physics properties of quantum gauge theories such as the renormalized electroweak SM, e.g., charge universality can be established based on both gauge choices [48,136]; see also Reference [137] for further discussions. As another example, the renormalization of the Higgs vacuum expectation values in spontaneously broken gauge theories can be controlled via a suitable Slavnov–Taylor identity [138,139].

6.3. *Algebraic Symmetry Restoration in the Context of DReg*

So far in this section, we have studied the role of symmetries in the process of renormalization. If the symmetry is respected by the regularization, this implies a great simplification for the UV counterterms. If it is not, algebraic renormalization constitutes a general setup that allows identifying symmetry violations and restoring the symmetry.

Here, we specialize the general procedure to the case of DReg. We use the BMHV scheme with non-anticommuting γ_5 in which gauge invariance may be broken.

6.3.1. Formulation of Symmetry and Symmetry Breaking in DReg

The ultimate symmetry requirement is the Slavnov–Taylor identity expressing the BRST invariance of the full renormalized theory, Equation (289). In the context of DReg, this requirement can be formulated as

$$\underset{D\to 4}{\text{LIM}}\left(\mathcal{S}_D(\Gamma_{\text{DRen}})\right) = 0. \tag{322}$$

As defined in Section 3.1, Γ_{DRen} denotes the renormalized effective action, still in D dimensions, but including all counterterms canceling $1/\epsilon$ divergences and restoring symmetries. The limit refers to the operation of letting $\epsilon \to 0$, as well as putting evanescent quantities such as $\hat{g}^{\mu\nu}$ to zero.

In order to discuss the inductive procedure, we considered some order n and supposed the theory has been renormalized and all counterterms have been constructed up to the previous order $n-1$. This provides us with

$$\Gamma^{(n)}_{\text{subren}}, \tag{323}$$

again using the notation of Section 3.1. At this point, we know from Section 5 that the divergences at the n-th order can be canceled by adding a local counterterm action S^n_{sct}. It may or may not be true that the divergences follow the simple pattern described in Section 6.1. In general, we can always write

$$S^n_{\text{sct}} = S^n_{\text{sct,inv}} + S^n_{\text{sct,non-inv}}, \tag{324}$$

where the first term corresponds to symmetric counterterms as described in Section 6.1 and the second term corresponds to whatever other divergent counterterms are required.

After subtracting these divergences, the theory is finite at the order n, and the Slavnov–Taylor identity may be written as

$$\mathcal{S}_D(\Gamma^{(n)}_{\text{subren}} + S^n_{\text{sct}}) = \hbar^n \Delta_D + \mathcal{O}(\hbar^{n+1}), \tag{325}$$

where Δ_D is a possible finite breaking term, still evaluated in D dimensions. The subrenormalized and finite effective action introduced for the algebraic analysis in Equation (312) is now given by $\text{LIM}_{D\to 4}(\Gamma^{(n)}_{\text{subren}} + S^n_{\text{sct}})$, and the counterpart of Equation (313) is given by the four-dimensional limit:

$$\underset{D\to 4}{\text{LIM}}\ \Delta_D = \Delta_{\text{from Eq. (313)}}. \tag{326}$$

This finite quantity Δ is the one constrained by algebraic renormalization and discussed after Equation (313). That is, it is a local breaking term that satisfies the Wess–Zumino consistency conditions and that can be canceled by adding suitable counterterms (we assumed that there is no genuine anomaly).

The practical question is then how to obtain, first, the breaking term Δ and, then, the symmetry-restoring counterterms. There are two strategies for this. The first, obvious option is to evaluate all Green functions appearing on the LHS of Equation (325) including their finite parts, plug them into the Slavnov–Taylor identity, and determine the potentially nonvanishing breaking. This straightforward procedure is convenient in that it operates on ordinary Green functions. Its drawback is that most finite parts of Green functions—in particular, parts that are non-polynomial in the momenta—will be in agreement with the symmetry and, hence, drop out of Equation (325), such that the calculation can become unnecessarily complicated. Nevertheless, this direct approach has been used in the literature, e.g., in References [140–144] on applications on chiral gauge theories and supersymmetric gauge theories. In the subsequent Section 7.4.1, we will also illustrate this approach with a concrete example.

A second, alternative approach is provided by using the regularized quantum action principle in DReg, described in Section 4. This theorem guarantees that we can rewrite the LHS of Equation (322) as

$$\mathcal{S}_D(\Gamma_{\mathrm{DRen}}) = (\widehat{\Delta} + \Delta_{\mathrm{ct}}) \cdot \Gamma_{\mathrm{DRen}}. \tag{327}$$

The possible breaking of the Slavnov–Taylor identity is, thus, rewritten as an operator insertion of the composite operator $\widehat{\Delta} + \Delta_{\mathrm{ct}}$, which is defined as

$$\widehat{\Delta} = \mathcal{S}_D(S_0)\,, \tag{328a}$$

$$\widehat{\Delta} + \Delta_{\mathrm{ct}} = \mathcal{S}_D(S_0 + S_{\mathrm{ct}})\,, \tag{328b}$$

In this approach, the breaking Δ may be computed in terms of the RHS of (327). The advantage lies in significantly restricting possible nonvanishing contributions. In particular, $\widehat{\Delta}$ is evanescent; hence, it can contribute in the $\mathrm{LIM}_{D\to 4}$ only in combination with the $1/\epsilon$ singularities of Feynman diagrams.

The RHS of (327) can be expanded in loop orders as

$$\widehat{\Delta} + \sum_{i=1}^{\infty} \hbar^i \left(\widehat{\Delta} \cdot \Gamma^i_{\mathrm{DRen}} + \sum_{k=1}^{i-1} \Delta^k_{\mathrm{ct}} \cdot \Gamma^{(i-k)}_{\mathrm{DRen}} + \Delta^i_{\mathrm{ct}} \right). \tag{329}$$

Plugging the previous definitions into Equation (322), we arrive at an equation expressing the symmetry requirement exactly at the order n:

$$\mathop{\mathrm{LIM}}_{D\to 4} \left(\widehat{\Delta} \cdot \Gamma^n_{\mathrm{DRen}} + \sum_{k=1}^{n-1} \Delta^k_{\mathrm{ct}} \cdot \Gamma^{n-k}_{\mathrm{DRen}} + \Delta^n_{\mathrm{ct}} \right) = 0, \tag{330}$$

for all $n \geq 1$. The individual terms in this equation have divergent and finite parts, but by construction, the entire expression is finite; hence, the cancellation of divergences may be used as a consistency check of practical calculations. For the determination of symmetry-restoring counterterms, Equation (330) should be viewed as follows. At the order n and after subrenormalization and adding divergent n-loop counterterms, everything in Equation (330) is already known except the finite counterterms of order n. They enter via Δ^n_{ct}, which in turn depends on the to-be-determined counterterms. The following subsubsection will make the dependence explicit. Hence, Equation (330) can be regarded as the optimized defining relation for the symmetry-restoring counterterms in DReg.

We close with the remark that Equation (330) does not fully determine all finite counterterms. It only determines the required form of counterterms in order to restore the symmetry. However, Equation (330) is blind to several types of counterterms: finite and symmetric counterterms (which often correspond to a renormalization transformation as described in Section 6.1) drop out; such counterterms can, therefore, still be adjusted at will, e.g., to satisfy the renormalization conditions corresponding to an on-shell or a different desirable renormalization scheme. In addition, evanescent and finite counterterms also drop out and may be added to optimize the counterterm action.

6.3.2. Practical Restoration of the Symmetry

Here, we illustrate the blueprint for the practical restoration of the symmetry, if Equation (330) is used as a basis.

We begin at the one-loop level and start from the regularized but unrenormalized effective action $\Gamma^{(1)}$.[14] At the one-loop level, the regularized action plus counterterms, as well as the symmetry breaking induced by the counterterms at one-loop order are given by

$$\Gamma^{(1)}_{\text{DRen}} = \Gamma^{(1)} + S^1_{\text{sct}} + S^1_{\text{fct}}, \tag{331a}$$

$$\Delta^1_{\text{ct}} = \mathcal{S}_D(S_0 + S_{\text{ct}})^1 = b_D S^1_{\text{sct}} + b_D S^1_{\text{fct}}, \tag{331b}$$

where the last part of the last equation is a specific rearrangement possible at the one-loop level and where the linearized Slavnov–Taylor operator b_D is defined in analogy to b in Equation (298). The general equations establishing the cancellation of divergences and symmetry restoration, (330), become

$$S^1_{\text{sct}} + \Gamma^1_{\text{div}} = 0, \tag{332a}$$

$$(\widehat{\Delta} \cdot \Gamma^1 + \Delta^1_{\text{ct}})_{\text{div}} = 0, \tag{332b}$$

$$\underset{D\to 4}{\text{LIM}} \, (\widehat{\Delta} \cdot \Gamma^1 + \Delta^1_{\text{ct}})_{\text{fin}} = 0. \tag{332c}$$

Compared to the general Equation (330), terms that vanish at one-loop order were dropped. The quantities that need to be explicitly computed here are the one-loop divergences Γ^1_{div} and the one-loop diagrams with one insertion of the evanescent operator $\widehat{\Delta}$, $\widehat{\Delta} \cdot \Gamma^1$. The first of these equations then determines the divergent one-loop counterterms S^1_{sct}, and the second equation provides a consistency check of the divergences. In view of Equation (331b), the last line contains $b_D S^1_{\text{fct}}$ and, thus, determines the symmetry-restoring one-loop counterterms.

Next, we consider the two-loop order. At the two-loop level, the corresponding equations for the effective action and the symmetry breaking of counterterms are

$$\Gamma^{(2)}_{\text{DRen}} = \Gamma^{(2)}_{\text{subren}} + S^2_{\text{sct}} + S^2_{\text{fct}}, \tag{333a}$$

$$\Delta^2_{\text{ct}} = \mathcal{S}_D(S_0 + S_{\text{ct}})^2 = \mathcal{S}_D(S_0 + S^1_{\text{ct}})^2 + b_D S^2_{\text{sct}} + b_D S^2_{\text{fct}}, \tag{333b}$$

where the upper index ^{2} corresponds to extracting the two-loop terms. The last equation exhibits the appearance of the genuine two-loop counterterms in a way specific to the two-loop level. The equations corresponding to finiteness and symmetry restoration read

$$S^2_{\text{sct}} + \Gamma^2_{\text{subren,div}} = 0, \tag{334a}$$

$$(\widehat{\Delta} \cdot \Gamma^2_{\text{subren}} + \Delta^1_{\text{ct}} \cdot \Gamma^1 + \Delta^2_{\text{ct}})_{\text{div}} = 0, \tag{334b}$$

$$\underset{D\to 4}{\text{LIM}} \, (\widehat{\Delta} \cdot \Gamma^2_{\text{subren}} + \Delta^1_{\text{ct}} \cdot \Gamma^1 + \Delta^2_{\text{ct}})_{\text{fin}} = 0. \tag{334c}$$

Here, we have to calculate, first, the two-loop divergences to obtain the two-loop divergent counterterms. Then, we have to calculate diagrams with insertions of $\widehat{\Delta}$ up to the two-loop level (and including one-loop subrenormalization), as well as one-loop diagrams with insertions of b_D-transformed one-loop counterterms. The second equation must automatically hold and provides a check. The third equation then determines the genuine finite two-loop symmetry-restoring counterterms $b_D S^2_{\text{fct}}$, which appear via Equation (333b) in Δ^2_{ct}.

In summary, the recipe is as follows:

- UV-renormalize the theory, previously renormalized up to order $n-1$, at order n to obtain the singular counterterms;

[14] We slightly simplify the notation and use $\Gamma^{(1)}$ in the following equations of this subsubsection to denote the unrenormalized effective action up to one-loop order. According to the general notational scheme defined in Section 3.1, this could also be called $\Gamma^{(1)}_{\text{subren}}$.

- Calculate genuine n-loop Green functions with one-time insertion of $\widehat{\Delta}$ for their divergent and finite part;
- Calculate the k-loop order insertion into $(n-k)$-loop order graphs, and determine their divergent and finite contributions;
- Check that the divergences thus obtained sum up to zero;
- Collect the finite contributions, and choose monomials X such that $b_D X$ cancels them. This is always possible, as discussed in the previous subsections.

6.3.3. The Counterterm Lagrangian in the BMHV Scheme

The output of the regularization/renormalization program is the renormalized effective action and the required counterterm action consisting of singular and finite counterterms. In the context of the BMHV scheme, the previous subsections showed that the counterterm action can, in general, contain five different kinds of terms:

$$S_{\text{ct}} = S_{\text{sct,inv}} + S_{\text{sct,non-inv}} + S_{\text{fct,inv}} + S_{\text{fct,restore}} + S_{\text{fct,evan}}. \tag{335}$$

This equation is a more detailed version of the generic decomposition explained in Section 3.1 into singular and finite counterterms. For both the singular and the finite counterterms, we may isolate a symmetry-invariant piece, which has the pattern of symmetric counterterms discussed in Section 6.1 and typically corresponds to counterterms generated by a renormalization transformation as

$$S_0 \stackrel{\text{ren. transf. (291)}}{\longrightarrow} S_0 + S_{\text{sct,inv}} + S_{\text{fct,inv}}. \tag{336}$$

In general, the conditions of Section 6.1 are not met, and symmetry-violating counterterms are required. Accordingly, the next type of counterterms

$$S_{\text{sct,non-inv}}$$

corresponds to additional singular counterterms needed to cancel additional $1/\epsilon$ poles of loop diagrams that cannot be canceled by symmetry-invariant counterterms. They may be evanescent and, starting from the two-loop order, also four-dimensional (non-evanescent). They cannot be obtained by renormalization transformations. We note that the subtraction of evanescent $1/\epsilon$ poles is a necessity for the consistency of higher orders (see also Reference [5] for a review discussing this point).

Next,

$$S_{\text{fct,restore}}$$

corresponds to finite counterterms needed to restore the Slavnov–Taylor identity and, thus, the underlying gauge invariance. They are the central objects of the present discussion and the outcome of the practical recipe of Section 6.3.2. Determining these counterterms is one of the key tasks in the usage of the BMHV scheme. Once those counterterms are found, the theory can be considered to be renormalized.

As mentioned before, the symmetry-restoring counterterms are not unique. Clearly, they may be modified by shifting around any symmetry-invariant counterterm between $S_{\text{fct,inv}}$ and $S_{\text{fct,non-inv}}$, since invariant terms would drop out of Equations (330), (332c) and (334c). The overall sum of $S_{\text{fct,inv}} + S_{\text{fct,non-inv}}$ can only be fixed by imposing a renormalization scheme (such as, e.g., the on-shell scheme), and the split into $S_{\text{fct,inv}}$ and $S_{\text{fct,non-inv}}$ can only be fixed by picking a convention. To illustrate this point, let us assume the counterterm Lagrangian must contain a non-gauge-invariant term $zA^\mu \Box A_\mu$, where z is a coefficient and A^μ a gauge field. Two different options for the counterterm Lagrangians would then be

$$\mathcal{L}_{\text{fct,non-inv}} = zA^\mu \Box A_\mu, \qquad \mathcal{L}_{\text{fct,inv}} = \delta Z (A^\mu \Box A_\mu + (\partial A)^2), \tag{337a}$$

$$\mathcal{L}_{\text{fct,non-inv}} = -z(\partial A)^2, \qquad \mathcal{L}_{\text{fct,inv}} = (\delta Z + z)(A^\mu \Box A_\mu + (\partial A)^2). \tag{337b}$$

The invariant counterterm here corresponds to an invariant counterterm generated by a field renormalization from the usual gauge-invariant kinetic term $F^{\mu\nu}F_{\mu\nu}$. According to the assumption, both options restore the symmetry, and they lead to the identical renormalized theory. The field renormalization constant δZ can be used to adopt a desired renormalization condition.

Finally,

$$S_{\text{fct,evan}}$$

corresponds to additional counterterms that are both finite and evanescent. Adding or changing such counterterms can change, e.g., a purely four-dimensional counterterm $A^\mu \bar{\psi}\bar{\gamma}_\mu\psi$ to a fully D-dimensional counterterm $A^\mu\bar{\psi}\gamma_\mu\psi$. These counterterms vanish in the four-dimensional limit, but they can affect calculations at higher orders. They also drop out of Equations (330), (332c) and (334c). Hence, one viable option is that the symmetry-restoring counterterms $S_{\text{fct,restore}}$ are always defined by using strictly four-dimensional quantities only. However, this is not the only option; in concrete cases, elevating four-dimensional terms to fully D-dimensional ones may simplify the expressions appearing at higher orders. At any rate, each such choice generates a different, valid, renormalized theory. From a practical point of view, it is desirable to make a computationally simple choice.

7. Practical Treatment of Chiral Gauge Theories in the BMHV Scheme of DReg

In recent years, the treatment of chiral gauge theories with the non-anticommuting γ_5 BMHV scheme has received increasing interest. Applications in the SM at the multiloop level and in effective field theories with additional operators involving chiral fermions have become more important; see, e.g., the discussions in [137,145–147]. Accordingly, the usefulness of regularization/renormalization schemes for which ultimate consistency is fully established is becoming more appreciated. After the pioneering one-loop discussion of gauge theories with chiral fermions in References [131,148], Reference [25] extended the analysis to general chiral gauge theories including scalar fields and Yukawa couplings to chiral fermions. Reference [26] pioneered the application of the BMHV scheme to chiral gauge theories at the two-loop level with a first, Abelian example. Reference [27] extended the one-loop analysis to the case of the background field gauge fixing and to the full gauge–fermion sector of the electroweak SM.

In this section, we give concrete illustrations of how to treat chiral gauge theories in the BMHV scheme with non-anticommuting γ_5. The discussion is based on our results in [25,26]. The following Section 7.1 provides an extended overview of the procedure and a guide for the present section.

7.1. Overview and Guide to the Present Section

In Section 2, we discussed the basic defining gauge invariance of gauge theories and reformulated it in terms of BRST symmetry and the Slavnov–Taylor and Ward identities. In Section 6, we explained how these symmetry identities are elevated to defining properties of the renormalized theory at higher orders. For the gauge theories we study here, it is known that these defining symmetry identities can be fulfilled in any consistent regularization/renormalization procedure, by appropriately defining the counterterms. In Section 3, we explained the definition of dimensional regularization and the BMHV scheme for γ_5, which in general breaks gauge invariance in the presence of chiral fermions. In Section 5, we explained the proof that dimensional regularization constitutes one of the consistent regularization/renormalization procedures.

As a result, it was in principle established that the dimensional regularization including the BMHV scheme for γ_5 may be used for chiral gauge theories. Further, Section 6.3 also provided a blueprint for how to determine the required counterterm structure in concrete calculations. In this section, we carry out such concrete calculations and illustrate all required steps in detail.

In the Abelian chiral gauge theory defined below, we expect the validity of simple QED-like Ward identities; the simplest one corresponds to the transversality of the photon self-energy. It turns out that, in the BMHV scheme, the actual one-loop self-energy violates this transversality (see Equation (372)). The violation affects both the divergent and the finite part in the BMHV scheme of dimensional regularization. The breaking, however, is a polynomial in the momentum; hence, it can be canceled by adding a local counterterm to the Lagrangian—in line with the general existence statement mentioned above. After adding this counterterm, the required transversality is fulfilled. The concrete required form of the counterterm can be found in Equations (376) and (378).

A question is then what is the most-efficient way to determine such symmetry breakings in general. Answers were given in Section 6.3.1 and can be illustrated as follows. One way in principle is to explicitly evaluate all Green functions and test the validity of all Ward and Slavnov–Taylor identities between all Green functions. The explicit computation of the non-transverse terms in Equation (372) provides an example. Given that there are, in principle, infinitely many identities between Green functions and given that the computation of Green functions involves also complicated non-local terms that cannot contribute to the symmetry violation, this strategy is not the most efficient.

Section 6.3.1 also explained a shortcut that is based on the regularized quantum action principle of dimensional regularization discussed in Section 4. Staying with the example of the photon self-energy, the terms violating the transversality in Equation (372) and then in Equation (374) may be equivalently obtained by computing one special Feynman diagram, shown in Equations (381) and (382). This diagram involves an insertion of the operator $\widehat{\Delta}$, which reflects the breaking of chiral gauge invariance in D dimensions, and the quantum action principle guarantees that the evaluation of this diagram reproduces directly the breaking of the transversality of the photon self-energy. The simplification is threefold: First and foremost, since $\widehat{\Delta}$ is evanescent, only the ultraviolet divergent part of the diagram can contribute—hence, the evaluation is simpler (the degree of simplification dramatically increases for more complicated Green functions and at higher orders). Second, in the general case, there are much fewer diagrams with insertions of $\widehat{\Delta}$ than ordinary diagrams. Third, since only divergent parts contribute, it is clear that the symmetry breaking/restoration procedure requires only the computation of power-counting divergent diagrams with insertions of $\widehat{\Delta}$.

This more efficient, but less obvious strategy based on the quantum action principle was applied to chiral gauge theories at the one-loop level in References [25,27,131,148] with and without the scalar sector and to an Abelian chiral gauge theory at the two-loop level in Reference [26]. It was also applied to the case of supersymmetric gauge theories in the context of dimensional reduction at the two- and three-loop level in References [102,105].

In the largest part of the present section, we focus on the simpler case of an Abelian chiral gauge theory. We begin in Section 7.2 by defining the considered model and collecting all relevant symmetry identities. Then, we discuss the subtleties in the continuation to D dimensions and determine the insertion operator $\widehat{\Delta}$. Section 7.3 provides a more technical overview of the procedure to determine the symmetry-restoring counterterms than the previous remarks. In Section 7.4, we then discuss the explicit computations in the Abelian model in detail. We begin with the case of the photon self-energy mentioned above and illustrate both strategies to determine the symmetry-restoring counterterms, then we progress to other Green functions and to the two-loop level. Thereafter, Section 7.5 discusses the case of non-Abelian Yang–Mills theories and presents explicit calculations and results at the one-loop level.

7.2. *Definition of an Abelian Chiral Gauge Theory*

Here, we define a concrete Abelian chiral gauge theory, which will be used in explicit calculations. It is first defined in four dimensions along with its symmetry requirements in Section 7.2.1; then, the definition is extended to D dimensions within the framework of the BMHV γ_5 scheme, and the resulting BRST symmetry breaking is exhibited in Section 7.2.2.

7.2.1. Chiral Electrodynamics in Four Dimensions

Following Section 2.6, the four-dimensional classical Lagrangian for quantum electrodynamics (QED) is given by

$$\mathcal{L}_{\mathrm{QED}} = i\overline{\psi}_i \not{D}_{ij}\psi_j - \frac{1}{4}F^{\mu\nu}F_{\mu\nu} - \frac{1}{2\xi}(\partial_\mu A^\mu)^2 - \bar{c}\partial^2 c + \rho^\mu s A_\mu + \bar{R}^i s\psi_i + R^i s\overline{\psi}_i, \tag{338}$$

with the U(1) ghost and external BRST sources included. In contrast to Section 2.6, we already integrated out the Nakanishi–Lautrup field $B(x)$, i.e., we used $B = -(\partial_\mu A^\mu)/\xi$ already in the Lagrangian. The only generator in this theory is the real and diagonal charge $Q_{ij} = Q_i \delta_{ij}$, so that the covariant derivative reads

$$D^\mu_{ij} = \partial^\mu \delta_{ij} + ieA^\mu Q_{ij}\,. \tag{339}$$

We now define a similar, but chiral Abelian gauge theory. We separated the fermionic content into left-handed and right-handed chirality parts:

$$\psi_{R/L} = \mathbb{P}_{R/L}\psi, \quad \mathbb{P}_{R/L} = \frac{\mathbb{1} \pm \gamma_5}{2}, \tag{340}$$

and allowed only purely right-handed fermions to appear as dynamical fields. This was a choice made to simplify the discussion, e.g., the $\mathrm{U}(1)_Y$ sector of the SM contains both left-handed and right-handed fermions with different gauge quantum numbers. It could be treated similarly. The four-dimensional and purely right-handed classical Lagrangian of the model then reads

$$\mathcal{L}_{\chi\mathrm{QED}} = i\overline{\psi_{R}}_i \not{D}_{ij}\psi_{Rj} - \frac{1}{4}F^{\mu\nu}F_{\mu\nu} - \frac{1}{2\xi}(\partial_\mu A^\mu)^2 - \bar{c}\partial^2 c + \rho^\mu s A_\mu + \bar{R}^i s\psi_{Ri} + R^i s\overline{\psi_{R}}_i, \tag{341}$$

where the interaction, coupling only to the right-handed fermions, is defined by the covariant derivative as

$$D^\mu_{ij} = \partial^\mu \delta_{ij} + ieA^\mu \mathcal{Y}_{Rij}\,. \tag{342}$$

Emphasizing the similarity with the $\mathrm{U}(1)_Y$ sector of the Standard Model, we call the generator $\mathcal{Y}_{Rij} = \mathcal{Y}_{Ri}\delta_{ij}$ the hypercharge. It can be seen that the left-handed fermions ψ_L are now decoupled from the theory. In order to avoid triangle anomalies, we need to impose the following additional anomaly cancellation condition to the hypercharge:

$$\mathrm{Tr}(\mathcal{Y}_R^3) = 0\,. \tag{343}$$

Following Section 2, the nonvanishing BRST transformations for this model are

$$sA_\mu = \partial_\mu c\,, \tag{344a}$$
$$s\psi_i = s\psi_{Ri} = -ie\,c\,\mathcal{Y}_{Rij}\psi_{Rj}\,, \tag{344b}$$
$$s\overline{\psi}_i = s\overline{\psi_R}_i = -ie\,\overline{\psi_R}_j c\mathcal{Y}_{Rji} \tag{344c}$$
$$s\bar{c} = B \equiv -\frac{1}{\xi}\partial A, \tag{344d}$$

where s is the nilpotent generator of the BRST transformations, which acts as a fermionic differential operator. This four-dimensional tree-level action:

$$S_0^{(4D)} = \int d^4x\, \mathcal{L}_{\chi\mathrm{QED}} \tag{345}$$

satisfies the following Slavnov–Taylor identity:

$$\mathcal{S}(S_0^{(4D)}) = 0\,, \tag{346}$$

where the Slavnov–Taylor operator, with the field content we considered, was already given in Equation (106).

At this point, we emphasize two additional functional identities that hold in four dimensions and that were derived and discussed in Section 2.6. The first is the ghost equation:

$$\left(\frac{\delta}{\delta \bar{c}} + \partial_\mu \frac{\delta}{\delta \rho_\mu}\right) S_0^{(4D)} = 0 \,. \tag{347}$$

The second is the functional form of the Abelian Ward identity:

$$\left(\partial^\mu \frac{\delta}{\delta A^\mu(x)} + ie\mathcal{Y}_R^j \sum_\Psi (-1)^{n_\Psi} \Psi(x) \frac{\delta}{\delta \Psi(x)}\right) S_0^{(4D)} = -\Box B(x)\,, \tag{348}$$

suitably adapted to the present theory χQED and its field content. The summation extends over the charged fermions and their sources, $\Psi \in \{\psi_{R_j}, \overline{\psi_{R_j}}, R^j, \bar{R}^j\}$ and $n_\Psi \in \{0,1,0,1\}$. Here, we kept the Nakanishi–Lautrup field $B(x)$ explicitly. However, one could integrate it out here as well using $B = -(\partial_\mu A^\mu)/\xi$.

Functional relations such as the ghost equation and the local Ward identity are part of the definition of our theory in four dimensions. Once we perform the regularization and renormalization procedure, the requirement that those identities still hold imposes important restrictions, as we will soon see in the explicit loop calculations. However, first, we extend the model to D dimensions and examine the consequences of this extension.

7.2.2. Definition of Chiral Electrodynamics in DReg

We can immediately see that the extension of χQED to D dimensions is not unique due to the right-handed chiral current $\overline{\psi_{R_i}}\gamma^\mu \psi_{R_j}$. The extension to D dimensions of this term has three *inequivalent choices*, each of them *equally correct*:

$$\overline{\psi}_i \gamma^\mu \mathbb{P}_R \psi_j\,, \qquad \overline{\psi}_i \mathbb{P}_L \gamma^\mu \psi_j\,, \qquad \overline{\psi}_i \mathbb{P}_L \gamma^\mu \mathbb{P}_R \psi_j\,. \tag{349}$$

They are different because $\mathbb{P}_L \gamma^\mu \neq \gamma^\mu \mathbb{P}_R$ in D dimensions. Each of these choices leads to a valid D-dimensional extension of the model that is renormalizable using dimensional regularization and the BMHV scheme and is expected to produce the same final results in physical four dimensions after the renormalization procedure is performed. However, the intermediate calculations and the D-dimensional results will differ, depending on the choice for this interaction term. The third option, which is equal to

$$\overline{\psi} \mathbb{P}_L \gamma^\mu \mathbb{P}_R \psi = \overline{\psi} \mathbb{P}_L \overline{\gamma}^\mu \mathbb{P}_R \psi = \overline{\psi_R} \overline{\gamma}^\mu \psi_R\,, \tag{350}$$

is the most symmetric one and leads to the simplest intermediate expressions. Notice that this choice is actually the most-straightforward one since it preserves the information that right-handed fermions were present on the left and on the right sides of the interaction term before the extension, see also the review [15].

The second, more serious problem, is that, as it stands, the pure fermionic kinetic term $i\overline{\psi_{R_i}} \not{\partial} \psi_{R_i} = i\overline{\psi}_i \mathbb{P}_L \not{\partial} \mathbb{P}_R \psi_i$ projects only the purely four-dimensional derivative, leading to a purely four-dimensional propagator:

$$\frac{i \mathbb{P}_R \not{\bar{p}} \mathbb{P}_L}{\bar{p}^2}\,, \tag{351}$$

and to unregularized loop diagrams. As discussed in Section 3.4, the *only* valid choice for the propagator in the D-dimensional theory in the context of dimensional regularization is

$$\frac{i \not{p}}{p^2}\,, \tag{352}$$

so we are thus led to consider the full Dirac fermion ψ with both a left- and right-handed component and used instead the fully D dimensional covariant kinetic term $i\overline{\psi}_i\partial\!\!\!/\psi_i$. It can be re-expressed in terms of projectors as follows:

$$\begin{aligned} i\overline{\psi}_i\partial\!\!\!/\psi_i &= i\overline{\psi}_i\overline{\partial\!\!\!/}\psi_i + i\overline{\psi}_i\widehat{\partial\!\!\!/}\psi_i \\ &= i(\overline{\psi}_i\mathbb{P}_L\partial\!\!\!/\mathbb{P}_R\psi_i + \overline{\psi}_i\mathbb{P}_R\partial\!\!\!/\mathbb{P}_L\psi_i) + i(\overline{\psi}_i\mathbb{P}_L\partial\!\!\!/\mathbb{P}_L\psi_i + \overline{\psi}_i\mathbb{P}_R\partial\!\!\!/\mathbb{P}_R\psi_i) \end{aligned} \tag{353}$$

Notice that the fictitious, *sterile* left-chiral field ψ_L is introduced, which appears only within the kinetic term and nowhere else; it does not interact, so it does not couple in particular to the gauge bosons of the theory, and we enforced it to be invariant under gauge transformations.

Unfortunately, the choice of the D-dimensional propagator, crucial for loop regularization, that led to the introduction to the left-handed component in the kinetic term breaks the gauge invariance of the fermionic part of the Lagrangian, which is evident if we separate it in this way:

$$\mathcal{L}_{\text{fermions}} = \mathcal{L}_{\text{fermions,inv}} + \mathcal{L}_{\text{fermions,evan}}\,, \tag{354a}$$

$$\mathcal{L}_{\text{fermions,inv}} = i\overline{\psi}_i\overline{\partial\!\!\!/}\psi_i - e\mathcal{Y}_{Rij}\overline{\psi_R}_i\overline{A\!\!\!/}\psi_{Rj}\,, \tag{354b}$$

$$\mathcal{L}_{\text{fermions,evan}} = i\overline{\psi}_i\widehat{\partial\!\!\!/}\psi_i\,, \tag{354c}$$

where the first term contains purely four-dimensional derivatives and gauge fields and preserves the gauge and BRST invariance, since the fictitious left-chiral field ψ_L is a gauge singlet. The invariant term can also be written as a sum of purely left-chiral and purely right-chiral terms involving the four-dimensional covariant derivative as

$$\mathcal{L}_{\text{fermions,inv}} = i\overline{\psi_L}_i\overline{\partial\!\!\!/}\psi_{Li} + i\overline{\psi_R}_i\overline{\partial\!\!\!/}\psi_{Ri} - e\mathcal{Y}_{Rij}\overline{\psi_R}_i\overline{A\!\!\!/}\psi_{Rj} \tag{355a}$$

$$= i\overline{\psi_L}_i\overline{\partial\!\!\!/}\psi_{Li} + i\overline{\psi_R}_i\overline{D\!\!\!\!/}\psi_{Ri}\,, \tag{355b}$$

where the gauge invariance is obvious. The second term in Equation (354a) is purely evanescent, i.e., it vanishes in the four-dimensional limit. If we rewrite the evanescent term as

$$\mathcal{L}_{\text{fermions,evan}} = i\overline{\psi_L}_i\widehat{\partial\!\!\!/}\psi_{Ri} + i\overline{\psi_R}_i\widehat{\partial\!\!\!/}\psi_{Li}\,, \tag{356}$$

it can be easily seen that it mixes left- and right-chiral fields with different gauge transformation properties. This causes *the breaking of gauge and BRST invariance*—the central difficulty of the BMHV scheme.[15]

We can summarize this symmetry property and the symmetry breaking as

$$s_D\mathcal{L}_{\text{fermions,inv}} = 0\,, \tag{357a}$$

$$s_D\mathcal{L}_{\text{fermions,evan}} \neq 0\,, \tag{357b}$$

where s_D is the obvious extension of the BRST operator (344) to D dimensions.

[15] We remark that the problem is not specific to the case where the left-handed fermion is sterile. As Equation (356) shows, the problem generally exists if the left-handed and right-handed fermions have different gauge quantum numbers. References [27,131] considered this case and ended up with essentially the same breaking of BRST invariance in D dimensions and the same further consequences.

Since the extension of BRST transformation of fields in D dimensions is straightforward, our D-dimensional action is then

$$\begin{aligned} S_0 = \int d^D x \Big(& i\overline{\psi}_i \not{\partial} \psi_i + e\mathcal{Y}_{Rij}\overline{\psi_R}_i \not{A} \psi_{Rj} - \frac{1}{4}F^{\mu\nu}F_{\mu\nu} - \frac{1}{2\xi}(\partial_\mu A^\mu)^2 \\ & - \bar{c}\partial^2 c + \rho^\mu(\partial_\mu c) + i\,e\,\bar{R}^i c\,\mathcal{Y}_{Rij}\psi_{Rj} + i\,e\,\overline{\psi_R}_i c\mathcal{Y}_{Rij}R^j \Big) \\ \equiv & \sum_i S^i_{\overline{\psi}\psi} + \sum_i \overline{S^i_{\overline{\psi}_R A \psi_R}} + S_{AA} + S_{\text{g-fix}} + S_{\bar{c}c} + S_{\rho c} + S_{\bar{R}c\psi_R} + S_{Rc\overline{\psi_R}}, \end{aligned} \tag{358}$$

where, also, useful abbreviations for the individual terms were introduced. Similar to the fermion Lagrangian, the full D-dimensional action may be written as the sum of two parts, an "invariant" and an "evanescent" part:

$$S_0 = S_{0,\text{inv}} + S_{0,\text{evan}}, \tag{359a}$$

$$S_{0,\text{evan}} = \int d^D x\, i\overline{\psi}_i \hat{\not{\partial}} \psi_i . \tag{359b}$$

The second part $S_{0,\text{evan}}$ consists solely of one single, evanescent fermion kinetic term, the remnant of the D-dimensional propagator.

Now, we quantify the symmetry breaking caused by the BMHV scheme, the non-anticommuting γ_5, and the resulting evanescent term in the action. Acting with the D-dimensional BRST operator on the D-dimensional tree-level action, Equation (358) gives

$$s_D S_0 = s_D S_{0,\text{inv}} + s_D S_{0,\text{evan}} = 0 + s_D \int d^D x\, i\overline{\psi}_i \hat{\not{\partial}} \psi_i \equiv \hat{\Delta}, \tag{360}$$

where the nonvanishing integrated breaking term $\hat{\Delta}$ is given by

$$\hat{\Delta} = -\int d^D x e\, \mathcal{Y}_{Rij}\, c \left\{ \overline{\psi}_i \left(\overleftarrow{\hat{\not{\partial}}} \mathbb{P}_R + \overrightarrow{\hat{\not{\partial}}} \mathbb{P}_L \right) \psi_j \right\} \equiv \int d^D x\, \hat{\Delta}(x). \tag{361}$$

Acting with the D-dimensional Slavnov–Taylor operator $\mathcal{S}_D$ on the tree-level action, we obtain

$$\mathcal{S}_D(S_0) = s_D S_{0,\text{inv}} + s_D S_{0,\text{evan}} = 0 + \hat{\Delta}; \tag{362}$$

hence, the Slavnov–Taylor identity in D dimensions is violated by the same BRST breaking term at the tree level.

The simpler linear Equations (100)–(103) specific for Abelian theories are manifestly valid also in D dimensions. We will not discuss them further, but they have the consequence that higher-order corrections, including counterterm actions, cannot depend on the ghost/antighost and source fields. For this reason, the linearized Slavnov–Taylor operator here reduces to BRST transformations, $b_D = s_D$.

As mentioned in the overview Section 7.1, this breaking term will be a crucial tool in practical calculations. This breaking will be used as a composite operator insertion in Feynman diagrams. It generates an interaction vertex whose Feynman rule (with all momenta incoming and derived from the combination $i\hat{\Delta}$) is:

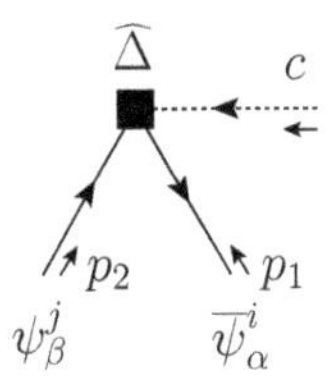

$$\begin{aligned} &= -\frac{e}{2}\mathcal{Y}_{Rij}((\hat{\not{p}}_1 + \hat{\not{p}}_2) + (\hat{\not{p}}_1 - \hat{\not{p}}_2)\gamma_5)_{\alpha\beta} \\ &= -e\,\mathcal{Y}_{Rij}(\hat{\not{p}}_1 \mathbb{P}_R + \hat{\not{p}}_2 \mathbb{P}_L)_{\alpha\beta} . \end{aligned} \tag{363}$$

As discussed in the context of Equation (80), in this way, the functional derivatives of $i\widehat{\Delta} \cdot \Gamma$ correspond to 1PI Feynman diagrams with one insertion of the Feynman rule (363). An analogous Feynman rule is derived for charge-conjugated fermions.

It is important to notice that this breaking $\widehat{\Delta}$ is evanescent, i.e., it vanishes in the four-dimensional limit. This results from the evanescent original term (359b) and has the consequence that insertions of $\widehat{\Delta}$ can only contribute in power-counting divergent Feynman diagrams.

7.3. Symmetry Restoration Requirements

Before beginning the explicit calculations, we recall and collect the required symmetry identities and the strategy for symmetry restoration in a more technical way than in the overview Section 7.1. We begin by collecting the required symmetry identities.

Symmetry identities expressing gauge/BRST invariance are considered part of the definition of the theory. Hence, they are required to be fulfilled at all orders; see Sections 2.6 and 6.2 for detailed discussions.

The symmetry requirements are defined for the renormalized and finite four-dimensional effective action of the form

$$\Gamma_{\text{ren}} = S_0^{(4D)} + \mathcal{O}(\hbar), \tag{364}$$

where we again highlight that the effective action coincides with the classical action up to higher-order corrections and that loop corrections are of higher order in $\hbar$; see Equation (76) and Section 3. The first symmetry requirement is BRST (and underlying gauge) invariance, which is expressed as the Slavnov–Taylor identity:

$$\mathcal{S}(\Gamma_{\text{ren}}) = 0, \tag{365}$$

for the renormalized theory. Notice that, in χQED, the fields c, $\bar{c}$, and ρ^μ do not have higher-order corrections, so relations

$$\frac{\delta\Gamma_{\text{ren}}}{\delta c(x)} = \frac{\delta S_0^{(4D)}}{\delta c(x)}, \qquad \frac{\delta\Gamma_{\text{ren}}}{\delta \bar{c}(x)} = \frac{\delta S_0^{(4D)}}{\delta \bar{c}(x)}, \qquad \frac{\delta\Gamma_{\text{ren}}}{\delta \rho^\mu(x)} = \frac{\delta S_0^{(4D)}}{\delta \rho^\mu(x)}. \tag{366}$$

hold trivially, since the respective derivatives of the tree-level action are linear in the dynamical fields as described in Section 2.6. The fact that the ghost does not have higher loop corrections will play a part in reducing the number of diagrams appearing in higher orders, compared to an analogous Yang–Mills theory. The local Ward identity:

$$\left(\partial^\mu \frac{\delta}{\delta A^\mu(x)} + ie\mathcal{Y}_R^j \sum_\Psi (-1)^{n_\Psi} \Psi(x) \frac{\delta}{\delta \Psi(x)}\right)\Gamma_{\text{ren}} = -\Box B(x), \tag{367}$$

is an automatic consequence of the Slavnov–Taylor identity, as we have shown in Section 2.6.

We record here the application of the Ward identity to the photon self-energy as an example that will later be illustrated in explicit computations. If we rewrite the Ward identity in the momentum-space representation and take a variation with the respect to photon field, we obtain the requirement:

$$ip_\nu \frac{\delta^2 \widetilde{\Gamma}_{\text{ren}}}{\delta A_\mu(p) \delta A_\nu(-p)} = 0, \tag{368}$$

which corresponds to the transversality of the photon self-energy.

All previous symmetry identities must hold after regularization and renormalization at each loop order. If the symmetries are broken in the intermediate regularization procedure, as is the case when we used the BMHV scheme, they must be restored order by order in perturbation theory, by adding suitable counterterms.

The symmetry identities are covered by the general analysis of algebraic renormalization discussed in Section 6.2.3, and the theory has no gauge anomaly; see Equation (343). This guarantees that the procedure of symmetry restoration works at all orders.

Now, we recapitulate the practical strategies for the concrete determination of symmetry-restoring counterterms, following the detailed outline given in Section 6.3. The application will be discussed in the subsequent subsections, where we treat not only the chiral model χQED, but also compare it with the familiar case of ordinary QED to highlight the features of the BMHV treatment of γ_5.

The first obvious difference is that ordinary QED is a vector-like gauge theory, and DReg preserves all relevant symmetry identities manifestly at each step: the counterpart to the tree-level breaking $\widehat{\Delta}$ in Equation (361) vanishes as already discussed in Section 4.3. Hence, generating counterterms by a renormalization transformation is sufficient; see the discussion in Section 6.1 and Equation (291).

For the case of χQED, the existence of a tree-level symmetry breaking, $\widehat{\Delta} \neq 0$, necessitates symmetry-restoring counterterms. Hence, generating counterterms by a renormalization transformation is not sufficient, and the general structure is the one discussed in Section 6.3.3, i.e., the combination:

$$S_{\text{sct,inv}} + S_{\text{sct,non-inv}} + S_{\text{fct,inv}} + S_{\text{fct,restore}} + S_{\text{fct,evan}}\,. \tag{369}$$

Section 6.3 presented two basic strategies to carry out the required computations of the crucial symmetry-restoring counterterms $S_{\text{fct,restore}}$. The first is based on the explicit computation of ordinary Green functions and explicitly checking symmetry identities. Its essential equation is Equation (325), which requires computing

$$\mathcal{S}_D(\Gamma^{(n)}_{\text{subren}} + S^n_{\text{sct}})$$

at each new order n. If this expression is nonzero, finite counterterms have to be found and added to the action such that the symmetry breaking is canceled.

The second strategy is based on using the regularized quantum action principle and represented by Equation (330):

$$\underset{D\to 4}{\text{LIM}}\left(\widehat{\Delta}\cdot\Gamma^n_{\text{DRen}} + \sum_{k=1}^{n-1}\Delta^k_{\text{ct}}\cdot\Gamma^{n-k}_{\text{DRen}} + \Delta^n_{\text{ct}}\right) = 0.$$

The computation of full Green functions and evaluating Slavnov–Taylor identities is replaced by the computation of Green functions with insertions of breaking operators such as $\widehat{\Delta}$. This equation is specialized to Equations (332) and (334) at the one- and two-loop level.

In the following subsections, we illustrate Feynman diagrammatic computations for both strategies. The more efficient second strategy is illustrated also at the two-loop level. We then see how the desired symmetry-restoring counterterms are determined.

7.4. Explicit Calculations and Results in the Abelian Chiral Gauge Theory

In this section, explicit calculations in the Abelian chiral gauge theory defined above in Section 7.2 are performed in the BMHV scheme of DReg, and all necessary counterterms are provided up to the two-loop level. In particular, the evaluation of the photon self-energy at the one-loop (Sections 7.4.1 and 7.4.2) and the two-loop level (Section 7.4.3) is highlighted, and the results are compared to the ordinary QED. As announced in Section 7.1, there are two different ways of determining symmetry-restoring counterterms. While the method in Section 7.4.1 amounts to the explicit evaluation of the full photon self-energy, i.e., a full Green function, including its finite part, Section 7.4.2 employs the direct method based on the regularized quantum action principle, where the symmetry breaking is determined via special Feynman diagrams with an insertion of the $\widehat{\Delta}$-operator, which reflects the breaking of chiral gauge invariance. Section 7.4.2 then concludes by providing the full one-loop counterterm action for chiral QED in the BMHV scheme. Similarly, in Section 7.4.3, the two-loop counterterms for the photon self-energy are obtained using the latter method based on the regularized quantum action principle, but are verified by comparing with the explicit result for the full-photon self-energy including its finite part. Concluding, Section 7.4.4 provides the full two-loop renormalization of the chiral QED in the BMHV scheme.

7.4.1. One-Loop Photon Self-Energy and Symmetry-Restoring Counterterms

To better understand the features of the BMHV scheme, we now focus on explicit loop calculations. We take the photon self-energy and compare its results in ordinary QED and chiral QED. The photon self-energy is subject to the simplest Ward identity (368); it must be transverse, to guarantee the correct physical interpretation of the theory describing a massless spin one particle with two transverse polarizations.

The photon self-energy is denoted as

$$i\widetilde{\Gamma}^{\nu\mu}_{AA}(p) =$$ 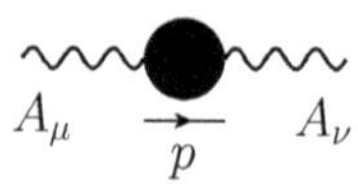 .

We used the notation explained in Section 2.4, corresponding to the one-particle irreducible diagrams with external fields and momentum as indicated.[16]

We begin by recalling the well-known one-loop result of the ordinary QED with massless fermions as defined in Equation (338),

$$i\widetilde{\Gamma}^{\nu\mu}_{AA}(p)|^{1}_{\text{div,QED}} = \frac{ie^2}{16\pi^2\epsilon}\frac{4\,\text{Tr}(Q^2)}{3}(p^\mu p^\nu - p^2 g^{\mu\nu}), \tag{370a}$$

$$i\widetilde{\Gamma}^{\nu\mu}_{AA}(p)|^{1}_{\text{fin,QED}} = \frac{ie^2}{16\pi^2}\frac{2\,\text{Tr}(Q^2)}{3}\left[\left(\frac{10}{3} - 2\ln\left(-p^2\right)\right)(p^\mu p^\nu - p^2 g^{\mu\nu})\right]. \tag{370b}$$

Here and in all following results, we set $D = 4 - 2\epsilon$ and suppress the dimensional regularization scale $\bar{\mu}^2 = \mu^2 4\pi e^{-\gamma_E}$ in dimensionful logarithms. We see that the result is transverse and satisfies the Ward identity (368), both in its divergent and finite parts. Adding the counterterm action:

$$S^{1}_{\text{sct,QED}} = \frac{-\hbar\, e^2}{16\pi^2\epsilon}\frac{4\text{Tr}(Q^2)}{3}S_{AA} + \ldots, \tag{371}$$

where the dots denote terms unrelated to the photon self-energy, cancels the divergences and preserves the validity of the Ward identity. The factor $\hbar$ was explicitly restored to highlight that the counterterm action is of one-loop order. As is well known, this counterterm action can be generated via a photon field renormalization transformation.

In comparison, the result for the one-loop photon self-energy diagram in χQED with massless fermions as defined in Equation (341) reads

$$i\widetilde{\Gamma}^{\nu\mu}_{AA}(p)|^{1}_{\text{div},\chi\text{QED}} = \frac{ie^2}{16\pi^2\epsilon}\frac{2\,\text{Tr}(\mathcal{Y}_R^2)}{3}\left[(\overline{p}^\mu\overline{p}^\nu - \overline{p}^2\overline{g}^{\mu\nu}) - \frac{1}{2}\widehat{p}^2\overline{g}^{\mu\nu}\right], \tag{372a}$$

$$\begin{aligned} i\widetilde{\Gamma}^{\nu\mu}_{AA}(p)|^{1}_{\text{fin},\chi\text{QED}} = \frac{ie^2}{16\pi^2}\frac{\text{Tr}(\mathcal{Y}_R^2)}{3}\Bigg[&\left(\frac{10}{3} - 2\ln\left(-p^2\right)\right)(\overline{p}^\mu\overline{p}^\nu - \overline{p}^2\overline{g}^{\mu\nu}) \\ &- \left(\overline{p}^2 + \widehat{p}^2\left(\frac{8}{3} - \ln\left(-p^2\right)\right)\right)\overline{g}^{\mu\nu}\Bigg]. \end{aligned} \tag{372b}$$

[16] However, in this subsection, we use a slightly simpler notation than in Section 3.1 for unrenormalized/subrenormalized expressions. We drop the subscript $_{\text{subren}}$ and simply write Γ^1 for the unrenormalized one-loop effective action and Γ^2 for the subrenormalized two-loop effective action. Accordingly, the following equations correspond to the unrenormalized one-loop photon self-energy.

From this illustrative example, we can extract several interesting comments. First, and most obviously, transversality is violated by the last terms in Equations (372). This will be our main focus. However, also the transverse part shows two differences compared to the ordinary QED. Since the interaction vertex in χQED differs from the one given in the standard QED by

$$V_{\text{QED}} \to -ie\gamma^{\mu}Q_{ij}, \qquad V_{\chi\text{QED}} \to -ie\bar{\gamma}^{\mu}\mathbb{P}_R\mathcal{Y}_{R,ij}, \tag{373}$$

it projects the fermion loop content, so the transverse part becomes purely four-dimensional, explaining the appearance of the covariants $\bar{g}^{\mu\nu}$ and $\bar{p}^{\mu}$ in Equations (372). Further, due to this projection, only half the number of fermionic degrees of freedom appear in the loop for the chiral case, resulting in the relative factor of two with respect to the ordinary QED.

Let us now focus on the breaking of transversality in the photon self-energy. The divergent breaking term in Equation (372a) is proportional to $\widehat{p}^2$, i.e., it is evanescent. In contrast, the finite breaking term in Equation (372b) contains finite expressions that do not vanish in the four-dimensional limit. The finite breaking also contains evanescent terms that vanish in $\text{LIM}_{D\to 4}$; these will be ignored in the following.

We can exhibit the breaking explicitly by plugging the photon self-energy into the Ward identity (368); we obtain

$$ip_{\nu}\widetilde{\Gamma}^{\nu\mu}_{AA}(p)|^1_{\text{div+fin},\chi\text{QED}} = \frac{ie^2}{16\pi^2}\frac{\text{Tr}(\mathcal{Y}_R^2)}{3}\left[-\frac{1}{\epsilon}\widehat{p}^2\bar{p}^{\mu} - \bar{p}^2\bar{p}^{\mu}\right] \neq 0\,. \tag{374}$$

Here, we ignored the finite, evanescent term, as announced. In line with the derivation of the Ward identity from the Slavnov–Taylor identity via derivatives with respect to a ghost field (see Equation (107)), the result is equivalent to the violation of the Slavnov–Taylor identity:

$$[\mathcal{S}(\Gamma)]^1_{A^{\mu}c} = \frac{ie^2}{16\pi^2}\frac{\text{Tr}(\mathcal{Y}_R^2)}{3}\left[-\frac{1}{\epsilon}\widehat{p}^2\bar{p}^{\mu} - \bar{p}^2\bar{p}^{\mu}\right], \tag{375}$$

where the left-hand side denotes functional derivatives in momentum-space, similarly to the notation of Γ_{AA}.

A decisive feature of the breaking terms is their locality: the breaking terms in all the previous equations are polynomials of the momentum in momentum-space, and this translates into local expressions on the level of the (effective) action. This locality is in line with the general statement discussed in Section 6.2.3, which forms the basis of algebraic renormalization. This means that a local counterterm can be defined that cancels the symmetry breaking.

In view of the explicit results, the required counterterms for the sector of the photon self-energy can be read off as follows. We first discuss the divergent counterterms. The divergent counterterms can be split into an invariant and a non-invariant part as in Equation (369) as $S_{\text{sct}} = S_{\text{sct,inv}} + S_{\text{sct,non-inv}}$ such that the one-loop parts relevant for the photon self-energy in χQED read

$$S^1_{\text{sct,inv},\chi\text{QED}} = \frac{-\hbar\, e^2}{16\pi^2\epsilon}\frac{2\text{Tr}(\mathcal{Y}_R^2)}{3}\overline{S_{AA}} + \dots, \tag{376a}$$

$$S^1_{\text{sct,non-inv},\chi\text{QED}} = \frac{-\hbar\, e^2}{16\pi^2\epsilon}\frac{\text{Tr}(\mathcal{Y}_R^2)}{3}\int d^Dx\,\frac{1}{2}\bar{A}_{\mu}\widehat{\partial}^2\bar{A}^{\mu} + \dots, \tag{376b}$$

where the dots denote terms unrelated to the photon self-energy. As in the case of the ordinary QED, the divergences are canceled, and the invariant counterterm can be generated via a photon field renormalization transformation. In contrast to the ordinary QED, however, the non-invariant term is required, and it cannot be obtained from a renormalization transformation, but must be read off by hand.

Obviously, adding these counterterms does not only cancel the divergences of the photon self-energy, but it also cancels the divergences in the breaking of the Ward/Slavnov–Taylor identities (374) and (375). Specifically, adding the counterterms to the action modifies the Slavnov–Taylor identity $\mathcal{S}(\Gamma)$ to $\mathcal{S}(\Gamma + S^1_{\text{sct},\chi\text{QED}}) = \mathcal{S}(\Gamma) + s_D S^1_{\text{sct},\chi\text{QED}} + \ldots$, where the dots denote higher-order terms and where

$$s_D S^1_{\text{sct},\chi\text{QED}} = \Delta^1_{\text{ct}}\big|_{\text{div}} = -\frac{\hbar}{16\pi^2\epsilon}\frac{e^2\text{Tr}(\mathcal{Y}_R^2)}{3}\int d^D x\,(\bar{\partial}_\mu c)\,(\widehat{\partial}^2 \bar{A}^\mu)\,. \tag{377}$$

In momentum-space, with incoming A^μ momentum p, this is precisely the negative of the divergent term in Equation (375). This is an automatic consequence of the finiteness.

Now, we discuss the required finite counterterms to the photon self-energy. The explicit result (372b) shows that the transversality is restored by the following finite counterterm:

$$S^1_{\text{fct},\chi\text{QED}} = \frac{\hbar}{16\pi^2}\int d^4x \frac{-e^2\text{Tr}(\mathcal{Y}_R^2)}{6}\bar{A}_\mu\bar{\partial}^2\bar{A}^\mu + \ldots \tag{378}$$

In momentum-space, this counterterm cancels the non-transverse $\overline{p}^2$-term of (372b) (we recall that the remaining non-transverse finite terms are evanescent and vanish in the $\text{LIM}_{D\to 4}$). On the level of the Slavnov–Taylor identity, adding the finite counterterm modifies the Slavnov–Taylor identity $\mathcal{S}(\Gamma)$ by the term:

$$s_D S^1_{\text{fct},\chi\text{QED}} = -\frac{\hbar}{16\pi^2}\int d^D x\,\frac{e^2\text{Tr}(\mathcal{Y}_R^2)}{3}(\bar{\partial}_\mu c)(\bar{\partial}^2\bar{A}^\mu)\,. \tag{379}$$

In momentum-space, this is the negative of the finite term in Equation (375).

In total, after adding all counterterms (376) and (378) to the photon self-energy and taking $\text{LIM}_{D\to 4}$, the renormalized one-loop photon self-energy is

$$i\widetilde{\Gamma}^{\nu\mu}_{AA}(p)\big|^1_{\text{ren},\,\chi\text{QED}} = \frac{ie^2}{16\pi^2}\frac{\text{Tr}(\mathcal{Y}_R^2)}{3}\left[\left(\frac{10}{3} - 2\ln\left(-p^2\right)\right)(\overline{p}^\mu\overline{p}^\nu - \overline{p}^2\overline{g}^{\mu\nu})\right]. \tag{380}$$

It is finite, defined in four dimensions, and it is properly transverse. One may still add further, finite, symmetric counterterms. These can be derived from usual field and parameter renormalization, but are not our focus here.

7.4.2. One-Loop Photon Self-Energy—Direct Computation of Symmetry Breaking

In the previous subsection, we determined the required counterterms (376) and (378) by carrying out an explicit computation of a Green function, including its finite part, and by explicitly evaluating the breaking of the relevant symmetry identity. We now show how the determination of the counterterms can be performed in a simpler way. We still illustrate it for the one-loop photon self-energy, but the advantage of that simplification will become more and more prominent for higher orders and more complicated Green functions.

Instead of evaluating the full-photon self-energy including its finite part (372), the following is sufficient: First, we need the divergent part of the photon self-energy, i.e., only (372a). This, of course, determines the divergent counterterms (376) unambiguously.

Second, we need the violation of the symmetry, expressed in terms of Equation (375). This violation can be obtained in a more direct way, by using the regularized quantum action principle discussed in Section 4. This tells us that the violation $\mathcal{S}(\Gamma) \neq 0$ is directly given by diagrams with insertions of the composite operator $\widehat{\Delta}$, corresponding to the tree-level violation of the Slavnov–Taylor identity in D dimensions. For the photon self-energy, the violation (375) can be obtained directly by computing the Green function $[\widehat{\Delta}\cdot\widetilde{\Gamma}^\mu_{Ac}]$, i.e., the one-particle irreducible Green function with an insertion of $\widehat{\Delta}$ and external A^μ and c fields.

At one-loop order, there is only one diagram.

$$i[\widehat{\Delta}\cdot\widetilde{\Gamma}^{\mu}_{Ac}]^{(1)} = \quad (381)$$

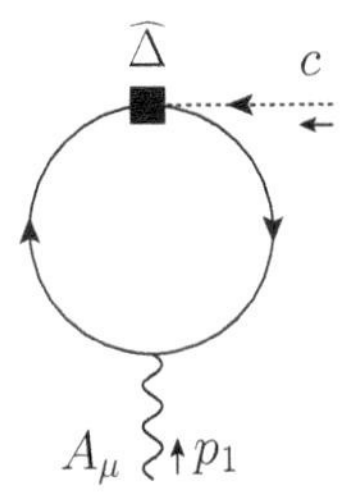

The result of this single diagram is

$$i[\widehat{\Delta}\cdot\widetilde{\Gamma}^{\mu}_{Ac}]^{1}_{\mathrm{div}} = \frac{e^2}{16\pi^2\epsilon}\frac{\mathrm{Tr}(\mathcal{Y}_R^2)}{3}\widehat{p_1}^{\,2}\overline{p_1}^{\mu}, \quad (382a)$$

$$i\,[\widehat{\Delta}\cdot\widetilde{\Gamma}^{\mu}_{Ac}]^{1}_{\mathrm{fin}} = \frac{e^2}{16\pi^2}\frac{\mathrm{Tr}(\mathcal{Y}_R^2)}{3}\overline{p}_1^{\,2}\overline{p}_1^{\mu}\,. \quad (382b)$$

We see that the result of this diagram indeed agrees with the right-hand side of Equation (375), as it is guaranteed by the regularized quantum action principle.

The important point is the technical simplification: the computation of this diagram is technically easier than the computation of the finite part of the photon self-energy since only power-counting divergent parts of the loop integrals are relevant. We reiterate that the technical advantage is much more dramatic at higher orders and for more complicated Green functions.

It is instructive to rewrite the result in coordinate space:

$$[\widehat{\Delta}\cdot\Gamma]^{(1)}_{\mathrm{div}} = \frac{e^2}{16\pi^2\epsilon}\frac{\mathrm{Tr}(\mathcal{Y}_R^2)}{3}\int d^Dx(\overline{\partial}_\mu c)(\widehat{\partial}^2\bar{A}^\mu)+\dots, \quad (383a)$$

$$[\widehat{\Delta}\cdot\Gamma]^{(1)}_{\mathrm{fin}} = \frac{e^2}{16\pi^2}\frac{\mathrm{Tr}(\mathcal{Y}_R^2)}{3}\int d^Dx\,(\overline{\partial}_\mu c)(\overline{\partial}^2\bar{A}^\mu)+\dots. \quad (383b)$$

The dots denote terms unrelated to the photon self-energy.

The divergent part provides no independent information, but a check. As discussed after Equation (377), the expression $s_D S^1_{\mathrm{sct},\chi\mathrm{QED}}$ must automatically cancel the divergent part of the symmetry breaking. Using our new result, this means that $s_D S^1_{\mathrm{sct},\chi\mathrm{QED}} + [\widehat{\Delta}\cdot\Gamma]^{(1)}_{\mathrm{div}} = 0$ must automatically hold. Clearly, this is true, and the check is passed.

The important new information is in the finite part of the $\widehat{\Delta}$-insertion diagram Equations (382) and (383). Its result is equal to the finite part of the violation of the Slavnov–Taylor identity (375), thus eliminating the need to explicitly evaluate the Slavnov–Taylor identity.

The finite, symmetry-restoring counterterm may now be obtained from solving the equation:

$$s_D S^1_{\mathrm{fct},\chi\mathrm{QED}} = -[\widehat{\Delta}\cdot\Gamma]^{(1)}_{\mathrm{fin}}\,. \quad (384)$$

For the sector of the photon self-energy, the result is the one given in Equation (378). In summary, there, the result was obtained from inspecting the finite part of the photon self-energy; here, the result can be obtained from evaluating Equation (382) and then solving the defining condition (384).

To conclude the section, we summarize the full one-loop results for the counterterm structure of χQED. First, all divergences of all one-loop diagrams need to be evaluated, generalizing Equation (372a). The negative of these results defines unambiguously the one-loop divergent counterterms, generalizing Equation (376). The result reads

$$S^1_{\text{sct},\chi\text{QED}} = \frac{-\hbar\, e^2}{16\pi^2\epsilon}\left(\frac{2\text{Tr}(\mathcal{Y}_R^2)}{3}\overline{S_{AA}} + \xi\sum_j(\mathcal{Y}_R^j)^2\left(\overline{S^j_{\overline{\psi}\psi_R}} + \overline{S^j_{\overline{\psi_R}A\psi_R}}\right) + \frac{\text{Tr}(\mathcal{Y}_R^2)}{3}\int d^Dx\,\frac{1}{2}\bar{A}_\mu\widehat{\partial}^2\bar{A}^\mu\right). \tag{385}$$

Most terms are similar to their counterparts in the ordinary QED and can be obtained by a renormalization transformation of the fields and parameters as in Equation (291), where it is noteworthy that only the physical, right-handed fermion is renormalized, while the sterile, left-handed fermion is not. However, this renormalization transformation does not generate the last term involving the $\widehat{\partial}^2$ operator, and it generates the full D-dimensional photon kinetic term S_{AA} instead of its four-dimensional version $\overline{S_{AA}}$. Hence, the $\widehat{\partial}^2$-term and the difference $\overline{S_{AA}} - S_{AA}$ correspond to symmetry-breaking singular counterterms. These counterterms become particularly important in the context of two-loop calculations, where they are necessary for the proper subrenormalization.

Second, all one-loop symmetry breakings need to be determined, generalizing either Equation (375) or Equation (383). We used the method based on the regularized quantum action principle. In this case, the full symmetry breaking is given by the complete set of all one-loop diagrams with a $\widehat{\Delta}$ insertion. Since only power-counting divergent diagrams can provide nonvanishing contributions, there are only precisely four contributing diagrams: with external fields cA, cAA, $cAAA$, or $c\overline{\psi}\psi$. One of them vanishes due to the anomaly cancellation condition (343). The full result of the symmetry breaking is

$$\begin{aligned}\widehat{\Delta}\cdot\Gamma^1 = \frac{1}{16\pi^2}\int d^Dx\Bigg[&\frac{e^2\text{Tr}(\mathcal{Y}_R^2)}{3}\left(\frac{1}{\epsilon}(\bar{\partial}_\mu c)\,(\widehat{\partial}^2\bar{A}^\mu) + (\bar{\partial}_\mu c)(\bar{\partial}^2\bar{A}^\mu)\right)\\ &+\frac{e^4\text{Tr}(\mathcal{Y}_R^4)}{3}\,c\,\bar{\partial}_\mu(\bar{A}^\mu\bar{A}^2)\\ &-\frac{(\xi+5)e^3}{6}\sum_j(\mathcal{Y}_R^j)^3\,c\,\bar{\partial}^\mu(\overline{\psi}_j\bar{\gamma}_\mu\mathbb{P}_R\psi_j)\Bigg].\end{aligned} \tag{386}$$

Using the defining condition (384) for the finite, symmetry-restoring counterterms, we obtain

$$S^1_{\text{fct}} = \frac{\hbar}{16\pi^2}\int d^4x\left\{\frac{-e^2\text{Tr}(\mathcal{Y}_R^2)}{6}\bar{A}_\mu\bar{\partial}^2\bar{A}^\mu + \frac{e^4\text{Tr}(\mathcal{Y}_R^4)}{12}\bar{A}_\mu\bar{A}^\mu\bar{A}_\nu\bar{A}^\nu + \frac{5+\xi}{6}e^2\sum_j(\mathcal{Y}_R^j)^2 i\overline{\psi}_j\bar{\gamma}^\mu\bar{\partial}_\mu\mathbb{P}_R\psi_j\right\}. \tag{387}$$

This is the complete result for the symmetry-restoring counterterms of the χQED model at the one-loop level. Each of the terms has a clear and simple interpretation. The first finite counterterm restores the transversality of the photon self-energy as discussed before. The second term restores a similar transversality identity for the photon four-point function. The last term restores the QED-like Ward identity relating the fermion self-energy with the fermion–photon three-point function.

These three counterterms must be inserted in higher-order calculations. They give additional contributions to loop diagrams compared to the renormalization in vector-like theories or to a naive γ_5 treatment, where gauge invariance is manifestly preserved.

7.4.3. Two-Loop Photon Self-Energy and Corresponding Breaking Diagram

Now, we illustrate the determination of two-loop counterterms in χQED using the BMHV scheme. We immediately follow the more direct strategy explained in Section 7.4.2 based on diagrams with $\widehat{\Delta}$-insertions.

At the 2-loop level, diagrams contributing to the subrenormalized photon self-energy are, on the one hand, genuine 2-loop diagrams and, on the other hand, 1-loop diagrams with counterterm insertions. Both the singular counterterms (385), as well as finite symmetry-restoring counterterms (387) must be used. The result for the divergent part of the sub-renormalized two-loop photon self-energy is given by (we still used the simplified notation described in footnote 16, where Γ^2 denotes the subrenormalized two-loop effective action.)

$$i\widetilde{\Gamma}^{\nu\mu}_{AA}(p)|^2_{\text{div},\chi\text{QED}} = \frac{ie^4}{256\pi^4}\frac{\text{Tr}(\mathcal{Y}_R^4)}{3}\left[\frac{2}{\epsilon}(\overline{p}^\mu\overline{p}^\nu - \overline{p}^2\overline{g}^{\mu\nu}) + \left(\frac{17}{24\epsilon} - \frac{1}{2\epsilon^2}\right)\widehat{p}^2\overline{g}^{\mu\nu}\right], \tag{388a}$$

which can be compared to the result in ordinary QED

$$i\widetilde{\Gamma}^{\nu\mu}_{AA}(p)|^2_{\text{div,QED}} = \frac{ie^4}{256\pi^4\epsilon}2\,\text{Tr}(Q^4)(p^\mu p^\nu - p^2 g^{\mu\nu})\,. \tag{388b}$$

Notice again that the transverse part for QED is fully D-dimensional, but projected to four dimensions in the chiral case, and in the chiral case, an evanescent term is present, again spoiling gauge and BRST invariance. Unlike at the 1-loop level, the global factor in front of the chiral transversal part is not half of the QED case, since the additional diagram with finite 1-loop counterterm insertion spoils this relationship.

From this singular part of the two-loop diagrams, we reconstruct an equivalent result in coordinate space:

$$\Gamma^{2,AA}_{\text{div}} = \frac{e^4}{256\pi^4}\frac{\text{Tr}(\mathcal{Y}_R^4)}{3}\left[\frac{1}{\epsilon}\overline{A}_\mu(\overline{\partial}^2\overline{g}^{\mu\nu} - \overline{\partial}^\mu\overline{\partial}^\nu)\overline{A}_\nu + \overline{A}_\mu\widehat{\partial}^2\overline{A}^\mu\left(\frac{1}{4\epsilon^2} - \frac{17}{48\epsilon}\right)\right], \tag{389}$$

which results in the required singular counterterm of the form:

$$S^2_{\text{sct}} = -\left(\frac{\hbar e^2}{16\pi^2}\right)^2\frac{\text{Tr}(\mathcal{Y}_R^4)}{3}\left[\frac{2}{\epsilon}\overline{S_{AA}} + \left(\frac{1}{4\epsilon^2} - \frac{17}{48\epsilon}\right)\int d^D x\overline{A}_\mu\widehat{\partial}^2\overline{A}^\mu\right] + \dots, \tag{390}$$

which cancels the divergences. Clearly, this counterterm also breaks BRST symmetry at the two-loop level by

$$\Delta^2_{\text{sct}} = s_D S^2_{\text{sct}} = \frac{-\hbar^2 e^4}{256\pi^4}\frac{\text{Tr}(\mathcal{Y}_R^4)}{6}\left(\frac{1}{\epsilon^2} - \frac{17}{12\epsilon}\right)\int d^D x(\overline{\partial}_\mu c)(\widehat{\partial}^2\overline{A}^\mu) + \dots. \tag{391}$$

Now, we use the regularized quantum action principle and determine the symmetry breaking at the two-loop level in the photon self-energy sector. Hence, we need to evaluate the Green function $\left([\widehat{\Delta} + \Delta^1_{\text{ct}}]\cdot\widetilde{\Gamma}\right)^2_{A_\mu c}$ at the two-loop level.

Compared to the one-loop level, there are several new features. There are four types of two-loop level diagrams; see Figure 4. The diagrams in the first column of the figure are genuine two-loop diagrams with one insertion of the tree-level breaking $\widehat{\Delta}$. The diagrams in the second column are one-loop diagrams with one insertion of a one-loop singular counterterm, denoted as a circled cross. The third column contains a one-loop diagram with an insertion of a one-loop symmetry-restoring counterterm obtained from the fermion self-energy operator, denoted by a boxed F, and a one-loop diagram with an insertion of the one-loop breaking Δ^1_{ct}.

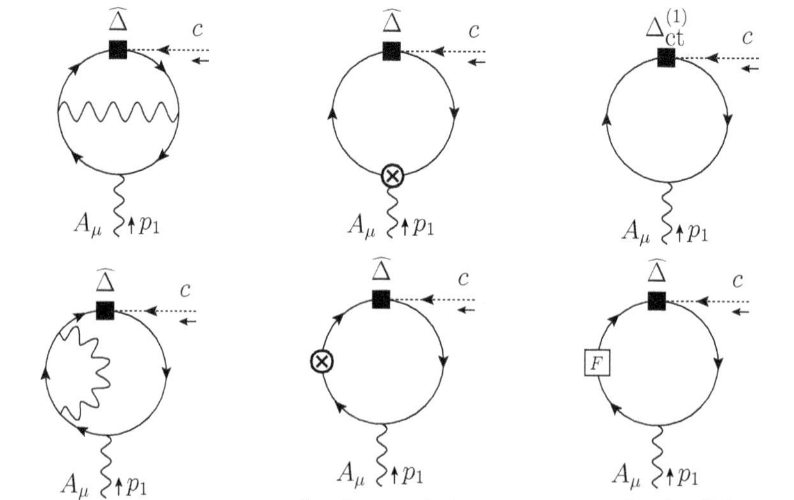

+ loop on the other fermion propagator.

+ fermion counterterm on the other fermion propagator.

+ fermion finite counterterm on the other fermion propagator.

Figure 4. List of Feynman diagrams for the ghost–photon breaking contribution given in Equation (392).

The total two-loop breaking in this sector, i.e., the result of the diagrams in Figure 4, is

$$i\left([\widehat{\Delta}+\Delta^{1}_{\text{ct}}]\cdot\widetilde{\Gamma}\right)^{2}_{A_\mu c} = \frac{1}{256\pi^4}\frac{e^4\text{Tr}(\mathcal{Y}_R^4)}{6}\left[\left(\frac{1}{\epsilon^2}-\frac{17}{12\epsilon}\right)\hat{p}_1^2\bar{p}_1^\mu - \frac{11}{4}\bar{p}_1^2\bar{p}_1^\mu + \mathcal{O}(\hat{\cdot})\right]. \quad (392)$$

The result contains $1/\epsilon^2$ poles and $1/\epsilon$ poles with local, evanescent coefficients and a finite, non-evanescent term.

Like at the one-loop level, we first use the result to check the cancellation of the UV divergences as prescribed by Equation (334).[17] Using this simplification, we can confirm that the expected cancellation of UV divergences with $s_D S^2_{\text{sct}}$ given in Equation (391) indeed occurs as

$$\Delta^2_{\text{sct}} = s_D S^2_{\text{sct}} = -\left([\widehat{\Delta}+\Delta^1_{\text{ct}}]\cdot\Gamma\right)^2_{\text{div}}. \quad (393)$$

The remaining finite part can then be evaluated in strictly four dimensions:

$$\begin{aligned}\Delta^2_{\text{fct}} &= -\underset{D\to4}{\text{LIM}}\left\{\left([\widehat{\Delta}+\Delta^{(1)}_{\text{ct}}]\cdot\Gamma\right)^{(2)} + s_D S^{(2)}_{\text{sct}}\right\}\\ &= \frac{e^4}{256\pi^4}\text{Tr}(\mathcal{Y}_R^4)\, s\left(\frac{11}{48}\int d^4x \bar{A}_\mu \bar{\partial}^2\bar{A}^\mu\right) + \ldots.\end{aligned} \quad (394)$$

The defining relation for the finite, symmetry-restoring counterterm is then

$$\underset{D\to4}{\text{LIM}}\, s_D S^2_{\text{fct}} = -\Delta^2_{\text{fct}}. \quad (395)$$

From this, we reconstruct the corresponding finite counterterm as

$$S^2_{\text{fct}} = \left(\frac{\hbar}{16\pi^2}\right)^2\int d^4x\, e^4\text{Tr}(\mathcal{Y}_R^4)\frac{11}{48}\bar{A}_\mu\bar{\partial}^2\bar{A}^\mu + \ldots. \quad (396)$$

17 As mentioned above in Section 7.2.2, in the Abelian case considered here we have $b_D = s_D$, and therefore in Equation (334) we can simplify and use $\Delta^2_{\text{ct}} = s_D S^2_{\text{ct}}$.

As before, we only display terms related to the photon self-energy. Adding this counterterm restores the photon self-energy transversality at the two-loop level.

At this point, the determination of the two-loop counterterms of this sector is complete, and the counterterms of other sectors can be determined analogously. The required computations were the ones of the divergent part of the photon self-energy and of the finite part of the diagrams of Figure 4.

Nevertheless, we now confirm the result by comparing with the explicit result for the finite part of the photon self-energy. The finite part of the photon self-energy at the two-loop level (including one-loop counterterms, but excluding two-loop counterterms) is given by

$$i\widetilde{\Gamma}^{\mu\nu}_{AA}(p)\Big|^{2}_{\text{fin}} = \frac{ie^4}{256\pi^4}\frac{\text{Tr}(\mathcal{Y}_R^4)}{3} \left[\left(\frac{673}{23} - 6\log\left(-\overline{p}^2\right) - 24\zeta(3)\right)(\overline{p}^\mu\overline{p}^\nu - \overline{p}^2\overline{g}^{\mu\nu}) + \frac{11}{8}\overline{p}^\mu\overline{p}^\nu\right]. \tag{397}$$

Similar to the one-loop result (372), the non-local $\log\left(-\overline{p}^2\right)$ and transcendental $\zeta(3)$ parts are by themselves transversal and, so, do not break the gauge invariance. The last term breaks the transversality, but this breaking term is local.

Plugging the result into the Ward or Slavnov–Taylor identity, we obtain

$$i\,p_\nu\,\widetilde{\Gamma}^{\mu\nu}_{A(-p)A(p)}\Big|^{2}_{\text{fin}} = \frac{ie^4}{256\pi^4}\frac{\text{Tr}(\mathcal{Y}_R^4)}{6}\frac{11}{4}\overline{p}^2\overline{p}^\mu \tag{398a}$$

$$= -\left([\widehat{\Delta} + \Delta^1_{\text{ct}}]\cdot\widetilde{\Gamma}\right)^{2}_{\text{fin},\,A_\mu(-p)c(p)}. \tag{398b}$$

The first of these equations is obtained by direct computation using the finite parts in Equation (397). The second equation is then observed by comparison with Equation (392). Hence, we confirmed that the violation of the symmetry is restored by our finite counterterm evaluated from breaking diagrams.

7.4.4. Full Two-Loop Renormalization of Chiral QED

In the previous sections, we performed the full one-loop renormalization with singular and finite, symmetry-restoring counterterms (385) and (387), respectively, and studied the photon self-energy and the corresponding breaking at the two-loop level; cf. Section 7.4.3. In this section, we present the full two-loop renormalization of chiral QED based on our results in Reference [26].

A list of all divergent 1PI two-loop Green functions together with the individual results is to be found in Chapter 7 of Reference [26]. From the singular part of these Green functions, we obtain the singular counterterm action at the two-loop level:

$$\begin{aligned}
S^2_{\text{sct}} = &-\left(\frac{\hbar e^2}{16\pi^2}\right)^2\frac{\text{Tr}(\mathcal{Y}_R^4)}{3}\left[\frac{2}{\epsilon}\overline{S_{AA}} + \left(\frac{1}{4\epsilon^2} - \frac{17}{48\epsilon}\right)\int d^Dx\,\bar{A}_\mu\widehat{\partial}^2\bar{A}^\mu\right] \\
&+\left(\frac{\hbar e^2}{16\pi^2}\right)^2\sum_j(\mathcal{Y}_R^j)^2\left[\left(\frac{1}{2\epsilon^2} + \frac{17}{12\epsilon}\right)(\mathcal{Y}_R^j)^2 - \frac{1}{9\epsilon}\text{Tr}(\mathcal{Y}_R^2)\right]\left(\overline{S^j_{\overline{\psi}\psi_R}} + \overline{S^j_{\overline{\psi}_R A\psi_R}}\right) \\
&-\left(\frac{\hbar e^2}{16\pi^2}\right)^2\sum_j\frac{(\mathcal{Y}_R^j)^2}{3\epsilon}\left(\frac{5}{2}(\mathcal{Y}_R^j)^2 - \frac{2}{3}\text{Tr}(\mathcal{Y}_R^2)\right)\overline{S^j_{\overline{\psi}\psi_R}}
\end{aligned} \tag{399}$$

which cancels the divergences. Comparing (399) with its one-loop counterpart in Equation (385), we see that its structure is the same up to the term in the last line, which breaks the BRST invariance by a non-evanescent amount and is, thus, a new feature emerging at the two-loop level.

This two-loop counterterm action (399) generates the BRST breaking:

$$\begin{aligned}\Delta^2_{\text{sct}} &= s_D S^2_{\text{sct}}\\ &= -\frac{\hbar^2 e^4}{256\pi^4}\frac{\text{Tr}(\mathcal{Y}_R^4)}{6}\left(\frac{1}{\epsilon^2}-\frac{17}{12\epsilon}\right)\int d^D x\,(\bar{\partial}_\mu c)(\widehat{\partial}^2 \bar{A}^\mu)\\ &\quad -\frac{\hbar^2 e^5}{256\pi^4}\frac{1}{3\epsilon}\sum_j(\mathcal{Y}_R^j)^3\left(\frac{5}{2}(\mathcal{Y}_R^j)^2-\frac{2}{3}\text{Tr}(\mathcal{Y}_R^2)\right)\int d^D x\, c\,\bar{\partial}_\mu(\overline{\psi}\bar{\gamma}^\mu \mathbb{P}_R \psi).\end{aligned} \tag{400}$$

Compared to the previous Section 7.4.3, we this time provided the full two-loop result explicitly and see that, in contrast to the one-loop case (377), this BRST breaking contains a non-evanescent contribution given by the last line of (400).

Following the restoration procedure described in Sections 6.3 or 7.3 and analogous to the ghost–gauge boson contribution (392) in the previous Section 7.4.3, we additionally need to calculate $([\widehat{\Delta}+\Delta^1_{\text{ct}}]\cdot\widetilde{\Gamma})^2$ for the ghost–fermion–fermion, the ghost–double gauge boson, and the ghost–triple gauge boson contributions (i.e. with external fields $c\psi\overline{\psi}$, cAA, $cAAA$, respectively). It turns out that the ghost–double gauge boson contribution vanishes and the ghost–triple gauge boson contribution does not contain UV divergences, but only finite terms. In total the result is

$$\begin{aligned}\left([\widehat{\Delta}+\Delta^1_{\text{ct}}]\cdot\Gamma\right)^2 = \frac{e^4}{256\pi^4}\int d^D x\\ \Bigg\{ -\frac{\text{Tr}(\mathcal{Y}_R^4)}{6}\left[\left(\frac{1}{\epsilon^2}-\frac{17}{12\epsilon}\right)c\,\bar{\partial}_\mu\widehat{\partial}^2\bar{A}^\mu - \frac{11}{4}c\,\bar{\partial}_\mu\bar{\partial}^2\bar{A}^\mu\right]\\ +e\sum_j\frac{(\mathcal{Y}_R^j)^3}{3}\left[\frac{1}{\epsilon}\left(\frac{5}{2}(\mathcal{Y}_R^j)^2-\frac{2}{3}\text{Tr}(\mathcal{Y}_R^2)\right)\right.\\ \left. +\frac{127}{12}(\mathcal{Y}_R^j)^2-\frac{1}{9}\text{Tr}(\mathcal{Y}_R^2)\right]c\,\bar{\partial}_\mu(\overline{\psi}_j\bar{\gamma}^\mu\mathbb{P}_R\psi_j)\\ +\frac{3e^2\text{Tr}(\mathcal{Y}_R^6)}{2}c\,\bar{\partial}_\mu(\bar{A}^\mu\bar{A}_\nu\bar{A}^\nu)\Bigg\} + \mathcal{O}(\hat{\cdot})\end{aligned} \tag{401}$$

for the full two-loop breaking of the Slavnov–Taylor identity of two-loop subrenormalized 1PI Green functions. Comparing this with the corresponding one-loop contribution (386), we see that the structure of the terms is the same.

For the symmetry restoration at the two-loop level, we first note that Δ^2_{sct} in Equation (400) completely cancels the UV divergent terms in Equation (401). In addition to that, we need to determine the finite, symmetry-restoring counterterms at the two-loop as indicated in Equation (394). Thus, our choice for the full finite counterterm action, which restores the Slavnov–Taylor identity at the two-loop level, is

$$\begin{aligned}S^2_{\text{fct}} = \left(\frac{\hbar}{16\pi^2}\right)^2\int d^D x\, e^4\Bigg\{\text{Tr}(\mathcal{Y}_R^4)\frac{11}{48}\bar{A}_\mu\bar{\partial}^2\bar{A}^\mu + 3e^2\frac{\text{Tr}(\mathcal{Y}_R^6)}{8}\bar{A}_\mu\bar{A}^\mu\bar{A}_\nu\bar{A}^\nu\\ -\sum_j(\mathcal{Y}_R^j)^2\left(\frac{127}{36}(\mathcal{Y}_R^j)^2-\frac{1}{27}\text{Tr}(\mathcal{Y}_R^2)\right)\left(\overline{\psi}_j i\bar{\partial\!\!\!/}\mathbb{P}_R\psi_j\right)\Bigg\}.\end{aligned} \tag{402}$$

Similar to its one-loop counterpart in Equation (387), S^2_{fct} consists of three kinds of terms, or in other words, the same three field monomials are involved. These three terms correspond to the restoration of the Ward identity relations for the photon self-energy, the photon four-point function, and the fermion self-energy/photon–fermion–fermion interaction. Reference [26] also gave a discussion of the explicit results for these three Ward identity relations, similar to the discussion at the end of Section 7.4.3. In all cases, the breaking terms of the Ward identity are explicitly exhibited, and the cancellation with the symmetry-restoring counterterms (402) is made manifest.

7.5. Non-Abelian Chiral Yang–Mills Theory and Comparison with the Abelian Chiral Theory at the One-Loop Level

In this section, we review the application of the BMHV scheme to non-Abelian chiral gauge theories and present the differences to the Abelian chiral QED discussed above. In particular, we study a massless chiral Yang–Mills theory at the one-loop level based on References [25,131]. Note that, in our publication [25], the considered theory also contained real scalar fields. Here, similar to Reference [131], scalar fields were omitted in order to focus on the key points of the BMHV scheme in the framework of chiral gauge theories and the differences compared to the Abelian case discussed above.

As discussed in Section 2.1, the group generators of the Yang–Mills theories satisfy the nontrivial commutation relations (1); in particular, they are not simultaneously diagonalizable. These algebraic structures of the non-Abelian gauge group of Yang–Mills theories lead to new effects, such as more interaction terms and nonlinear BRST transformations of the gauge fields and the ghosts, compared to the Abelian case; cf. Section 2.6. Especially, gauge boson self-interactions, interactions of the Faddeev–Popov ghosts with the rest of the theory, and the renormalization of the BRST transformations distinguish non-Abelian Yang–Mills theories from the Abelian case above.

The outline of this section is analogous to the Abelian case discussed above. First, we briefly introduce the Lagrangian of the theory and the BRST transformations using the notations from Section 2. Second, we discuss the analytical continuation of the theory to D dimensions in DReg treating γ_5 with the BMHV scheme and comment on the BRST breaking induced by this scheme. Finally, we present the results for the singular and the symmetry restoring counterterms at the one-loop level (cf. [25,131]) necessary to consistently renormalize the theory, and discuss the differences to the Abelian theory.

7.5.1. Definition of the Non-Abelian Chiral Yang–Mills Theory

Following the conventions of Section 2.3, the Lagrangian in four dimensions can be written as

$$\mathcal{L}_{\chi\text{YM}} = \mathcal{L}_{\text{inv}} + \mathcal{L}_{\text{fix,gh}} + \mathcal{L}_{\text{ext}}\,. \tag{403}$$

The physical part of the Yang–Mills Lagrangian reads

$$\mathcal{L}_{\text{inv}} = -\frac{1}{4}G^a_{\mu\nu}G^{a,\mu\nu} + i\,\overline{\psi_R}_i \not{D}_{ij}\psi_{Rj}\,, \tag{404}$$

with covariant derivative $D^\mu_{ij} = \partial^\mu\delta_{ij} + igG^{a,\mu}\,T^a_{Rij}$ and field strength tensor $G^a_{\mu\nu} = \partial_\mu G^a_\nu - \partial_\nu G^a_\mu - gf^{abc}G^b_\mu G^c_\nu$, leading to three- and four-point gauge boson self-interactions. The gauge-fixing and ghost Lagrangian, already presented in Equation (55), is

$$\mathcal{L}_{\text{fix,gh}} = s\left[\bar{c}^a\left((\partial^\mu G^a_\mu) + \frac{\xi}{2}B^a\right)\right] = B^a(\partial^\mu G^a_\mu) + \frac{\xi}{2}B^aB^a - \bar{c}^a\partial^\mu D^{ab}_\mu c^b, \tag{405}$$

with $D^{ab}_\mu = \partial_\mu\delta^{ab} + gf^{abc}G^c_\mu$, implying ghost–antighost–gauge boson interactions, which is a consequence of the nonlinear gauge transformations of the gauge fields G^a_μ, as shown below in (407). The Lagrangian of the external sources, as introduced in Section 2.3, is

$$\mathcal{L}_{\text{ext}} = \rho^{a,\mu}sG^a_\mu + \zeta^a sc^a + \bar{R}^i s\psi_{Ri} + R^i s\overline{\psi_R}_i\,. \tag{406}$$

The BRST transformations are given by

$$sG^a_\mu(x) = D^{ab}_\mu c^b(x) = \partial_\mu c^a(x) + g f^{abc} c^b(x) G^c_\mu(x)\,, \tag{407a}$$
$$s\psi_i(x) = s\psi_{R_i}(x) = -igT^a_{Rij} c^a(x)\psi_{R_j}(x)\,, \tag{407b}$$
$$s\overline{\psi}_i(x) = s\overline{\psi_{R_i}}(x) = -ig\overline{\psi_{R_j}}(x) c^a(x) T^a_{Rji}\,, \tag{407c}$$
$$sc^a(x) = \frac{1}{2} g f^{abc} c^b(x) c^c(x)\,, \tag{407d}$$
$$s\bar{c}^a(x) = B^a(x)\,, \tag{407e}$$
$$sB^a(x) = 0\,. \tag{407f}$$

In contrast to the Abelian case, the BRST transformations of the gauge boson G^a_μ and the Faddeev–Popov ghost c^a are nonlinear, which means that nontrivial quantum corrections are expected.

Hence, the tree-level action of the considered chiral Yang–Mills theory in four dimensions is given by

$$S_0^{(4D)} = \int d^4x\, \mathcal{L}_{\chi\text{YM}} \tag{408}$$

and satisfies the tree-level Slavnov–Taylor identity:

$$\begin{aligned} 0 &= \mathcal{S}(S_0^{(4D)}) \\ &= \int d^4x \left(\frac{\delta S_0^{(4D)}}{\delta\rho^{a\mu}(x)} \frac{\delta S_0^{(4D)}}{\delta G^a_\mu(x)} + \frac{\delta S_0^{(4D)}}{\delta\zeta^a(x)} \frac{\delta S_0^{(4D)}}{\delta c^a(x)} \right. \\ &\qquad \left. + \frac{\delta S_0^{(4D)}}{\delta\bar{R}^i(x)} \frac{\delta S_0^{(4D)}}{\delta\psi_{R_i}(x)} + \frac{\delta S_0^{(4D)}}{\delta R^i(x)} \frac{\delta S_0^{(4D)}}{\delta\overline{\psi_{R_i}}(x)} + B^a(x) \frac{\delta S_0^{(4D)}}{\delta\bar{c}^a(x)} \right), \end{aligned} \tag{409}$$

which just manifests the BRST invariance of $S_0^{(4D)}$.

The different group invariants, which will be employed in the following results below, (follow the notations of [25]) and are provided by

$$C_2(R)\,\mathbb{1} = T^a_R T^a_R\,, \qquad\qquad S_2(R)\,\delta^{ab} = \text{Tr}(T^a_R T^b_R), \tag{410}$$

with an irreducible representation R of the gauge group for the right-handed fermions with corresponding Hermitian group generators T^a_R. The adjoint representation of the gauge group is denoted by G, and its Casimir index is $C_2(G)$.

7.5.2. Chiral Yang–Mills Theory in DReg

To regularize the theory, we employ dimensional regularization, treating γ_5 with the BMHV scheme. Analogous to the Abelian case above, there are two problems regarding the continuation of the chiral Yang–Mills theory (408) to D dimensions, as already discussed in Section 7.2.2 for the Abelian case and extensively discussed in [25] for chiral Yang–Mills theories.

First, there is an ambiguity in extending the fermion–gauge interaction term in Equation (404), which involves the right-handed chiral current $\overline{\psi_{R_i}}\gamma^\mu\psi_{R_j}$, to D dimensions. Again, there are three inequivalent choices for the D-dimensional version of this chiral current (cf. Equation (349)), which are all equally correct. Analogous to the Abelian case above, we resolved this problem by choosing the most-symmetric version; cf., Equation (350).

Second, the purely fermionic kinetic term $i\overline{\psi_{R_i}}\partial\!\!\!/\psi_{R_i}$ projects only the purely 4-dimensional derivative, leading to a purely 4-dimensional propagator and, thus, to unregularized loop diagrams, as explained above in Section 7.2.2. Hence, we again introduce a gauge-singlet left-chiral field ψ_L with trivial BRST transformations:

$$s\psi_{L_i}(x) = 0, \qquad s\overline{\psi_{L_i}}(x) = 0, \tag{411}$$

which appears solely in the fermionic kinetic term and nowhere else and which is thus completely decoupled from the rest of the theory. Using it we obtain a fully D-dimensional covariant kinetic term $i\overline{\psi}_i \partial\!\!\!/ \psi_i$.

Finally, we can again separate the D-dimensional fermionic Lagrangian into an invariant and an evanescent part, analogous to Equations (354a)–(357b). Hence, we may write the D-dimensional action as

$$\begin{aligned} S_0 &= S_{0,\text{inv}} + S_{0,\text{evan}} \\ &= (S_{GG} + S_{GGG} + S_{GGGG}) + \sum_i \left(S^i_{\overline{\psi}\psi} + \overline{S^i_{\overline{\psi}_R G \psi_R}} \right) + S_{\text{g-fix}} \\ &\quad + (S_{\bar{c}c} + S_{\bar{c}Gc}) + (S_{\rho c} + S_{\rho Gc} + S_{\zeta cc} + S_{\bar{R}c\psi_R} + S_{Rc\overline{\psi_R}}), \end{aligned} \tag{412}$$

having it separated into an "invariant" and an "evanescent" part in the first line (cf. Equation (359) in Section 7.2.2) and having used the notation of [25,26] and of Equation (358) to present the D-dimensional action as a sum of its integrated field monomials in the last two lines.

Similar to the Abelian case in Section 7.2.2, we quantify the symmetry breaking caused by the BMHV scheme, the non-anticommuting γ_5, and the evanescent term $S_{0,\text{evan}}$ by acting with the D-dimensional BRST operator s_D on the D-dimensional tree-level action S_0. Thus, for the BRST breaking, we obtain

$$s_D S_0 = s_D S_{0,\text{inv}} + s_D S_{0,\text{evan}} = 0 + s_D \int d^D x\, i\overline{\psi}_i \widehat{\partial\!\!\!/} \psi_i \equiv \widehat{\Delta}, \tag{413}$$

which leads to a breaking of the Slavnov–Taylor identity of the form:

$$\mathcal{S}_D(S_0) = \widehat{\Delta}, \tag{414}$$

with the nonvanishing integrated breaking:

$$\widehat{\Delta} = -\int d^D x g\, T^a_{Rij}\, c^a \left\{ \overline{\psi}_i \left(\overleftarrow{\widehat{\partial\!\!\!/}} \mathbb{P}_R + \overrightarrow{\widehat{\partial\!\!\!/}} \mathbb{P}_L \right) \psi_j \right\} \equiv \int d^D x\, \widehat{\Delta}(x). \tag{415}$$

As in the Abelian case, this breaking term will be a crucial tool in practical calculations and will be used as a composite operator insertion in the Feynman diagrams. It generates an interaction vertex whose Feynman rule (with all momenta incoming) is

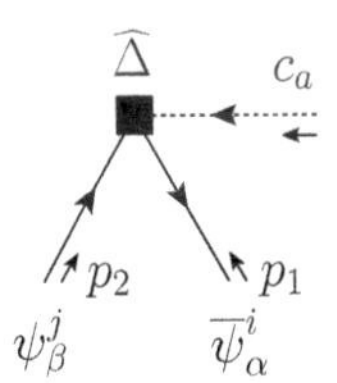

$$\begin{aligned} &= -\frac{g}{2} T^a_{Rij} \left((\widehat{p\!\!\!/_1} + \widehat{p\!\!\!/_2}) + (\widehat{p\!\!\!/_1} - \widehat{p\!\!\!/_2})\gamma_5 \right)_{\alpha\beta} \\ &= -g\, T^a_{Rij} (\widehat{p\!\!\!/_1} \mathbb{P}_R + \widehat{p\!\!\!/_2} \mathbb{P}_L)_{\alpha\beta}. \end{aligned} \tag{416}$$

For charge-conjugated fermions, an analogous Feynman rule can be derived.

7.5.3. One-Loop Singular Counterterm and Symmetry-Restoring Counterterm Action in Chiral Yang–Mills Theory

In this subsection, we present the results of the one-loop renormalization of the above introduced chiral Yang–Mills theory based on the results of [25], but also already discussed in [131].[18]

The basic renormalization procedure is the same as in the Abelian theory discussed above. The difference is that there are more interaction terms; in particular, the gauge bosons interact with themselves and the Faddeev–Popov ghosts. The fact that the ghosts now participate in the interactions, and thus may propagate as internal particles in loop diagrams, leads to a nontrivial renormalization of the field monomials including external sources. Besides this, the renormalization procedure is also more demanding than in an Abelian theory, due to the larger number of loop diagrams and the more complicated algebraic structures of the non-Abelian gauge group.

After computing all UV divergent one-loop 1PI Feynman diagrams, which can be found in Section 5 of [25] with detailed individual results, the singular one-loop counterterm action is given by

$$\begin{aligned} S_{\text{sct}}^{(1)} = \frac{\hbar g^2}{16\pi^2\epsilon}\bigg\{ & -\frac{2S_2(R)}{3}\big(\overline{S_{GG}} + \overline{S_{GGG}} + \overline{S_{GGGG}}\big) - \xi C_2(R)\big(\overline{S_{\overline{\psi}\psi_R}} + \overline{S_{\overline{\psi}G\psi_R}}\big) \\ & + \frac{13-3\xi}{6} C_2(G) S_{GG} + \frac{17-9\xi}{12} C_2(G) S_{GGG} + \frac{2-3\xi}{3} C_2(G) S_{GGGG} \\ & - \frac{3+\xi}{4} C_2(G)\overline{S_{\overline{\psi}G\psi_R}} + \frac{3-\xi}{4} C_2(G)\big(S_{\bar{c}c} + S_{\rho c}\big) \\ & - \frac{\xi C_2(G)}{2}\big(S_{\bar{c}Gc} + S_{\rho Gc} + S_{\zeta cc} + S_{\bar{R}c\psi_R} + S_{Rc\overline{\psi}_R}\big)\bigg\} \\ & - \frac{\hbar g^2}{16\pi^2\epsilon}\frac{S_2(R)}{3}\int d^4x\, \frac{1}{2}\bar{G}^{a,\mu}\hat{\partial}^2\bar{G}^a_\mu, \end{aligned} \tag{417}$$

such that it cancels all UV divergences. The structure has similarities with the Abelian counterpart, Equation (385). Again, most terms can be obtained by a renormalization transformation of the kind (291), and only the right-handed fermions renormalize. However, again, also non-symmetric singular counterterms appear.

Comparing Equations (417) and (385) in detail, we can see many additional contributions. Only the $\overline{S_{GG}}$, $\overline{S_{\overline{\psi}\psi_R}}$ and $\overline{S_{\overline{\psi}G\psi_R}}$ terms in the first line of the RHS of (417), as well as the explicit evanescent operator in last line of (417) have Abelian counterparts. All other terms in (417) do not appear in the Abelian theory and are, thus, new effects of the non-Abelian Yang–Mills theory due to additional interaction terms, as mentioned above. In particular, we can see new contributions to the field monomials including the Faddeev–Popov ghosts and the external sources in the last term of the third line and the penultimate line of (417), as announced at the beginning of this subsection.

Similar to the Abelian result (385), we have just one explicit evanescent operator in the last line of (417) in the considered Yang–Mills theory, generating the Feynman rule $-i\hat{p}^2\hat{g}_{\mu\nu}\delta^{ab}$. This is specific to our choice for the fermion-gauge interaction term, corresponding to the most symmetric version of Equation (350). We would have obtained many more evanescent operators if we used another D-dimensional choice instead.

Following the algebraic renormalization procedure described in Section 6, as well as in Section 6 of [25], specifically for the considered case, we need to check that

$$0 = \underset{D\to 4}{\text{LIM}}\left(\big[\hat{\Delta}\cdot\Gamma^{(1)}\big]^{(1)}_{\text{div}} + b_D S^{(1)}_{\text{sct}} + \big[\hat{\Delta}\cdot\Gamma^{(1)}\big]^{(1)}_{\text{fin}} + b_D S^{(1)}_{\text{fct,restore}}\right). \tag{418}$$

[18] Note the different sign convention with respect to the covariant derivative D^μ_{ij} in this review compared to [25]. This influences some signs, such as the relative sign in the forthcoming Equation (420) and the relative sign in the brackets of the last term of Equation (421).

In other words, we need to check that the b_D-variation of the singular counterterms (417) cancels the divergent part of the symmetry breaking $[\widehat{\Delta}\cdot\Gamma^{(1)}]^{(1)}_{\text{div}}$, and we need to determine finite symmetry-restoring counterterms $S^{(1)}_{\text{fct,restore}}$ whose b_D-variation cancels the finite part of the symmetry breaking $[\widehat{\Delta}\cdot\Gamma^{(1)}]^{(1)}_{\text{fin}}$.

The b_D-variation of the singular counterterms (417), calculated in [25], is provided by

$$b_D S^{(1)}_{\text{sct}} = \frac{-\hbar}{16\pi^2\epsilon}\left\{g^2\frac{\xi C_2(G)}{2}\widehat{\Delta} + g^2\frac{S_2(R)}{3}b_D\int d^Dx\,\frac{1}{2}\bar{G}^{a,\mu}\widehat{\partial}^2\bar{G}^a_\mu\right\}, \tag{419}$$

where, in the last term, b_D acts like the BRST transformation, leading to

$$\begin{aligned} b_D\int d^Dx\,\frac{1}{2}\bar{G}^{a,\mu}\widehat{\partial}^2\bar{G}^a_\mu &= \int d^Dx\,(s_D\bar{G}^{a,\mu})\widehat{\partial}^2\bar{G}^a_\mu \\ &= \int d^Dx\,\left(\bar{\partial}^\mu c^a - g f^{abc}\bar{G}^{b,\mu}c^c\right)\widehat{\partial}^2\bar{G}^a_\mu. \end{aligned} \tag{420}$$

Indeed, (419) is a pure $1/\epsilon$ singular term and perfectly cancels the nonvanishing contribution:

$$[\widehat{\Delta}\cdot\Gamma]^{(1)}_{\text{div}} = \frac{1}{16\pi^2\epsilon}\left\{g^2\frac{\xi C_2(G)}{2}\widehat{\Delta} + g^2\frac{S_2(R)}{3}\int d^Dx\,\left(\bar{\partial}^\mu c^a - g f^{abc}\bar{G}^{b,\mu}c^c\right)\widehat{\partial}^2\bar{G}^a_\mu\right\}, \tag{421}$$

as explicitly shown in [25].

Now, the finite symmetry-restoring counterterms $S^{(1)}_{\text{fct,restore}}$ need to be determined following (418) in order to cancel the remaining finite part of the symmetry breaking, which was explicitly performed in Section 6 of [25] with the result:

$$\begin{aligned} S^{(1)}_{\text{fct, restore}} = \frac{\hbar}{16\pi^2}\bigg\{ & g^2\frac{S_2(R)}{6}\left(5S_{GG} - \int d^4x\,G^{a,\mu}\partial^2G^a_\mu\right) \\ & + g^2\frac{(T_R)^{abcd}}{3}\int d^4x\,\frac{g^2}{4}G^a_\mu G^{b,\mu}G^c_\nu G^{d,\nu} + g^2\left(1+\frac{\xi-1}{6}\right)C_2(R)S_{\overline{\psi}\psi} \\ & + g^2\frac{S_2(R)}{6}S_{GGG} - g^2\frac{\xi C_2(G)}{4}\left(S_{\bar{R}c\psi_R} + S_{Rc\overline{\psi_R}}\right)\bigg\}, \end{aligned} \tag{422}$$

where $(T_R)^{a_1\cdots a_n} \equiv \text{Tr}[T_R^{a_1}\cdots T_R^{a_n}]$. Comparing (422) with the Abelian result (387), we can again see that only the first two lines of (422) have Abelian counterparts, whereas the terms in the last line of (422) do not appear in an Abelian theory. The new terms in the last line of (422) are due to triple gauge boson contributions and contributions including external sources. The latter implies that, again, Green functions with external sources have to been evaluated, this time with a $\widehat{\Delta}$-vertex insertion, which stands in contrast to the Abelian case.

These finite counterterms (422) are necessary and sufficient to restore the BRST symmetry at the one-loop level in the BMHV scheme, if the (non-spurious) anomalies cancel, which are given by [25]

$$\begin{aligned} -\frac{g^2}{16\pi^2}\bigg(& -\frac{S_2(R)}{3}d_R^{abc}\int d^4x\,g\epsilon^{\mu\nu\rho\sigma}c^a(\partial_\rho G^b_\mu)(\partial_\sigma G^c_\nu) \\ & + \frac{\mathcal{D}_R^{abcd}}{3\times 3!}\int d^4x\,g^2c^a\epsilon^{\mu\nu\rho\sigma}\partial_\sigma\left(G^b_\mu G^c_\nu G^d_\rho\right)\bigg), \end{aligned} \tag{423}$$

with fully symmetric $d_R^{abc} \equiv \text{Tr}[T_R^a\{T_R^b, T_R^c\}]$ and fully antisymmetric $\mathcal{D}_R^{abcd} \equiv (-i)3!\text{Tr}[T_R^aT_R^{[b}T_R^cT_R^{d]}]$ for the R-representation. This result, of course, agrees with the general result (318) obtained by the analysis of algebraic renormalization, and it provides an explicit result for the coefficient L appearing there. To ensure the renormalizability of the theory, the fermionic content and their associated group representations have to be chosen such that these anomalies cancel, i.e., such that the expression (321) vanishes, which equivalently means that d_R^{abc} vanishes. This then also implies the vanishing of $\mathcal{D}_R^{abcd}$; see Equation (320). It becomes apparent that also the possible anomalies are more complex than in the Abelian model.

These finite counterterms (422), purely four-dimensional and non-evanescent, are not gauge-invariant. They modify all self-energies, as well as some specific interactions: the gauge boson self-interactions and the interactions between gauge bosons and fermions.

Concluding, we see that the resulting counterterm action, not only for the Abelian case at the one- and two-loop level, but also for non-Abelian Yang–Mills theories, may be written in a relatively compact way. Thus, treating γ_5 rigorously in the BMHV scheme does not lead to extraordinarily lengthy or complicated results, but, in fact, to counterterms, which can easily be implemented in computer algebra systems.

Author Contributions: Conceptualization, H.B.-M., A.I., P.K., M.M.-B., D.S. and M.W.; writing, H.B.-M., A.I., P.K., M.M.-B., D.S. and M.W. All authors have read and agreed to the published version of the manuscript.

Funding: This research was partially funded by the Croatian Science Foundation (HRZZ) under the project "PRECIOUS" ("Precise Computations of Physical Observables in Supersymmetric Models") number | HRZZ-IP-2016-06-7460 |, and by the German Science Foundation DFG, Grant STO 876/8-1.

Institutional Review Board Statement: Not applicable.

Informed Consent Statement: Not applicable.

Data Availability Statement: Data sharing not applicable.

Acknowledgments: A.I., H.B., and M.M. acknowledge financial support from the Croatian Science Foundation (HRZZ) under the project "PRECIOUS" ("Precise Computations of Physical Observables in Supersymmetric Models") number | HRZZ-IP-2016-06-7460 |. P.K., D.S., and M.W. acknowledge financial support by the German Science Foundation DFG, Grant STO 876/8-1.

Conflicts of Interest: The authors declare no conflict of interest.

References

1. 't Hooft, G.; Veltman, M.J.G. Regularization and Renormalization of Gauge Fields. *Nucl. Phys. B* **1972**, *44*, 189–213. [CrossRef]
2. Bollini, C.G.; Giambiagi, J.J. Dimensional Renormalization: The Number of Dimensions as a Regularizing Parameter. *Nuovo Cim. B* **1972**, *12*, 20–26. [CrossRef]
3. Ashmore, J.F. A Method of Gauge Invariant Regularization. *Lett. Nuovo Cim.* **1972**, *4*, 289–290. [CrossRef]
4. Breitenlohner, P.; Maison, D. Dimensional Renormalization and the Action Principle. *Commun. Math. Phys.* **1977**, *52*, 11–38. [CrossRef]
5. Gnendiger, C.; Signer, A.; Stöckinger, D.; Broggio, A.; Cherchiglia, A.L.; Driencourt-Mangin, F.; Fazio, A.R.; Hiller, B.; Mastrolia, P.; Peraro, T.; Pittau, R. To *d*, or not to *d*: Recent developments and comparisons of regularization schemes. *Eur. Phys. J. C* **2017**, *77*, 471. [CrossRef]
6. Adler, S.L. Axial vector vertex in spinor electrodynamics. *Phys. Rev.* **1969**, *177*, 2426–2438. [CrossRef]
7. Bell, J.S.; Jackiw, R. A PCAC puzzle: $\pi^0 \to \gamma\gamma$ in the σ model. *Nuovo Cim. A* **1969**, *60*, 47–61. [CrossRef]
8. Adler, S.L.; Bardeen, W.A. Absence of higher order corrections in the anomalous axial vector divergence equation. *Phys. Rev.* **1969**, *182*, 1517–1536. [CrossRef]
9. Bouchiat, C.; Iliopoulos, J.; Meyer, P. An Anomaly Free Version of Weinberg's Model. *Phys. Lett. B* **1972**, *38*, 519–523. [CrossRef]
10. Gross, D.J.; Jackiw, R. Effect of anomalies on quasirenormalizable theories. *Phys. Rev. D* **1972**, *6*, 477–493. [CrossRef]
11. Geng, C.Q.; Marshak, R.E. Uniqueness of Quark and Lepton Representations in the Standard Model from the Anomalies Viewpoint. *Phys. Rev. D* **1989**, *39*, 693. [CrossRef]
12. Fujikawa, K. Path Integral Measure for Gauge Invariant Fermion Theories. *Phys. Rev. Lett.* **1979**, *42*, 1195–1198. [CrossRef]
13. Fujikawa, K. Path Integral for Gauge Theories with Fermions. *Phys. Rev. D* **1980**, *21*, 2848, Erratum in *Phys. Rev. D* **1980**, *22*, 1499. [CrossRef]
14. Chanowitz, M.S.; Furman, M.; Hinchliffe, I. The Axial Current in Dimensional Regularization. *Nucl. Phys. B* **1979**, *159*, 225–243. [CrossRef]
15. Jegerlehner, F. Facts of life with gamma(5). *Eur. Phys. J. C* **2001**, *18*, 673–679. [CrossRef]
16. Kreimer, D. The $\gamma(5)$ Problem and Anomalies: A Clifford Algebra Approach. *Phys. Lett. B* **1990**, *237*, 59–62. [CrossRef]
17. Korner, J.G.; Kreimer, D.; Schilcher, K. A Practicable gamma(5) scheme in dimensional regularization. *Z. Phys. C* **1992**, *54*, 503–512. [CrossRef]
18. Kreimer, D. The Role of gamma(5) in dimensional regularization. *arXiv* **1993**, arXiv:hep-ph/9401354.

19. Cicuta, G.M.; Montaldi, E. Analytic renormalization via continuous space dimension. *Lett. Nuovo Cim.* **1972**, *4*, 329–332. [CrossRef]
20. Akyeampong, D.A.; Delbourgo, R. Dimensional regularization and PCAC. *Nuovo Cim. A* **1973**, *18*, 94–104. [CrossRef]
21. Akyeampong, D.A.; Delbourgo, R. Dimensional regularization, abnormal amplitudes and anomalies. *Nuovo Cim. A* **1973**, *17*, 578–586. [CrossRef]
22. Speer, E.R. Renormalization and ward identities using complex spacetime dimension. *J. Math. Phys.* **1974**, *15*, 1–6. [CrossRef]
23. Breitenlohner, P.; Maison, D. Dimensionally Renormalized Green's Functions for Theories with Massless Particles. 1. *Commun. Math. Phys.* **1977**, *52*, 39. [CrossRef]
24. Breitenlohner, P.; Maison, D. Dimensionally Renormalized Green's Functions for Theories with Massless Particles. 2. *Commun. Math. Phys.* **1977**, *52*, 55. [CrossRef]
25. Bélusca-Maïto, H.; Ilakovac, A.; Mađor-Božinović, M.; Stöckinger, D. Dimensional regularization and Breitenlohner–Maison/'t Hooft–Veltman scheme for γ_5 applied to chiral YM theories: Full one-loop counterterm and RGE structure. *J. High Energy Phys.* **2020**, *8*, 24. [CrossRef]
26. Bélusca-Maïto, H.; Ilakovac, A.; Kühler, P.; Mađor-Božinović, M.; Stöckinger, D. Two-loop application of the Breitenlohner–Maison/'t Hooft-Veltman scheme with non-anticommuting γ_5: Full renormalization and symmetry-restoring counterterms in an Abelian chiral gauge theory. *J. High Energy Phys.* **2021**, *11*, 159. [CrossRef]
27. Cornella, C.; Feruglio, F.; Vecchi, L. Gauge Invariance and Finite Counterterms in Chiral Gauge Theories. *arXiv* **2022**, arXiv:hep-ph/2205.10381.
28. Bélusca-Maïto, H. Renormalisation Group Equations for BRST-Restored Chiral Theory in Dimensional Renormalisation: Application to Two-Loop Chiral-QED. *arXiv* **2022**, arXiv:hep-th/2208.09006.
29. Bélusca-Maïto, H.; Ilakovac, A.; Mađor-Božinović, M.; Kühler, P.; Stöckinger, D. Gamma5 in dimensional regularization - a no-compromise approach using the BMHV scheme. *PoS* **2022**, *LL2022*, 11. [CrossRef]
30. Cheng, T.P.; Li, L.F. *Gauge theory of elementary particle physics*; Oxford University Press: Oxford, UK, 1984.
31. Weinberg, S. *The Quantum Theory of Fields*; Cambridge University Press: Cambridge, UK, 2005; Volume 1: Foundations.
32. Weinberg, S. *The Quantum Theory of Fields*; Cambridge University Press: Cambridge, UK, 2013; Volume 2: Modern Applications.
33. Peskin, M.E.; Schroeder, D.V. *An Introduction to Quantum Field Theory*; Addison-Wesley: Reading: Boston, MA, USA, 1995.
34. Bohm, M.; Denner, A.; Joos, H. *Gauge Theories of the Strong and Electroweak Interaction*; Springer: Berlin/Heidelberg, Germany, 2001. [CrossRef]
35. Srednicki, M. *Quantum Field Theory*; Cambridge University Press: Cambridge, UK, 2007.
36. Schwartz, M.D. *Quantum Field Theory and the Standard Model*, 10th ed.; Cambridge University Press: Cambridge, UK, 2018.
37. Wigner, E.P. On Unitary Representations of the Inhomogeneous Lorentz Group. *Ann. Math.* **1939**, *40*, 149–204. [CrossRef]
38. Ryder, L.H. *Quantum Field Theory*; Cambridge University Press: Cambridge, UK, 1996.
39. Dreiner, H.K.; Haber, H.E.; Martin, S.P. Two-component spinor techniques and Feynman rules for quantum field theory and supersymmetry. *Phys. Rept.* **2010**, *494*, 1–196. [CrossRef]
40. Faddeev, L.D.; Popov, V.N. Feynman Diagrams for the Yang–Mills Field. *Phys. Lett. B* **1967**, *25*, 29–30. [CrossRef]
41. Becchi, C.; Rouet, A.; Stora, R. The Abelian Higgs-Kibble Model. Unitarity of the S Operator. *Phys. Lett. B* **1974**, *52*, 344–346. [CrossRef]
42. Becchi, C.; Rouet, A.; Stora, R. Renormalization of the Abelian Higgs-Kibble Model. *Commun. Math. Phys.* **1975**, *42*, 127–162. [CrossRef]
43. Becchi, C.; Rouet, A.; Stora, R. Renormalization of Gauge Theories. *Ann. Phys.* **1976**, *98*, 287–321. [CrossRef]
44. Tyutin, I.V. Gauge Invariance in Field Theory and Statistical Physics in Operator Formalism. *arXiv* **1975**, arXiv:hep-th/0812.0580.
45. Kugo, T.; Ojima, I. Local Covariant Operator Formalism of NonAbelian Gauge Theories and Quark Confinement Problem. *Prog. Theor. Phys. Suppl.* **1979**, *66*, 1–130. [CrossRef]
46. Henneaux, M.; Teitelboim, C. *Quantization of Gauge Systems*; Princeton University Press: Princeton, NJ, USA, 1992.
47. Piguet, O.; Sorella, S.P. *Algebraic Renormalization: Perturbative Renormalization, Symmetries and Anomalies*; Springer: Berlin/Heidelberg, Germany, 1995; Volume 28. [CrossRef]
48. Denner, A.; Weiglein, G.; Dittmaier, S. Application of the background field method to the electroweak standard model. *Nucl. Phys. B* **1995**, *440*, 95–128. [CrossRef]
49. Duncan, A. *The Conceptual Framework of Quantum Field Theory*; Oxford University Press: Oxford, UK, 2012. [CrossRef]
50. Zinn-Justin, J. *Quantum Field Theory and Critical Phenomena*; International Series of Monographs on Physics; Oxford University Press: Oxford, UK, 2021; Volume 77.
51. Itzykson, C.; Zuber, J.B. *Quantum Field Theory*; International Series in Pure and Applied Physics; McGraw-Hill: New York, NY, USA, 1980.
52. Brown, L.S. *Quantum Field Theory*; Cambridge University Press: Cambridge, UK, 1992. [CrossRef]
53. Kraus, E.; Sibold, K. Rigid invariance as derived from BRS invariance: The Abelian Higgs model. *Z. Phys. C* **1995**, *68*, 331–344. [CrossRef]
54. Haussling, R.; Kraus, E. Gauge parameter dependence and gauge invariance in the Abelian Higgs model. *Z. Phys. C* **1997**, *75*, 739–750. [CrossRef]
55. Haussling, R.; Kraus, E.; Sibold, K. Gauge parameter dependence in the background field gauge and the construction of an invariant charge. *Nucl. Phys. B* **1999**, *539*, 691–719. [CrossRef]
56. Grassi, P.A. The Abelian antighost equation for the standard model in the 't Hooft background gauge. *Nucl. Phys. B* **1999**, *537*, 527–548. [CrossRef]

57. Kraus, E. Renormalization of the Electroweak Standard Model to All Orders. *Ann. Phys.* **1998**, *262*, 155–259. [CrossRef]
58. Grassi, P.A. Renormalization of nonsemisimple gauge models with the background field method. *Nucl. Phys. B* **1999**, *560*, 499–550. [CrossRef]
59. Hollik, W.; Kraus, E.; Roth, M.; Rupp, C.; Sibold, K.; Stockinger, D. Renormalization of the minimal supersymmetric standard model. *Nucl. Phys. B* **2002**, *639*, 3–65. [CrossRef]
60. Bohm, M.; Spiesberger, H.; Hollik, W. On the One Loop Renormalization of the Electroweak Standard Model and Its Application to Leptonic Processes. *Fortsch. Phys.* **1986**, *34*, 687–751. [CrossRef]
61. Wilson, K.G. Quantum field theory models in less than four-dimensions. *Phys. Rev. D* **1973**, *7*, 2911–2926. [CrossRef]
62. Speer, E.R.; Westwater, M.J. Generic Feynman Amplitudes. *Ann. IHP Phys. Theor.* **1970**, *14*, 1–55.
63. Collins, J.C. *Renormalization: An Introduction to Renormalization, the Renormalization Group and the Operator-Product Expansion*; Cambridge Monographs on Mathematical Physics; Cambridge University Press: Cambridge, UK, 1984. [CrossRef]
64. Smirnov, V.A. Evaluating Feynman integrals. *Springer Tracts Mod. Phys.* **2004**, *211*, 1–244.
65. Hepp, K. Proof of the Bogolyubov-Parasiuk theorem on renormalization. *Commun. Math. Phys.* **1966**, *2*, 301–326. [CrossRef]
66. Anikin, S.A.; Polivanov, M.K.; Zavyalov, O.I. Simple Proof of the Bogolyubov-Parasiuk Theorem. *Theor. Math. Phys.* **1973**, *17*, 1082. [CrossRef]
67. Bergere, M.C.; Zuber, J.B. Renormalization of feynman amplitudes and parametric integral representation. *Commun. Math. Phys.* **1974**, *35*, 113–140. [CrossRef]
68. Speer, E.R. On the structure of analytic renormalization. *Commun. Math. Phys.* **1971**, *23*, 23–36, Erratum in *Commun. Math. Phys.* **1972**, *25*, 336. [CrossRef]
69. Nakanishi, N. *Graph Theory and Feynman Integrals*; Mathematics and Its applications: A Series of Monographs and Texts; Gordon and Breach: Philadelphia, PA, USA, 1971.
70. Larin, S.A. The Renormalization of the axial anomaly in dimensional regularization. *Phys. Lett. B* **1993**, *303*, 113–118. [CrossRef]
71. Trueman, T.L. Spurious anomalies in dimensional renormalization. *Z. Phys. C* **1996**, *69*, 525–536. [CrossRef]
72. Chetyrkin, K.G.; Misiak, M.; Munz, M. $|\Delta F| = 1$ nonleptonic effective Hamiltonian in a simpler scheme. *Nucl. Phys. B* **1998**, *520*, 279–297. [CrossRef]
73. Bonneau, G. Trace and Axial Anomalies in Dimensional Renormalization Through Zimmermann Like Identities. *Nucl. Phys. B* **1980**, *171*, 477–508. [CrossRef]
74. Bonneau, G. Zimmermann Identities and Renormalization Group Equation in Dimensional Renormalization. *Nucl. Phys. B* **1980**, *167*, 261–284. [CrossRef]
75. Gnendiger, C.; Signer, A. γ_5 in the four-dimensional helicity scheme. *Phys. Rev. D* **2018**, *97*, 096006. [CrossRef]
76. Bruque, A.M.; Cherchiglia, A.L.; Pérez-Victoria, M. Dimensional regularization vs methods in fixed dimension with and without γ_5. *JHEP* **2018**, *8*, 109. [CrossRef]
77. Neubert, M. Renormalization Theory and Effective Field Theories. In *Effective Field Theory in Particle Physics and Cosmology: Lecture Notes of the Les Houches Summer School*; Oxford University Press: Oxford, UK, 2019. . [CrossRef]
78. Stockinger, D. Regularization by dimensional reduction: Consistency, quantum action principle, and supersymmetry. *J. High Energy Phys.* **2005**, *3*, 76. [CrossRef]
79. Signer, A.; Stockinger, D. Using Dimensional Reduction for Hadronic Collisions. *Nucl. Phys. B* **2009**, *808*, 88–120. [CrossRef]
80. Siegel, W. Supersymmetric Dimensional Regularization via Dimensional Reduction. *Phys. Lett. B* **1979**, *84*, 193–196. [CrossRef]
81. Capper, D.M.; Jones, D.R.T.; van Nieuwenhuizen, P. Regularization by Dimensional Reduction of Supersymmetric and Non-supersymmetric Gauge Theories. *Nucl. Phys. B* **1980**, *167*, 479–499. [CrossRef]
82. Siegel, W. Inconsistency of Supersymmetric Dimensional Regularization. *Phys. Lett. B* **1980**, *94*, 37–40. [CrossRef]
83. Jack, I.; Jones, D.R.T.; Roberts, K.L. Dimensional reduction in nonsupersymmetric theories. *Z. Phys. C* **1994**, *62*, 161–166. [CrossRef]
84. Jack, I.; Jones, D.R.T.; Roberts, K.L. Equivalence of dimensional reduction and dimensional regularization. *Z. Phys. C* **1994**, *63*, 151–160. [CrossRef]
85. van Damme, R.; 't Hooft, G. Breakdown of Unitarity in the Dimensional Reduction Scheme. *Phys. Lett. B* **1985**, *150*, 133–138. [CrossRef]
86. Harlander, R.V.; Jones, D.R.T.; Kant, P.; Mihaila, L.; Steinhauser, M. Four-loop beta function and mass anomalous dimension in dimensional reduction. *J. High Energy Phys.* **2006**, *12*, 24. [CrossRef]
87. Kilgore, W.B. Regularization Schemes and Higher Order Corrections. *Phys. Rev. D* **2011**, *83*, 114005. [CrossRef]
88. Kunszt, Z.; Signer, A.; Trocsanyi, Z. One loop helicity amplitudes for all 2 —> 2 processes in QCD and N=1 supersymmetric Yang–Mills theory. *Nucl. Phys. B* **1994**, *411*, 397–442. [CrossRef]
89. Catani, S.; Seymour, M.H.; Trocsanyi, Z. Regularization scheme independence and unitarity in QCD cross-sections. *Phys. Rev. D* **1997**, *55*, 6819–6829. [CrossRef]
90. Catani, S.; Dittmaier, S.; Trocsanyi, Z. One loop singular behavior of QCD and SUSY QCD amplitudes with massive partons. *Phys. Lett. B* **2001**, *500*, 149–160. [CrossRef]
91. Beenakker, W.; Kuijf, H.; van Neerven, W.L.; Smith, J. QCD Corrections to Heavy Quark Production in p anti-p Collisions. *Phys. Rev. D* **1989**, *40*, 54–82. [CrossRef]
92. Beenakker, W.; Hopker, R.; Zerwas, P.M. SUSY QCD decays of squarks and gluinos. *Phys. Lett. B* **1996**, *378*, 159–166. [CrossRef]

93. Smith, J.; van Neerven, W.L. The Difference between n-dimensional regularization and n-dimensional reduction in QCD. *Eur. Phys. J. C* **2005**, *40*, 199–203. [CrossRef]
94. Signer, A.; Stockinger, D. Factorization and regularization by dimensional reduction. *Phys. Lett. B* **2005**, *626*, 127–138. [CrossRef]
95. Kilgore, W.B. The Four Dimensional Helicity Scheme Beyond One Loop. *Phys. Rev. D* **2012**, *86*, 014019. [CrossRef]
96. Broggio, A.; Gnendiger, C.; Signer, A.; Stöckinger, D.; Visconti, A. Computation of $H \to gg$ in $DRED$ and FDH: Renormalization, operator mixing, and explicit two-loop results. *Eur. Phys. J. C* **2015**, *75*, 418. [CrossRef]
97. Broggio, A.; Gnendiger, C.; Signer, A.; Stöckinger, D.; Visconti, A. SCET approach to regularization-scheme dependence of QCD amplitudes. *J. High Energy Phys.* **2016**, *1*, 78. [CrossRef]
98. Bern, Z.; Kosower, D.A. The Computation of loop amplitudes in gauge theories. *Nucl. Phys. B* **1992**, *379*, 451–561. [CrossRef]
99. Martin, S.P.; Vaughn, M.T. Regularization dependence of running couplings in softly broken supersymmetry. *Phys. Lett. B* **1993**, *318*, 331–337. [CrossRef]
100. Mihaila, L. Two-loop parameter relations between dimensional regularization and dimensional reduction applied to SUSY-QCD. *Phys. Lett. B* **2009**, *681*, 52–59. [CrossRef]
101. Stockinger, D.; Varso, P. FeynArts model file for MSSM transition counterterms from DREG to DRED. *Comput. Phys. Commun.* **2012**, *183*, 422–430. [CrossRef]
102. Hollik, W.; Stockinger, D. MSSM Higgs-boson mass predictions and two-loop non-supersymmetric counterterms. *Phys. Lett. B* **2006**, *634*, 63–68. [CrossRef]
103. Avdeev, L.V.; Chochia, G.A.; Vladimirov, A.A. On the Scope of Supersymmetric Dimensional Regularization. *Phys. Lett. B* **1981**, *105*, 272–274. [CrossRef]
104. Harlander, R.V.; Mihaila, L.; Steinhauser, M. The SUSY-QCD beta function to three loops. *Eur. Phys. J. C* **2009**, *63*, 383–390. [CrossRef]
105. Stöckinger, D.; Unger, J. Three-loop MSSM Higgs-boson mass predictions and regularization by dimensional reduction. *Nucl. Phys. B* **2018**, *935*, 1–16. [CrossRef]
106. Lowenstein, J.H. Normal product quantization of currents in Lagrangian field theory. *Phys. Rev. D* **1971**, *4*, 2281–2290. [CrossRef]
107. Lowenstein, J.H. Differential vertex operations in Lagrangian field theory. *Commun. Math. Phys.* **1971**, *24*, 1–21. [CrossRef]
108. Lam, Y.M.P. Perturbation Lagrangian theory for scalar fields: Ward-Takahasi identity and current algebra. *Phys. Rev. D* **1972**, *6*, 2145–2161. [CrossRef]
109. Lam, Y.M.P. Perturbation lagrangian theory for dirac fields—Ward-takahashi identity and current algebra. *Phys. Rev. D* **1972**, *6*, 2161–2167. [CrossRef]
110. Lam, Y.M.P. Equivalence theorem on Bogolyubov-Parasiuk-Hepp-Zimmermann renormalized Lagrangian field theories. *Phys. Rev. D* **1973**, *7*, 2943–2949. [CrossRef]
111. Clark, T.E.; Lowenstein, J.H. Generalization of Zimmermann's Normal-Product Identity. *Nucl. Phys. B* **1976**, *113*, 109–134. [CrossRef]
112. Popineau, G.; Stora, R. A pedagogical remark on the main theorem of perturbative renormalization theory. *Nucl. Phys. B* **2016**, *912*, 70–78. [CrossRef]
113. Bogoliubov, N.N.; Parasiuk, O.S. On the Multiplication of the causal function in the quantum theory of fields. *Acta Math.* **1957**, *97*, 227–266. [CrossRef]
114. Speer, E.R. The Convergence of BPH renormalization. *Commun. Math. Phys.* **1974**, *35*, 151–154. . [CrossRef]
115. Zimmermann, W. Convergence of Bogolyubov's method of renormalization in momentum space. *Commun. Math. Phys.* **1969**, *15*, 208–234. [CrossRef]
116. Epstein, H.; Glaser, V. The Role of locality in perturbation theory. *Ann. Inst. H. Poincare Phys. Theor. A* **1973**, *19*, 211–295.
117. Hepp, K. Renomalization theory. In Proceedings of the Les Houches Summer School of Theoretical Physics: Statistical Mechanics and Quantum Field Theory, Les Houches, France, 5 July–29 August 1971; pp. 429–500.
118. DeWitt, C.; Stora, R. (Eds.) *Mécanique statistique et théorie quantique des champs: Proceedings, Ecole d'Eté de Physique Théorique, Les Houches, France, 5 July–29 August, 1970*; Les Houches Summer School: New York, NY, USA; Gordon and Breach: New York, NY, USA, 1971; Volume 20,
119. Piguet, O.; Rouet, A. Symmetries in Perturbative Quantum Field Theory. *Phys. Rept.* **1981**, *76*, 1. [CrossRef]
120. Bogolyubov, N.; Shirkov, D. *Introduction to the Theory of Quantized Fields*; John Wiley & Sons: Hoboken, NJ, USA, 1980; p. 636.
121. Binoth, T.; Heinrich, G. An automatized algorithm to compute infrared divergent multiloop integrals. *Nucl. Phys. B* **2000**, *585*, 741–759. [CrossRef]
122. Binoth, T.; Heinrich, G. Numerical evaluation of multiloop integrals by sector decomposition. *Nucl. Phys. B* **2004**, *680*, 375–388. [CrossRef]
123. Bergere, M.C.; Lam, Y.M.P. Bogolyubov-Parasiuk Theorem in the alpha Parametric Representation. *J. Math. Phys.* **1976**, *17*, 1546–1557. [CrossRef]
124. Bogolyubov, N.; Shirkov, D. *Quantum Fields*; Benjamin-Cummings Publishing Company: San Francisco, CA, USA, 1982; p. 411.
125. 't Hooft, G. Renormalizable Lagrangians for Massive Yang–Mills Fields. *Nucl. Phys. B* **1971**, *35*, 167–188. [CrossRef]
126. 't Hooft, G. Renormalization of Massless Yang–Mills Fields. *Nucl. Phys. B* **1971**, *33*, 173–199. [CrossRef]
127. Lee, B.W.; Zinn-Justin, J. Spontaneously Broken Gauge Symmetries Part 1: Preliminaries. *Phys. Rev. D* **1972**, *5*, 3121–3137. [CrossRef]

128. Lee, B.W.; Zinn-Justin, J. Spontaneously Broken Gauge Symmetries Part 2: Perturbation Theory and Renormalization. *Phys. Rev. D* **1972**, *5*, 3137–3155. Erratum in *Phys. Rev. D* **1973**, *8*, 4654. [CrossRef]
129. Lee, B.W.; Zinn-Justin, J. Spontaneously Broken Gauge Symmetries Part 3: Equivalence. *Phys. Rev. D* **1972**, *5*, 3155–3160. [CrossRef]
130. Lee, B.W.; Zinn-Justin, J. Spontaneously Broken Gauge Symmetries Part 4: General Gauge Formulation. *Phys. Rev. D* **1973**, *7*, 1049–1056. [CrossRef]
131. Martin, C.P.; Sanchez-Ruiz, D. Action principles, restoration of BRS symmetry and the renormalization group equation for chiral nonAbelian gauge theories in dimensional renormalization with a nonanticommuting gamma(5). *Nucl. Phys. B* **2000**, *572*, 387–477. [CrossRef]
132. Witten, E. An SU(2) Anomaly. *Phys. Lett. B* **1982**, *117*, 324–328. [CrossRef]
133. Delbourgo, R.; Salam, A. The gravitational correction to pcac. *Phys. Lett. B* **1972**, *40*, 381–382. [CrossRef]
134. Eguchi, T.; Freund, P.G.O. Quantum Gravity and World Topology. *Phys. Rev. Lett.* **1976**, *37*, 1251. [CrossRef]
135. Alvarez-Gaume, L.; Witten, E. Gravitational Anomalies. *Nucl. Phys. B* **1984**, *234*, 269. [CrossRef]
136. Aoki, K.I.; Hioki, Z.; Konuma, M.; Kawabe, R.; Muta, T. Electroweak Theory. Framework of On-Shell Renormalization and Study of Higher Order Effects. *Prog. Theor. Phys. Suppl.* **1982**, *73*, 1–225. [CrossRef]
137. Denner, A.; Dittmaier, S. Electroweak Radiative Corrections for Collider Physics. *Phys. Rept.* **2020**, *864*, 1–163. [CrossRef]
138. Sperling, M.; Stöckinger, D.; Voigt, A. Renormalization of vacuum expectation values in spontaneously broken gauge theories. *J. High Energy Phys.* **2013**, *7*, 132. [CrossRef]
139. Sperling, M.; Stöckinger, D.; Voigt, A. Renormalization of vacuum expectation values in spontaneously broken gauge theories: Two-loop results. *J. High Energy Phys.* **2014**, *1*, 068. [CrossRef]
140. Grassi, P.A.; Hurth, T.; Steinhauser, M. Practical algebraic renormalization. *Ann. Phys.* **2001**, *288*, 197–248. [CrossRef]
141. Grassi, P.A.; Hurth, T.; Steinhauser, M. The Algebraic method. *Nucl. Phys. B* **2001**, *610*, 215–250. [CrossRef]
142. Hollik, W.; Kraus, E.; Stockinger, D. Renormalization and symmetry conditions in supersymmetric QED. *Eur. Phys. J. C* **1999**, *11*, 365–381. [CrossRef]
143. Hollik, W.; Stockinger, D. Regularization and supersymmetry restoring counterterms in supersymmetric QCD. *Eur. Phys. J. C* **2001**, *20*, 105–119. [CrossRef]
144. Fischer, I.; Hollik, W.; Roth, M.; Stockinger, D. Restoration of supersymmetric Slavnov–Taylor and Ward identities in presence of soft and spontaneous symmetry breaking. *Phys. Rev. D* **2004**, *69*, 015004. [CrossRef]
145. Blondel, A. Standard model theory for the FCC-ee Tera-Z stage. In *CERN Yellow Reports: Monographs, Proceedings of the Mini Workshop on Precision EW and QCD Calculations for the FCC Studies: Methods and Techniques, Geneva, Switzerland, 12–13 January 2018*; CERN: Geneva, Switzerland, 2018; Volume 3/2019. [CrossRef]
146. Fuentes-Martín, J.; König, M.; Pagès, J.; Thomsen, A.E.; Wilsch, F. Evanescent Operators in One-Loop Matching Computations. *arXiv* **2022**, arXiv:hep-ph/2211.09144.
147. Carmona, A.; Lazopoulos, A.; Olgoso, P.; Santiago, J. Matchmakereft: Automated tree-level and one-loop matching. *SciPost Phys.* **2022**, *12*, 198. [CrossRef]
148. Sanchez-Ruiz, D. BRS symmetry restoration of chiral Abelian Higgs-Kibble theory in dimensional renormalization with a nonanticommuting gamma(5). *Phys. Rev. D* **2003**, *68*, 025009. [CrossRef]

Disclaimer/Publisher's Note: The statements, opinions and data contained in all publications are solely those of the individual author(s) and contributor(s) and not of MDPI and/or the editor(s). MDPI and/or the editor(s) disclaim responsibility for any injury to people or property resulting from any ideas, methods, instructions or products referred to in the content.

MDPI AG
Grosspeteranlage 5
4052 Basel
Switzerland
Tel.: +41 61 683 77 34

Symmetry Editorial Office
E-mail: symmetry@mdpi.com
www.mdpi.com/journal/symmetry

Disclaimer/Publisher's Note: The title and front matter of this reprint are at the discretion of the Guest Editor. The publisher is not responsible for their content or any associated concerns. The statements, opinions and data contained in all individual articles are solely those of the individual Editor and contributors and not of MDPI. MDPI disclaims responsibility for any injury to people or property resulting from any ideas, methods, instructions or products referred to in the content.

www.ingramcontent.com/pod-product-compliance
Lightning Source LLC
Chambersburg PA
CBHW040835250926
51850CB00007BA/1325